Genomics of Pain and Co-Morbid Symptoms

Susan G. Dorsey • Angela R. Starkweather
Editors

Genomics of Pain and Co-Morbid Symptoms

 Springer

Editors
Susan G. Dorsey
Department of Pain and Translational
Symptom Science
University of Maryland School of Nursing
Baltimore, MD
USA

Angela R. Starkweather
University of Connecticut School
of Nursing
Storrs, CT
USA

ISBN 978-3-030-21656-6   ISBN 978-3-030-21657-3   (eBook)
https://doi.org/10.1007/978-3-030-21657-3

This Springer imprint is published by the registered company Springer Nature Switzerland AG
The registered company address is: Gewerbestrasse 11, 6330 Cham, Switzerland

Contents

History of Integrating Genomics in Nursing Research: The Importance of Omics in Symptom Science

1

Patricia A. Grady, Ann K. Cashion, and Louise M. Rosenbaum

Contents

The mission of the National Institute of Nursing Research (NINR) is to promote and improve the health of individuals, families, and communities. To achieve this mission, NINR supports and conducts clinical and basic research, and research training across the behavioral and biological sciences. NINR has a long history of promoting the integration of omics technologies and methodologies into the research it supports, with the understanding that omics would ultimately contribute to the scientific basis for clinical practice.

P. A. Grady · A. K. Cashion · L. M. Rosenbaum (✉)
National Institutes of Health, National Institutes of Nursing Research, Bethesda, MD, USA
e-mail: pgrady@mail.nih.gov, gradyp@mail.nih.gov

© Springer Nature Switzerland AG 2020
S. G. Dorsey, A. R. Starkweather (eds.), *Genomics of Pain and Co-Morbid Symptoms*, https://doi.org/10.1007/978-3-030-21657-3_1

1

1.1 The Unique Perspective Nurse Scientists Bring to Symptom Science Research

Nurses work in a variety of health care settings and are trained to function in teams with diverse practitioners, such as physicians, respiratory therapists, social workers, and psychologists. Because of this experience, nurse scientists are in pivotal positions in the transformation of health care science. They have become essential leaders and participants in cross-disciplinary team science and clinical care.

Nurse scientists address research questions that are informed from clinical experience and they interpret research results through a clinical lens. Rather than focusing on any particular disease or condition, nursing science addresses the needs of individual patients and their families. It plays a critical role in the health research enterprise: bridging the gaps between the bench, the research clinic, and communities, and translating the findings to clinical care. It provides a unique scientific perspective into both the clinical and biological features of symptoms and the negative effects of conditions and treatments. Nursing science develops and applies new knowledge, such as genomics and biomarkers, to advance our understanding of symptoms, including fatigue, sleep disturbance, and pain, as well as impaired cognition and disordered mood.

NINR has supported research on new and better ways to characterize adverse symptoms in the pursuit of improved clinical management for more than 30 years. As the US population lives longer with chronic conditions and their associated symptoms, a concerted and transformative approach to symptom science research will be needed to address the resulting increased health burden and enable people to live well over the entire lifespan. NINR is committed to supporting innovative symptom science research that will generate meaningful and practical clinical applications.

1.2 The History of Promoting Precision Health Research at NINR

Attending to the comfort and well-being of individual patients is a core value of nursing. At its inception in 1985, advocates for NINR's predecessor, the National Center for Nursing Research (NCNR), highlighted research advances by nurse scientists in patient-centered care, such as alleviating side effects of cancer chemotherapy and mitigating patient anxiety with communication interventions. The initial organization of NCNR's research programs addressed biological and behavioral factors affecting health and health outcomes from illness and treatments, with an emphasis on symptom management.

The second Conference on Nursing Research Priorities in 1992 marked a shift from behavioral to biobehavioral interventions involving interdisciplinary research that analyzed the interaction of psychological, biological, social, and clinical factors in health care issues. As nurse scientists embarked on this pathway of discovery, they adopted the latest techniques, initiatives, and innovations in science. At the

1996 commemoration of the tenth anniversary of nursing science at NIH, NINR reiterated its commitment to patient-oriented research and announced its leading role in an NIH-wide initiative in understanding molecular, genetic, and behavioral underpinnings of pain, as well as encouraging nurse scientists to incorporate molecular studies and genetics into future research.

1.2.1 Harnessing the Future with Genetics and Genomics: Education and Funding Initiatives

During this period of new directions, NINR leveraged a 1995 report on NIH's investment in gene therapy research (Stuart et al. 1995) to urge increased involvement of nurse scientists in genetics research. Based on the amount of meaningful contact and relationships of trust that nurses have with patients and families, nurse scientists were poised to translate genetics research to the public and to collaborate in the incorporation of genetic data into behavioral studies.

A 1996 genetics research workshop sponsored by NINR identified areas in which nurse scientists could have a significant impact, including the translation of basic research into clinical practice and investigation of the interaction of genetic, behavior, and environmental factors in health (Sigmon et al. 1996). Participants noted that nursing science efforts to identify and isolate disease markers and risk factors for individuals and families with inherited conditions would enable clinicians to tailor and personalize preventive strategies and treatments. The workshop also reinforced NINR's foundation in what would become known as precision health, with a push for nurse scientists to turn their attention from broad populations to individual phenotypes for the development of more effective interventions and health care delivery.

In addressing the recommendations from the workshop, NINR encouraged the incorporation of genetics research into ongoing nursing science projects through the use of research funding supplements. In addition, NINR created extramural funding opportunities in genetics research.

For almost 20 years, the Summer Genetics Institute (SGI) has provided training in genetic technologies and methodologies that can be applied to nursing research, curricula, and clinical practice. The SGI is an intensive program of classroom and laboratory training, aimed at building a foundation in molecular genetics. More than 400 SGI alumni are now advancing genetics and genomics research and integrating new knowledge into clinical practice. The SGI has also been credited with the continued scientific growth of the International Society of Nurses in Genetics (ISONG), which works toward use of genomic information and public understanding of genomic health.

An additional summer training opportunity was launched in 2010, to further develop expertise in the nursing science community, with a focus on symptom science. NINR began its annual, week-long Methodologies Boot Camp on the NIH campus in Bethesda, Maryland to bolster symptom science research training for graduate students and faculty. It features lectures by nationally and internationally known scientists from the NIH and universities across the country, as well as classroom discussion and laboratory training. Topics have included pain, fatigue and sleep, data science, precision health, and approaches using mHealth and digital health technologies.

1.3   Efforts to Develop Nursing Research Expertise in Genomics and Symptom Science

Genomics is transforming healthcare across the disease continuum and across the lifespan. Because nurses have prominent roles in the delivery of direct patient care, there is a clear need for research to support the incorporation of genomics within the science of nursing practice. The scope of genomic nursing care is broad, encompassing risk assessment, risk management, and treatment options, including pharmacogenomics. Hence, nurse scientists must keep pace with the latest techniques and innovations in science.

The "Blueprint for Genomic Nursing Science" (Calzone et al. 2013) appeared in the 2013 Genomic Special Issue of the *Journal of Nursing Scholarship* and was the primary outcome of the 2012 Genomic Nursing State of the Science Advisory Panel meeting. The participants described genomic nursing research as an important factor in developing evidence-based translation of genomics into practice to improve health outcomes.

The Blueprint provides a framework for furthering genomic nursing science, which was aligned with areas of the NINR strategic plan, as well as cross-cutting themes applicable to any genomic nursing research initiative, such as health disparities, cost, policy, and public education. Implementation recommendations for the Blueprint included infrastructure changes, health care, and building research capacity, acknowledging that NINR's SGI was an important element. Funding for interdisciplinary genomics team science was also cited as critical for advancing the Blueprint.

A 2013 Genomics Workshop convened by NINR brought together experts from diverse disciplines to identify a series of research questions to assist in guiding future research directions in genomic nursing science. The resulting questions are aligned with the four areas of scientific focus that evolved from NINR's 2011 strategic plan: symptom science, self-management, wellness, and end-of-life and palliative care, such as:

- How should omic discoveries be used to create and test technologies (such as clinical tools) that can be used to diagnose clinical problems, predict the clinical course, and promote optimal outcomes?
- In what ways can genomic information be used to promote adherence and improve self-management of chronic conditions?
- How does the social environment interact with gene expression to influence resilience in coping with life challenges?
- For high-risk patients who are at the end of life, how can genetic assessment and DNA banking be used to address familial risk?

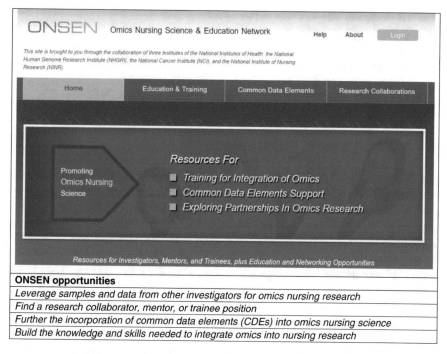

Fig. 1.1 The ONSEN website

An expanding array of tools for genomics and other omics research provides numerous opportunities to advance nursing science. NINR, the National Cancer Institute (NCI), and the National Human Genome Research Institute (NHGRI) have collaborated in the development of the Omics Nursing Science & Education Network (ONSEN) website (https://omicsnursingnetwork.net/) (Fig. 1.1). It is a central resource to assist those interested in including omics in their program of nursing research.

The model of the Genomics Workshop was expanded when NINR launched its Innovative Questions (IQ) initiative in November 2013 to develop lists of creative and results-oriented research questions that could assist in guiding future research directions in nursing science. The initiative was organized by NINR's areas of science and involved two components: IQ workshops brought together leading interprofessional scientists to identify and refine research questions through a consensus-building discussion format, followed by an opportunity for members of the scientific community, professional organizations, and the general public to submit research questions directly to NINR, and to comment on questions through the IQ website.

The collection of IQs (Table 1.1), along with other considerations, such as emerging research areas, portfolio balance, and trans-NIH research priorities, were used to develop a series of specific research emphases, outlined in NINR's 2016 Strategic Plan, to guide NINR-supported science over the next several years.

Table 1.1 Sample of Symptom Science Innovative Questions

How do lifestyle factors, environmental conditions, symptom clusters, and symptom treatments impact quality of life and symptom management in different chronic conditions?
How do symptom precursors (e.g., biomarkers or conditions such as obesity) contribute to the physiology of symptom risk, severity, duration, and response to treatment?
What are the omic, phenotypic, and state-dependent indicators related to the mechanisms, assessment, and management of high impact symptoms (e.g., pain, fatigue, dyspnea)?

1.4 NIH Symptom Science Model

The NIH Symptom Science Model (NIH SSM) (Cashion et al. 2016) was developed by NINR to guide the Institute's symptom science research process. Beginning with the characterization of symptom phenotypes, genomic and other omics approaches are used to identify biomarkers and pathways that affect outcomes, while taking into account environmental factors, such as diet, behavior, and lifestyle (Fig. 1.2). Subsequently, these data contribute to the development of clinical interventions, in areas such as cancer-related fatigue, gastrointestinal disorders, and traumatic brain injuries.

The NIH SSM is an opportunity to standardize symptom science research and offer NINR's expertise on symptom assessment to the entire NIH intramural community and, potentially, extramural nurse scientists. To facilitate expansion of this resource, NINR has developed a framework for symptom assessment and collection of patient-reported outcomes (PROs), which includes a computer adaptive testing

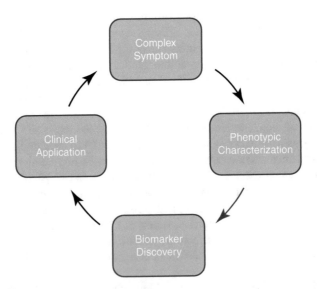

Fig. 1.2 The National Institutes of Health Symptom Science Model (NIH-SSM). Source: Cashion AK, Gill J, Hawes R, Henderson WA, Saligan L. National Institutes of Health Symptom Science Model sheds light on patient symptoms. Nurs Outlook. 2016;64(5):499–506, p. 500, Fig. 1

(CAT) format for collecting PROs. CAT provides improved efficiency through automated administration, scoring, and reporting. In addition, CAT reduces respondent burden, because symptom-oriented questions are administered based on prior responses and level of severity.

1.5 NINR's Division of Intramural Research and Symptom Science

NINR maintains a robust intramural research program on the NIH Campus in Bethesda, Maryland, dedicated to improving the understanding of the underlying biological mechanisms of a range of symptoms, their effect on patients, and the biological and behavioral bases for how patients respond to interventions. Nurse scientists in the Division of Intramural Research (DIR) use the NIH SSM to address complex research questions and improve clinical care.

Research on the biobehavioral mechanisms of fatigue from cancer and cancer treatments is driving the identification of biomarkers associated with functional pathways of cancer-related fatigue (Feng et al. 2017). Ongoing investigations on the etiology of cancer-related fatigue are geared towards the development of interventions for this debilitating symptom.

To understand the neurological symptoms and deficits in military personnel with traumatic brain injuries (TBIs), as well as in athletes with concussions, DIR scientists have linked omic biomarkers to neuronal damage detected with novel brain imaging methods (Livingston et al. 2017). They are also developing methods to identify brain-injured patients at risk for poor recovery who are candidates for early interventions to mitigate symptoms resulting from neurological damage (Gill et al. 2017).

DIR scientists are investigating underlying processes of the primary senses to understand the fundamental molecular, behavioral, and neural mechanisms associated with chemosensory symptoms (taste and smell alterations) in metabolic conditions such as obesity, type 2 diabetes, and related comorbidities. In addition, they are developing a framework for obesity risk based on genetic risk scores and gene expression profiles (Joseph et al. 2019).

1.6 NINR's Support of Symptom Science and Omics Research

NINR's support of extramural symptom science and omics research has surged in recent years. The Institute has issued diverse funding opportunity announcements to encourage the use of innovative technologies and approaches to address unmet needs in health care. These opportunities have facilitated interdisciplinary research to decrease symptom burden and enhance health-related quality of life in persons with chronic conditions. They have encouraged increased understanding of the biological mechanisms of symptoms and development of innovative, cost-effective, targeted interventions to prevent, manage, or ameliorate symptoms of chronic illness.

Nurse scientists are leading pain research efforts that apply genetics, genomics, and interdisciplinary approaches to the development of precision health interventions for chronic pain. For example, one project is identifying predictive phenotypic and gene expression patterns related to chronic pain in patients with lower extremity fractures. Researchers are also examining the role of genomics and other biological factors on the mechanisms and outcomes of pain self-management.

Nurse scientists have been dissecting chronic pain and other conditions associated with neurological function toward improving patient outcomes. Using preclinical cell culture and mouse models, they have characterized a neural cell receptor that modulates pain and muscle activity responses to injury. Studies with a genetically modified version of the receptor produced a potential therapeutic target for neuropathic pain (Matyas et al. 2017). Researchers have also manipulated cell signaling pathways that dampen pain sensitivity in mouse models (Wu et al. 2016). Research in unresolved lower back pain (LBP) has revealed changes in expression of genes coding for cellular receptors and signaling molecules associated with pain sensitivity in LBP patients (Starkweather et al. 2016).

A higher prevalence of pain associated irritable bowel syndrome (IBS) in female populations has stimulated biomarker discovery to classify and characterize variants in pediatric and adult patients. This research has yielded urinary protein biomarker patterns associated with IBS symptom profiles (Goo et al. 2012) and genetic polymorphisms that identify patients who might benefit from self-management interventions. Further research on IBS-associated symptoms has examined the interactions of gut microbiota with biological and psychological factors.

NINR has supported investigations on the contribution of the microbiome to preterm labor and delivery through enhanced knowledge of the microbiome, the changes that are associated with increased risk for preterm labor and delivery, and the influence of genetic and environmental factors.

This research has revealed differences between the placental microbiomes of a small cohort of women experiencing spontaneous preterm birth, in comparison with women who had term pregnancies. Inflammation in association with chorioamnionitis appeared to be a factor in spontaneous preterm birth (Prince et al. 2016). Additional studies found that newborn babies had reasonably homogeneous microbiota populations in most of their body parts. In contrast, their individual mothers—like most adults—exhibited significant differences in microbiota from different parts of their bodies. Babies born by Cesarean section had less diverse microbiota immediately after birth, representing their mothers' skin microbiota, in comparison with vaginally born babies, who had microbiota reflecting their mothers' skin and vaginal microbiota (greater diversity in the microbiome is associated with greater resiliency and better health). However, both groups of babies had similarly diverse microbiota 6 weeks after delivery, indicating that a Cesarean section delivery should not put babies at a disadvantage in terms of microbiota diversity (Chu et al. 2017).

NINR's efforts to understand the genomic mechanisms associated with development and repair of chronic wounds span basic, translational, and clinical research (Table 1.2).

Table 1.2 Research in genomic mechanisms of chronic wound development and repair

Genomic marker discovery and characterization
Development and testing of genomic-based interventions aimed at chronic wound prevention and expedited wound healing
The influence of the social environment (e.g., social engagement or isolation, socioeconomic status, chronic stress) on gene expression and how this gene expression may affect health outcomes

Research in chronic wounds in skin has identified microRNAs that regulate the activity of genes involved in wound healing processes, thus affecting molecular signaling and delaying the formation of new tissue (Pastar et al. 2012). Extension of these studies to tangible clinical challenges, such as diabetic foot ulcers, has led to discoveries of microRNAs expressed in skin cells that inhibit cellular processes that contribute to impaired wound healing (Liang et al. 2016). These findings help identify potential therapeutic targets.

Investigation of other factors affecting wound healing has led to novel observations, such as a slower rate of wound healing in patients who have taken higher doses of opioid pain killers (due to larger wound size or painful comorbidities) in comparison with those who had taken lower doses or no opioids (Shanmugam et al. 2017). Related research has supported molecular characterization of hidradenitis suppurativa, a chronic, inflammatory condition that involves abscesses and delayed wound healing (Jones et al. 2018).

NINR's funding opportunities have also fostered synergies between omics and symptom science to integrate diverse information from molecular and biological processes with patient-reported symptoms, functional status, and health-related quality of life. In this vein, nurse scientists have identified genetic variants associated with poorer outcomes and slower rates of recovery following TBI (Hoh et al. 2010). Further studies have found gene variants, associated gene products, and age-related factors that affect post-TBI mortality risk (Failla et al. 2016). In animal models of TBI, problems in mitochondrial function have been detected prior to the loss of neurons. Additional investigations have revealed mitochondrial gene variants that are linked to worse outcomes and mitochondrial dysfunction in the acute post-TBI period, which could serve as therapeutic targets (Conley et al. 2014).

Nurse scientists have conducted research on the interaction of genetic, psychological, and environmental factors, and their influence on health disparities in minority populations. For example, interest in hypertension has propelled investigations on the effects of lead exposure and psychological stress from social factors on blood pressure in African Americans (Taylor et al. 2017).

Symptom science research has revealed associations between genetic risk factors and molecular pathways of cancer-related symptoms and symptom clusters, in which symptoms occur together and are interrelated; treatment strategies designed for a group of symptoms may yield better outcomes than targeting individual symptoms. Common inflammation-related biological mechanisms were found in symptom clusters of cancer patients undergoing radiation therapy, as well as inflammatory cytokine gene polymorphisms associated with these symptom clusters (Miaskowski et al. 2017). Genetic and epigenetic variations that modulate the expression of cytokine genes have been linked to persistent pain following breast cancer surgery (Stephens et al. 2017). Research on symptom trajectories has identified inflammation-related gene variants associated specifically with either morning or evening fatigue during cancer chemotherapy (Wright et al. 2017). Advances in this field could inform personalized treatment strategies in the future.

The Study of Women's Health Across the Nation (SWAN) is a long-term, community-based, multi-site longitudinal study that was initiated with funding from NINR, the National Institute on Aging, and the NIH Office of Research on Women's Health. A multi-ethnic cohort of more than 3000 pre-menopausal, middle-aged women were enrolled in SWAN in the mid-1990s and a wealth of clinical and psychosocial data were collected annually for nearly two decades. The project has tracked the health changes of these women as they age and experience menopause. This rich and robust dataset has yielded numerous important findings, including identification of a urinary peptide biomarker of bone loss (Shieh et al. 2016), serum proteins and other molecular biomarkers that predict future physical functioning (Karvonen-Gutierrez et al. 2016), and trajectories of pain across the menopausal transition (Lee et al. 2017).

1.6.1 Extramural Symptom Science and Omics Research Training

The development of a strong cadre of nurse scientists has been a primary goal of NINR since its establishment. To continue to support advancements in science and improvements in health, it is essential that the scientific workforce of the future be innovative, multidisciplinary, and diverse. NINR supports a wide range of activities to ensure excellence in the next generation of nurse scientists. In addition to training opportunities in the Division of Intramural Research described earlier, pre-doctoral and post-doctoral research training programs have been supported in nursing schools across the nation.

Precision health has been a guiding principle for preparing the next generation of nurse scientists in harnessing advances in genetics, transcriptomics, metabolomics, and microbiome research for integration with clinical practice. Training in a wide range of methodologies in genomics targets the development of clinical applications of genomics. In addition, nurse scientists and their colleagues published guidance for methodological pathways in epigenetic analyses for clinical researchers (Wright et al. 2016).

1.7 Trans-NIH Initiatives

NINR is one of 27 Institutes and Centers that comprise the National Institutes of Health (NIH), the steward of biomedical and behavioral research for the USA. NINR has been involved in the development of several cross-disciplinary research initiatives associated with symptom science and omics.

1.7.1 The NIH Pain Consortium

The NIH Pain Consortium (https://www.painconsortium.nih.gov/) aims to enhance pain research and promote collaboration among researchers across the many NIH Institutes and Centers that have programs and activities addressing pain. NINR has been part of this NIH-wide initiative since its inception, with the NINR Director serving on the Pain Consortium's Executive Committee.

The Pain Consortium's goals are to:

- Develop a comprehensive and forward-thinking pain research agenda for the NIH, building on what was learned from past efforts.
- Identify key opportunities in pain research, particularly those that provide for multidisciplinary and trans-NIH participation.
- Increase the visibility of pain research both within the NIH intramural and extramural communities, as well as outside the NIH. The latter audiences include various pain advocacy and patient groups who have expressed their interests through scientific and legislative channels.
- Pursue the pain research agenda through public-private partnerships, wherever applicable.

The annual NIH Pain Consortium Symposium has frequently included cutting-edge research on topics germane to nursing science, including genetic and epigenetic effects on pain (2015), biological and psychological factors contributing to chronic pain (2014), and self-management of pain (2013).

1.7.2 The *All of Us* Research Program

The NIH-wide *All of Us* Research Program (https://allofus.nih.gov/) is a bold initiative to develop a new model of patient-centered and patient-driven research to accelerate biomedical discoveries and provide clinicians with new tools, knowledge, and therapies to select which treatments will work best for individual patients. *All of Us* is a historic effort to gather data from one million or more people living in the USA to accelerate research and improve health. By taking into account individual differences in lifestyle, environment, and biology, researchers will uncover paths toward delivering precision medicine.

The goal of *All of Us* is to set the foundation for a new way of doing research that fosters open, responsible data sharing with the highest regard to participant privacy, and that puts engaged participants at the center of research efforts. Because Americans are improving their health and participating in health research more than ever before, electronic health records have been widely adopted, genomic analysis costs have dropped significantly, data science has become increasingly sophisticated, and health technologies have become mobile. The time is right to embark on this monumental approach to improving the health of individuals.

Nursing science is aligned with many of the initiative's long-term goals that will stimulate research using the cohort data, such as identifying new targets for treatment and prevention, testing whether mobile devices can encourage healthy behaviors, and laying the scientific foundation for precision medicine for many diseases.

1.7.3   Common Data Elements

In smaller scale population-based studies involving genomics, one of the formidable challenges is obtaining data from sufficient numbers of research participants to yield adequate statistical power. Common data elements (CDEs) are a potential solution to the problem of small sample sizes in individual studies, by allowing the consolidation of several data sets. CDEs are data items that are common to multiple data sets across different studies, a type of standardized nomenclature. Several Institutes at NIH, such as the National Institute of Neurological Disorders and Stroke (NINDS), the National Institute on Drug Abuse (NIDA), and the National Cancer Institute (NCI), have blazed the trail for generating CDE libraries and resources; some are relevant to a number of conditions, but many are disease-specific.

Directors of NINR-supported research centers have collaborated to develop CDEs to support symptom science and self-management research, including sleep, fatigue, pain, self-regulation, and self-efficacy for managing chronic conditions (Redeker et al. 2015; Moore et al. 2016). Future efforts include finalizing and adding more CDEs; creating websites for CDE definitions and data collection protocols; and developing and implementing quality assurance and administrative processes for CDE data management and other activities.

1.7.4   Repurposing Omics Data

NIH has a long-standing policy to share and make available to the public the results and accomplishments of the activities that it funds (https://grants.nih.gov/grants/policy/data_sharing/), with the belief that data sharing is essential for expedited translation of research results into knowledge, products, and procedures to improve human health.

Nurse scientists have summarized opportunities for applying existing omics datasets to symptom science (Osier et al. 2017). This useful guidance notes that some studies with large, well-characterized cohorts have focused on a phenotype or outcome in which symptoms are not a primary focus. The proposed line of research could generate extensive symptom-related data that may be available for further analysis by nurse scientists, particularly if the project received NIH funding. The robust datasets can include genomic, transcriptomic, epigenomic, microbiomic data, and even clinical data from linked electronic medical records—a wide array of opportunities to address research questions in symptom science.

1.8   Conclusion

This chapter provides a summary of scientific initiatives and activities from NINR over several decades. The efforts to anticipate, stimulate, and capture opportunities for nursing science and the response of the community have been remarkable. As a result, nursing is poised to accelerate forward in scientifically exciting and beneficial ways. Through implementation of its strategic plan and continued support of symptom science and omics research, NINR looks forward to providing the leadership that guides the scientific advances that improve clinical care.

References

Calzone KA, Jenkins J, Bakos AD, Cashion AK, Donaldson N, Feero WG, et al. A blueprint for genomic nursing science. J Nurs Scholarsh. 2013;45(1):96–104.

Cashion AK, Gill J, Hawes R, Henderson WA, Saligan L. National Institutes of Health symptom science model sheds light on patient symptoms. Nurs Outlook. 2016;64(5):499–506.

Chu DM, Ma J, Prince AL, Antony KM, Seferovic MD, Aagaard KM. Maturation of the infant microbiome community structure and function across multiple body sites and in relation to mode of delivery. Nat Med. 2017;23(3):314–26.

Conley YP, Okonkwo DO, Deslouches S, Alexander S, Puccio AM, Beers SR, et al. Mitochondrial polymorphisms impact outcomes after severe traumatic brain injury. J Neurotrauma. 2014;31(1):34–41.

Failla MD, Conley YP, Wagner AK. Brain-derived Neurotrophic factor (BDNF) in traumatic brain injury-related mortality: interrelationships between genetics and acute systemic and central nervous system BDNF profiles. Neurorehabil Neural Repair. 2016;30(1):83–93.

Feng LR, Suy S, Collins SP, Saligan LN. The role of TRAIL in fatigue induced by repeated stress from radiotherapy. J Psychiatr Res. 2017;91:130–8.

Gill J, Merchant-Borna K, Jeromin A, Livingston W, Bazarian J. Acute plasma tau relates to prolonged return to play after concussion. Neurology. 2017;88(6):595–602.

Goo YA, Cain K, Jarrett M, Smith L, Voss J, Tolentino E, et al. Urinary proteome analysis of irritable bowel syndrome (IBS) symptom subgroups. J Proteome Res. 2012;11(12):5650–62.

Hoh NZ, Wagner AK, Alexander SA, Clark RB, Beers SR, Okonkwo DO, et al. BCL2 genotypes: functional and neurobehavioral outcomes after severe traumatic brain injury. J Neurotrauma. 2010;27(8):1413–27.

Jones D, Banerjee A, Berger PZ, Gross A, McNish S, Amdur R, et al. Inherent differences in keratinocyte function in hidradenitis suppurativa: evidence for the role of IL-22 in disease pathogenesis. Immunol Investig. 2018;47(1):57–70.

Joseph PV, Jaime-Lara RB, Wang Y, Xiang L, Henderson WA. Comprehensive and systematic analysis of gene expression patterns associated with body mass index. Sci Rep. 2019;9(1):7447.

Karvonen-Gutierrez CA, Zheng H, Mancuso P, Harlow SD. Higher Leptin and Adiponectin concentrations predict poorer performance-based physical functioning in midlife women: the Michigan Study of Women's Health Across the Nation. J Gerontol A Biol Sci Med Sci. 2016;71(4):508–14.

Lee YC, Karlamangla AS, Yu Z, Liu CC, Finkelstein JS, Greendale GA, et al. Pain severity in relation to the final menstrual period in a prospective multiethnic observational cohort: results from the Study of Women's Health Across the Nation. J Pain. 2017;18(2):178–87.

Liang L, Stone RC, Stojadinovic O, Ramirez H, Pastar I, Maione AG, et al. Integrative analysis of miRNA and mRNA paired expression profiling of primary fibroblast derived from diabetic foot ulcers reveals multiple impaired cellular functions. Wound Repair Regen. 2016;24(6):943–53.

Livingston WS, Gill JM, Cota MR, Olivera A, O'Keefe JL, Martin C, et al. Differential gene expression associated with meningeal injury in acute mild traumatic brain injury. J Neurotrauma. 2017;34(4):853–60.

Matyas JJ, O'Driscoll CM, Yu L, Coll-Miro M, Daugherty S, Renn CL, et al. Truncated TrkB. T1-mediated astrocyte dysfunction contributes to impaired motor function and neuropathic pain after spinal cord injury. J Neurosci. 2017;37(14):3956–71.

Miaskowski C, Conley YP, Mastick J, Paul SM, Cooper BA, Levine JD, et al. Cytokine gene polymorphisms associated with symptom clusters in oncology patients undergoing radiation therapy. J Pain Symptom Manag. 2017;54(3):305–16.e3.

Moore SM, Schiffman R, Waldrop-Valverde D, Redeker NS, McCloskey DJ, Kim MT, et al. Recommendations of common data elements to advance the science of self-Management of Chronic Conditions. J Nurs Scholarsh. 2016;48(5):437–47.

Osier ND, Imes CC, Khalil H, Zelazny J, Johansson AE, Conley YP. Symptom science: repurposing existing Omics data. Biol Res Nurs. 2017;19(1):18–27.

Pastar I, Khan AA, Stojadinovic O, Lebrun EA, Medina MC, Brem H, et al. Induction of specific microRNAs inhibits cutaneous wound healing. J Biol Chem. 2012;287(35):29324–35.

Prince AL, Ma J, Kannan PS, Alvarez M, Gisslen T, Harris RA, et al. The placental membrane microbiome is altered among subjects with spontaneous preterm birth with and without chorioamnionitis. Am J Obstet Gynecol. 2016;214(5):627 e1–e16.

Redeker NS, Anderson R, Bakken S, Corwin E, Docherty S, Dorsey SG, et al. Advancing symptom science through use of common data elements. J Nurs Scholarsh. 2015;47(5):379–88.

Shanmugam VK, Couch KS, McNish S, Amdur RL. Relationship between opioid treatment and rate of healing in chronic wounds. Wound Repair Regen. 2017;25(1):120–30.

Shieh A, Ishii S, Greendale GA, Cauley JA, Lo JC, Karlamangla AS. Urinary N-telopeptide and rate of bone loss over the menopause transition and early Postmenopause. J Bone Miner Res. 2016;31(11):2057–64.

Sigmon HD, Amende LM, Grady PA. Development of biological studies to support biobehavioral research at the National Institute of Nursing Research. Image J Nurs Sch. 1996;28(2):88.

Starkweather AR, Ramesh D, Lyon DE, Siangphoe U, Deng X, Sturgill J, et al. Acute low Back pain: differential somatosensory function and gene expression compared with healthy no-pain controls. Clin J Pain. 2016;32(11):933–9.

Stephens KE, Levine JD, Aouizerat BE, Paul SM, Abrams G, Conley YP, et al. Associations between genetic and epigenetic variations in cytokine genes and mild persistent breast pain in women following breast cancer surgery. Cytokine. 2017;99:203–13.

Stuart H, Orkin M.D, Arno G, Motulsky, M.D. Report and Recommendations of the Panel to Assess the NIH Investment in Research on Gene Therapy. 1995.

Taylor JY, Sun YV, de Mendoza VB, Ifatunji M, Rafferty J, Fox ER, et al. The combined effects of genetic risk and perceived discrimination on blood pressure among African Americans in the Jackson heart study. Medicine. 2017;96(43):e8369.

Wright ML, Dozmorov MG, Wolen AR, Jackson-Cook C, Starkweather AR, Lyon DE, et al. Establishing an analytic pipeline for genome-wide DNA methylation. Clin Epigenetics. 2016;8:45.

Wright F, Hammer M, Paul SM, Aouizerat BE, Kober KM, Conley YP, et al. Inflammatory pathway genes associated with inter-individual variability in the trajectories of morning and evening fatigue in patients receiving chemotherapy. Cytokine. 2017;91:187–210.

Wu J, Zhao Z, Zhu X, Renn CL, Dorsey SG, Faden AI. Cell cycle inhibition limits development and maintenance of neuropathic pain following spinal cord injury. Pain. 2016;157(2):488–503.

Introduction to Omics Approaches in Symptom Science

2

Yvette P. Conley and Monica A. Wagner

Contents

Omics-based approaches (i.e., genomics, transcriptomics, epigenomics, proteomics, metabolomics, microbiomics) have potential to advance pain and symptom science by connecting biology to symptom phenotypes. Knowing that a particular gene or biological pathway plays a role in symptom development provides information that could lead to interventions directed at preventing or reducing severity of symptoms. This knowledge could also assist with identifying individuals at risk for development of pain and other symptoms and explain variability in treatment response for pain and other symptoms. Additionally, using omics-based approaches provides a way to identify shared biology for co-occurring symptoms.

The terms genetic and genomic are often used interchangeably; however, they differ based on breadth with genetics focusing on a single gene and genomics focusing on many genes, often all genes within the genome. The suffix "omic" is used to indicate breadth for other approaches, such as transcriptomics where the focus is on

Y. P. Conley (✉) · M. A. Wagner
Health Promotion and Development, University of Pittsburgh School of Nursing, Pittsburgh, PA, USA
e-mail: yconley@pitt.edu; MOW26@pitt.edu

© Springer Nature Switzerland AG 2020
S. G. Dorsey, A. R. Starkweather (eds.), *Genomics of Pain and Co-Morbid Symptoms*, https://doi.org/10.1007/978-3-030-21657-3_2

the transcripts from many or all genes instead of just one (Fig. 2.1). The purpose of this chapter is to provide practical information about a variety of omics approaches that have utility in symptom science as well as provide online resources to allow the reader to access additional and up-to-date omics-based information (Table 2.1).

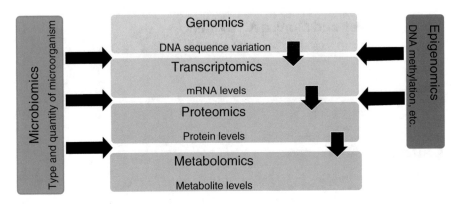

Fig. 2.1 Omics approaches

Table 2.1 Exemplar online omics resources for general knowledge and advanced inquiry

Name of site	Address of site	Description of site
National Human Genome Research Institute talking glossary	www.genome.gov/glossary	Defines genetic vocabulary terms many of which have substantial elaboration
National Human Genome Research Institute brief guide to genomics	https://www.genome.gov/18016863/a-brief-guide-to-genomics/	General information about DNA, genes, and genomes
National Human Genome Research Institute Transcriptomics fact sheet	https://www.genome.gov/13014330/transcriptome-fact-sheet/	General information about gene expression and the transcriptome
National Human Genome Research Institute epigenomics fact sheet	https://www.genome.gov/27532724/epigenomics-fact-sheet/	General information about epigenomics
National Institutes of Health common fund programs—Human microbiome project	https://commonfund.nih.gov/hmp/public	General information about microbiomics
National Institutes of Health common fund programs—Metabolomics	https://commonfund.nih.gov/Metabolomics/PublicHealthRelevance	General information about metabolomics

Table 2.1 (continued)

Name of site	Address of site	Description of site
National Human Genome Research Institute frequently asked questions about genetic and genomic science	https://www.genome.gov/19016904/	General information about genetics, genomics, and other omics including proteomics
National Human Genome Research Institute genome-wide association studies fact sheet	https://www.genome.gov/20019523/genomewide-association-studies-fact-sheet/	General information about genome-wide association studies (GWAS)
National Human Genome Research Institute frequently asked questions about pharmacogenomics	https://www.genome.gov/27530645/faq-about-pharmacogenomics/	General information about pharmacogenomics and its clinical utility
National Institute of General Medical Sciences pharmacogenomics fact sheet	https://www.nigms.nih.gov/education/pages/factsheet-pharmacogenomics.aspx	General information about pharmacogenomics.
Online Mendelian Inheritance in Man (OMIM)	https://www.ncbi.nlm.nih.gov/omim	Source for clinical information on conditions that have a genetic component. It will also inform you of any gene mapping efforts and mutations that are known for that disorder
Centers for Disease Control and Prevention (CDC) public health genomics	https://www.cdc.gov/genomics	Information for health professionals and general public regarding how genomics influences public health
ClinVar	https://www.ncbi.nlm.nih.gov/clinvar	Aggregates information about genomic variation and its relationship to human health
Genetics home reference	https://ghr.nlm.nih.gov/	Information about the effects of genetic variation on human health. Can be searched by health condition or gene of interest. Also contains resources to understand fundamental topics in human genetics
dbSNP	https://www.ncbi.nlm.nih.gov/snp	The single nucleotide polymorphism database. It allows you to search for SNPs within a gene of interest by using the gene as a key work or look up information about a specific SNP

(continued)

Table 2.1 (continued)

Name of site	Address of site	Description of site
dbGaP	https://www.ncbi.nlm.nih.gov/gap	The genotype and phenotype database that houses data from genome-wide association studies. Many cohorts with various phenotypes characterized have been subjected to a whole genome SNP scan and the data from these SNPs as well as the phenotype data is made publicly available
Gene Expression Omnibus (GEO)	https://www.ncbi.nlm.nih.gov/gds	A database for storage and retrieval of gene expression (transcriptomic) data
GeneCards suite	https://www.genecards.org/	A gene-centric database housing multi-omic information about your gene of interest
Genome	https://www.ncbi.nlm.nih.gov/genome/	Organizes information on genomes including sequences, maps, chromosomes, assemblies, and annotations for a variety of organisms
University of California Santa Cruz (UCSC) genome browser	genome.ucsc.edu	Working draft of the human genome → put in a region of a chromosome and it will give you all of the genes and variants within that region. Put in a gene and it gives you the flanking area and information about the gene including what variants are known

2.1 Genomics

The central dogma of molecular biology (Crick 1970) is a framework for understanding how the information housed in DNA sequence within the cell eventually results in protein production and provides the basis for omics research. There are two key events of the central dogma: transcription and translation. Transcription takes place in the nucleus of the cell and is defined as the synthesis of RNA using DNA sequence as a template. Translation is the production of a protein from the messenger RNA (mRNA) transcript and occurs in the cytoplasm. Genomics is the term used to describe the branch of omics research that considers the genome at the level of the DNA sequence. In comparison with the term genetics, which alludes to the study of heredity and inherited traits from a single gene, genomics refers to the entire genome of a species, including the structure, function, and evolution of the DNA sequence (Cheek 2013; Johnson et al. 2013). In a genomic study, variation in the DNA sequence, such as single nucleotide polymorphisms (SNPs), copy number variation (CNVs), and insertion/deletion polymorphisms, is characterized and used for analyses.

The experimental approach chosen for a genomic study depends on the type of knowledge the researcher is interested in gaining and the knowledge that the researcher brings to the experiment. A genomic study can be aimed toward investigating selected genes (a candidate gene or targeted approach) or exploring the entire genome (a genome-wide or discovery approach). A targeted, candidate gene association study investigates the correlation between a phenotype and genetic variation based on the known biological function of the specific pre-determined gene or group of genes. Alternatively, the discovery approach uses a genome-wide association study (GWAS) to interrogate the entire genome for a correlation between the phenotype of interest and common genetic variations across the entire genome. See Table 2.2 for exemplar targeted and discovery based genomic studies of pain.

Table 2.2 Exemplar pain-related research studies using omics approaches

Genomics				
Study design (method)	Polymorphism	Population/ sample	Findings	Citation
Discovery (GWAS)	SNP	Children with sickle cell disease	Evidence of an association between a variant SNP located upstream of the *KIAA1109* gene and acute, severe vaso-occlusive pain	Chaturvedi et al. (2018)
Targeted (Candidate gene)	*FAAH*	Men and women diagnosed with lower back pain	Significant association between FAAH SNP genotypes pain sensitivity in patients who developed chronic lower back pain	Ramesh et al. (2018)
Transcriptomics				
Study design (method)	Gene symbol	Population/ sample	Findings	Citation
Targeted (qPCR)	*FAAH* *CNR2* *TRPV1*	Men and women diagnosed with lower back pain	*CNR2* was significantly upregulated in all lower back pain patients; *FAAH* and TRPV1 are significantly upregulated in chronic lower back pain patients	Ramesh et al. (2018)
Discovery (microarray)		Caucasian females age 18 and older	A total of 421 genes are differentially expressed between fibromyalgia and healthy control groups, including genes important for pain processing	Jones et al. (2016)

(continued)

Table 2.2 (continued)

Proteomics

Study design (method)	Protein symbol	Population/ sample	Findings	Citations
Discovery (mass spectrometry)		Male and female patients undergoing surgical treatment for trigeminal neuralgia with vascular compression	Upregulated expression of TTR, RBP4, and AGP2 in human blood plasma of trigeminal neuralgia patients compared to control group	Farajzadeh and Bathaie (2018)
Discovery (mass spectrometry)		Fibromyalgia patients, rheumatoid arthritis patients, and patients with non-inflammatory neurological symptoms as controls	Out of 3787 proteins in cerebrospinal fluid, 176 were associated with pain and related to pathways of synaptic transmission, inflammatory responses, neuropeptide signaling, and hormonal activity	Khoonsari et al. (2019)
Targeted (Antibody detection)	Rab7	Spinal cord samples from C57BL/6N mice	Deficiency in Rab7 (an oxidized target of pain processing) reduces inflammatory and reactive oxygen species-induced pain behaviors	Kallenborn-Gerhardt et al. (2017)

Epigenomics

Study design (method)	Gene symbol	Population/ sample	Findings	Citations
Discovery (microarray)		Women with fibromyalgia and age matched healthy controls.	There are 1610 variations in methylation sites, 1042 hypomethylated, and 568 hypermethylated in cases compared to controls	Ciampi de Andrade et al. (2017)
Targeted (selected data from microarray)	TRPA1	Patients with either chronic back pain or postherpetic neuralgia	Results indicate a significant correlation between the increase in the DNA methylation level at the CpG island of the TRPA1 gene and increases in pain scores	Sukenaga et al. (2016)

Table 2.2 (continued)

Metabolomics

Study design (method)	Metabolites	Population/ sample	Findings	Citation
Targeted (metabolic profiling)	TRP serotonin kynurenine	Women with early-stage breast cancer before and after standard chemotherapy	Higher kynurenine levels and kynurenine/ TRP ratio post chemotherapy; symptoms of pain and fatigue strongly associated with targeted metabolites	Lyon et al. (2018)
Untargeted	Simultaneous measure of all possible metabolites	Women with early-stage breast cancer before and after standard chemotherapy	Compared to before chemotherapy, global metabolites after chemotherapy contained higher concentrations of acetyl-L-alanine and indoxyl sulfate and lower levels of 5-oxo-L-proline	Lyon et al. (2018)
Untargeted	Simultaneous measure of all possible metabolites	Patients with complex regional pain syndrome-related dystonia and healthy controls	Metabolic profile of cerebrospinal fluid in patients is significantly different compared to controls	Meissner et al. (2014)

Microbiomics

Study design (method)	Purpose	Population/ sample	Findings	Citation
16S rRNA	To explore the gut microbiome for alterations between patients with chronic pelvic pain syndrome and controls	Patients with chronic pelvic pain syndrome and healthy controls	The gut microbiome differed significantly in patients with chronic pain compared with controls	Shoskes et al. (2016)

GWAS genome-wide association study, *SNP* single nucleotide polymorphism, *KIAA1109* a predicted gene of unknown function, *FAAH* fatty acid amide hydrolase, *qPCR* real time quantitative polymerase chain reaction, *CNR2* cannabinoid type 2 receptor, *TRIPV* transient receptor potential cation channel subfamily V member 1, *TTR* transthyretin, *RBP4* retinol-binding protein 4, *AGP2* alpha-1-acid glycoprotein 2, *Rab7* Ras-associated protein 7, *TRPA1* transient receptor protein ankyrin 1, *TRP* tryptophan

2.2 Transcriptomics

Transcriptomics is the study of mRNA expression (i.e., gene expression, gene regulation) in a specific tissue or cell type at an indicated moment in time or in response to disease or symptom burden (Ferranti et al. 2017; Wang et al. 2009). A transcript is comprised of the mRNA sequence produced through transcription, base for base, from the original DNA sequence, and because gene expression is tightly regulated, genes are only expressed at times when the distinct function of the transcribed gene is useful. This is different from genomics because a genomic approach uses DNA, and the DNA sequence is the same in all cell and tissue types and does not change over time or under physiological stress. Transcriptome data can be used to identify alterations in a gene, set of genes, or biological pathways at the level of mRNA (Harrington et al. 2000; Kurella et al. 2001; Noordewier and Warren 2001).

Transcriptomics research is similar to genomics in that transcriptomic studies can be designed as a non-biased genome-wide analysis or a more specific targeted approach. It is not uncommon for researchers to perform a genome-wide analysis comparing the level of mRNA from genes across the genome in a sample population to a reference or control population to reveal previously unknown or novel genes related to a particular condition or disease. A more specific, targeted transcriptomic approach is used when the research question relates to the mRNA expression levels of certain pre-determined candidate genes or biological pathways. An example of a targeted approach is when a researcher is interested in the expression levels of a distinctive gene or gene set with implications in chronic pain development in subjects with pain versus healthy controls. See Table 2.2 for exemplar targeted and discovery based transcriptomic studies of pain.

2.3   Epigenomics

Epigenomics is the branch of omics science that considers those modifications to the DNA molecule that influence gene expression but do not alter the underlying DNA sequence (Baumgartel et al. 2011). The prefix "epi" implies "above," therefore it can be thought that epigenomics involve all those modifications that take place above the genetic code (Mccue et al. 2017). Epigenetic modifications are the results of epigenetic markers that modify gene regulation, which in turn has an effect on transcription and protein production and hence the function of the cell. These markers are histone modification impacting chromatin condensation, DNA methylation, and noncoding RNA (Fessele and Wright 2018).

The frequency and extent of epigenomic changes can be affected by numerous environmental factors (endogenous and exogenous) and are tissue or cell dependent. Researchers use an epigenomic approach to evaluate mechanisms of gene regulation. This approach is used in studies that investigate multifactorial diseases that can be influenced by environmental impact (Baumgartel et al. 2011). Like other omics approaches depending on the nature of the research question and a priori knowledge, epigenomic studies can be designed to interrogate the entire epigenome (a discovery approach) or to probe certain areas of the epigenome that are associated with the gene(s) of interest (a targeted approach). See Table 2.2 for exemplar targeted and discovery based epigenomic studies of pain.

2.4   Proteomics

Proteomics is the study of proteins and protein function (Lee et al. 2017). Similar to the transcriptome, the proteome is not consistent and will differ according to cell or tissue type and is subject to temporal changes (Mccue et al. 2017). Through the use of proteomics, researchers can discover the answers to questions involving the location and timing of protein expression, rates of protein production and degradation, posttranslational modifications and the results of alternative splicing, how proteins move within the cell and function in metabolic pathways, and what are the important protein–protein interactions (Mccue et al. 2017). The understanding of proteomics is important to the understanding gene function. Proteins are primarily responsible for the work of the cell, accounting for the structure, regulation, and function of tissues and organs in the body. The abundance of proteins is not only the result of mRNA transcription, but also a consequence of epigenetic modifications. Through the use of proteomics, researchers can learn more about physiology and pathophysiology of disease by considering a more extensive set of proteins rather than investigating each individual protein one by one.

There are many ways in which to design a proteomics study, but each depends on the research question of interest. An untargeted, discovery approach, such as mass spectrometry, is used to identify and distinguish the relative abundance of proteins based on the mass of the protein. For example, this method is useful when a researcher wishes to compare proteins found in samples from patients with chronic pain to a set of genes implicated in pain signaling.

Targeted approaches, on the other hand, apply methods such as antibody detection to separate and identify a candidate protein of interest. Researchers use this design when the question of inquiry involves identification or amount of a specific protein in the sample. See Table 2.2 for exemplar targeted and discovery based proteomic studies of pain.

2.5   Metabolomics

Metabolomics is the omics science that refers to the characterization of metabolism in biological systems including endogenous (gene-derived) and exogenous (environmentally derived) metabolites. The origins of metabolomics can be traced back to techniques and methods used in chemistry and biochemistry to characterize complex biochemical mixtures (Wishart 2016; Clish 2015). The term metabolite refers to the resulting product of protein expression, while metabolome refers to the totality of metabolites present in a given cell or tissue type at a given time under certain physiological conditions. The metabolomic approach is widely used in the discovery of new biomarkers and when combined with other genomic approaches can increase the understanding of biological systems and pathology of disease and provide more information in the genotype-phenotype gap (Bujak et al. 2015; Trivedi et al. 2017).

There are a variety of designs and uses of metabolomics that depend on the type of question that the researcher has put forward. Approaches to interrogating the metabolome included targeted approaches such as metabolic profiling. In metabolic profiling, groups of metabolites that either are part of the same biochemical pathways or have similar physiochemical properties are selected and hypothesized about prior to testing (Bujak et al. 2015). This type of approach is useful when a researcher is trying to determine the mechanism behind a certain condition or disease. Alternatively, metabolic fingerprinting uses an untargeted approach to specify changes throughout the entire metabolome of specimens collected at a certain time point and/or under a specific physiological condition. Metabolic fingerprinting is most often used as a comparative analysis between two groups such as healthy versus disease (Bujak et al. 2015). Untargeted metabolomic approaches intend to perform an unbiased simultaneous measurement of all possible metabolites in a biological sample. An untargeted approach is useful in characterizing the function of unknown genes and proteins or beginning to classify unknown cellular pathways and starting to link pathways to biological mechanisms (Patti et al. 2012). See Table 2.2 for exemplar targeted and untargeted metabolomic studies of pain.

2.6   Microbiomics

The microbiome consists of the population of microorganisms (e.g. bacteria, virus, and fungi) living on or in an organism (Mccue et al. 2017). For instance, the human microbiota is made up of anywhere from 10 to 100 trillion symbiotic microbial cells, each with their own genome (Ursell et al. 2012). The most frequently studied microbiome is the gut microbiome. A microbiome can be influenced by numerous factors such as environment, experimental condition, age, diet, and sex. The first phase of the Human Microbiome Project was initiated to observe diversity of the human microbiota structure to find common patterns associated with health and disease (Integrative 2014). A multitude of experiments using microbiota from healthy and diseased subjects have shown that the microbiome can be a foundation for disease (Goodrich et al. 2014). An understanding of microbiome research helps investigators to better understand the pathophysiological basis of disease, and results of microbiome studies can be used in diagnosis and treatment of altered biological states (Maki et al. 2018). Results of microbiome research studies can also be used to identify interventions using prebiotics, probiotics, or antibiotics (Mccue et al. 2017).

The most common study designs for a microbiome study use technology that is able to sequence specific genes from many different organisms living in the microbiome. The most common methods are the 16S ribosomal (rRNA) RNA gene sequencing and shotgun metagenome sequencing (Mccue et al. 2017; Maki et al. 2018). The 16S rRNA gene exists in all bacteria and sequencing targets regions of this gene allow for bacterial identification and quantification, but researchers must be aware that this method only identifies bacteria and not virus or fungi. The 16S rRNA method is beneficial in studies that compare samples from different environments or those that compare samples at different time intervals. Shotgun metagenomic sequencing can sample the entire genome of all organisms in a sample (Maki et al. 2018). An advantage to shotgun sequencing is that it can identify bacteria, virus, and fungi, which is more comprehensive than the 16S rRNA method. In general, due to the amount of sequencing accomplished with each method, if the study sample size is larger, 16S rRNA sequencing is suggested, while shotgun metagenomics is recommended for smaller sample sizes. See Table 2.2 for an exemplar 16S based microbiomic study of pain.

2.7   Overall Omics Study Design Considerations and Challenges

- Genomic studies involve the use of DNA. The DNA of an organism is consistent regardless of cell, tissue type, or time and is not altered under physiological challenge. The structure of DNA is a double stranded helix and is therefore a more stable molecule than single stranded RNA. The advantage of a genomic approach is that it uses stable DNA that is not as easily impacted by length of storage or storage conditions, it does not matter from what tissue(s) the DNA was extracted, and it does not matter when in a disease process, therapy, or age a sample was taken.
- Transcriptomic studies measure the level of single stranded mRNA to measure level of gene expression. Because genes are only expressed at times when gene function is useful to the cell, it is important to consider cell and tissue type as well as time points(s) of RNA collection. RNA is much less stable than DNA, therefore preservation of RNA integrity is also important.
- Epigenomic studies explore the highly controlled regulation of gene expression via epigenetic mechanisms; therefore, cell or tissue type as well as timing of sample collection need to be taken into consideration when designing the study. Researchers must also be aware of environmental factors that may influence changes in the epigenetic structure of the DNA molecule and that epigenomic modifications are dynamic and tissue specific. Therefore, consideration must be given to tissue type and timing of sample collection similarly to transcriptomic studies, but because the focus is on DNA or small RNA molecules stability of the sample is less of a concern.
- Proteomic studies require the consideration of cell or tissue type being analyzed as well as the timing of the sample collection. It is important to take steps to protect against degradation when storing the sample and extracting the protein. Also, depending on the technology used to explore proteomics, the protein structure may need to be unwound, or reduced to peptides.
- Metabolomic studies are time and tissue dependent. It is important to consider not only the protein that the metabolite is the product of, but also how that protein may have reacted with other proteins to produce the metabolites and how this interaction may affect the pathophysiology of disease.

- Microbiomic studies involve evaluating the genetic information of all of the microorganisms found within a sample and using that information to identify what organism and how much of it is in a sample. It is important to consider the organ or tissue type being explored as the microbiome will differ according to body structure being interrogated. It is also important to consider the type of microorganism (e.g. bacteria, virus, fungi, or all) under analysis.

For all approaches addressed in this chapter, it is possible to focus on a targeted number of genes or pathway of genes (candidate gene approach) where a priori knowledge is used to select genes on which to focus. Greater breadth and focus on the entire -ome (genome, transcriptome, epigenome, proteome, metabolome, microbiome) has the advantage that no a priori knowledge is implied and instead one evaluates all possibilities allowing for novel biological connections to be established. However, sample size and power are impacted by the number of analyses with greater breadth requiring larger sample sizes.

References

Baumgartel K, Zelazny J, Timcheck T, Snyder C, Bell M, Conley YP. Molecular genomic research designs. Annu Rev Nurs Res. 2011;29:1–26.

Bujak R, Struck-Lewicka W, Markuszewski MJ, Kaliszan R. Metabolomics for laboratory diagnostics. J Pharm Biomed Anal. 2015;113:108–20.

Chaturvedi S, Bhatnagar P, Bean CJ, Steinberg MH, Milton JN, Casella JF, DeBaun MR. Genome-wide association study to identify variants associated with acute severe vaso-occlusive pain in sickle cell anemia. Blood. 2018;130(5):686–8.

Cheek DJ. What you need to know about pharmacogenomics. Nursing. 2013;43:44–8. quiz 49.

Ciampi de Andrade D, Maschietto M, Galhardoni R, Gouveia G, Chile T, Victorino Krepischi AC, Brentani HP. Epigenetics insights into chronic pain. Pain. 2017;158(8):1473–80.

Clish CB. Metabolomics: an emerging but powerful tool for precision medicine. Mol Case Stud. 2015;1(1):a000588. 1–6.

Crick F. Central dogma of molecular biology. Nature. 1970;227:561–3.

Farajzadeh A, Bathaie SZ. Different pain states of trigeminal neuralgia make significant changes in the plasma proteome and some biochemical parameters: a preliminary cohort study. J Mol Neurosci. 2018;66:524–34.

Ferranti EP, Grossmann R, Starkweather A, Heitkemper M. Biological determinants of health: genes, microbes, and metabolism exemplars of nursing science. Nurs Outlook. 2017;65:506–14.

Fessele KL, Wright F. Primer in genetics and genomics, article 6 : basics of epigenetic control. Biol Res Nurs. 2018;20:103–10.

Goodrich JK, Di Rienzi SC, Poole AC, Koren O, Walters WA, Caporaso JG, Knight R, Ley RE. Primer conducting a microbiome study. Cell. 2014;158:250–62.

Harrington CA, Rosenow C, Retief J. Monitoring gene expression using DNA microarrays. Curr Opin Microbiol. 2000;3:285–91.

Integrative T. Perspective the integrative human microbiome project: dynamic analysis of microbiome-host Omics profiles during periods of human health and disease. Cell Host Microbe. 2014;16:276–89.

Johnson NL, Giarelli E, Lewis C, Rice CE. Genomics and autism spectrum disorder. J Nurs Scholarsh. 2013;45:69–78.

Jones KD, Gelbart T, Whisenant TC, Waalen J, Mondala TS, Iklé DN, Author R. Genome-wide expression profiling in the peripheral blood of patients with fibromyalgia. Clin Exp Rheumatol. 2016;34(2):89–98.

Kallenborn-Gerhardt W, Möser CV, Lorenz JE, Steger M, Heidler J, Scheving R, Schmidtko A. Rab7 - A novel redox target that modulates inflammatory pain processing. Pain. 2017;158(7):1354–65.

Khoonsari PE, Ossipova E, Lengqvist J, Svensson CI, Kosek E, Kadetoff D, Lampa J. The human CSF pain proteome. J Proteomics. 2019;190:67–76.

Kurella M, Hsiao LL, Yoshida T, Randall JD, Chow G, Sarang SS, Jensen RV, Gullans SR. DNA microarray analysis of complex biologic processes. J Am Soc Nephrol. 2001;12:1072–8.

Lee H, Gill J, Barr T, Yun S, Kim H. Primer in genetics and genomics, article 2-advancing nursing research with genomic approaches. Biol Res Nurs. 2017;19:229–39.

Lyon DE, Starkweather A, Yao Y, Garrett T, Kelly DL, Menzies V, et al. Pilot study of metabolomics and psychoneurological symptoms in women with early stage breast cancer. Biol Res Nurs. 2018;20(2):227–36.

Maki KA, Diallo AF, Lockwood MB, Franks AT, Green SJ, Joseph PV. Considerations when designing a microbiome study : implications for nursing science. Biol Res Nurs. 2018;21(2):125–41. https://doi.org/10.1177/1099800418811639.

Mccue ME, Mccoy AM, Mccue ME. The scope of big data in one medicine: unprecedented opportunities and challenges. Front Vet Sci. 2017;4:1–23.

Meissner A, Van Der Plas AA, Van Dasselaar NT, Deelder AM, Van Hilten JJ, Mayboroda OA. 1H-NMR metabolic profiling of cerebrospinal fluid in patients with complex regional pain syndrome-related dystonia. Pain. 2014;155(1):190–6.

Noordewier MO, Warren PV. Gene expression microarrays and the integration of biological knowledge. Trends Biotechnol. 2001;19:412–5.

Patti GJ, Yanes O, Siuzdak G. Innovation: metabolomics: the apogee of the omics trilogy. Nat Rev Mol Cell Biology. 2012;13:263–9.

Ramesh D, D'Agata A, Starkweather AR, Young EE. Contribution of endocannabinoid gene expression and genotype on low back pain susceptibility and chronicity. Clin J Pain. 2018;34(1):8–14.

Shoskes DA, Wang H, Polackwich AS, Tucky B, Altemus J, Eng C. Analysis of gut microbiome reveals significant differences between men with chronic prostatitis/chronic pelvic pain syndrome and controls. J Urol. 2016;196(2):435–41.

Sukenaga N, Ikeda-Miyagawa Y, Tanada D, Tunetoh T, Nakano S, Inui T, Hirose M. Correlation between DNA methylation of TRPA1 and chronic pain states in human whole blood cells. Pain Med. 2016;17(10):1906–10.

Trivedi DK, Hollywood KA, Goodacre R. New horizons in translational medicine metabolomics for the masses: the future of metabolomics in a personalized world. New Horizons Transl Med. 2017;3:294–305.

Ursell LK, Metcalf JL, Parfrey LW, Knight R. Defining the human microbiome. Nutr Rev. 2012;70(Suppl 1):S38–44.

Wang Z, Gerstein M, Snyder M. RNA-Seq: a revolutionary tool for transcriptomics. Nat Rev Genet. 2009;10:57–63.

Wishart DS. Emerging applications of metabolomics in drug discovery and precision medicine. Nat Rev Drug Discov. 2016;15:473–84.

Pain Physiology and the Neurobiology of Nociception

3

Cynthia L. Renn, Susan G. Dorsey, and Mari A. Griffioen

Contents

Pain is the primary symptom for many diseases and the number one reason that patients seek treatment from a healthcare provider (James 2013; Henschke et al. 2015; Kroenke 2003; Kroenke et al. 2013). In an effort to draw attention to the dilemma of pain and increase federal funding for research, treatment, and education, Congress declared the years 2000 to 2010 to be the Decade of Pain Control and Research (Nelson 2003). Yet, despite the many advances that have been made since the congressional declaration, pain continues to be a leading healthcare problem in the USA. According to the Institute of Medicine (IOM) report on Relieving Pain in America (Institute of Medicine 2011), chronic pain is a public health epidemic that impacts over 116 million Americans at a cost greater than $600 billion annually for

C. L. Renn (✉) · S. G. Dorsey
Department of Pain and Translational Symptom Science,
University of Maryland School of Nursing, Baltimore, MD, USA
e-mail: Renn@umaryland.edu; SDorsey@umaryland.edu

M. A. Griffioen
University of Delaware School of Nursing, Newark, DE, USA
e-mail: mgriffi@udel.edu

© Springer Nature Switzerland AG 2020
S. G. Dorsey, A. R. Starkweather (eds.), *Genomics of Pain and Co-Morbid Symptoms*, https://doi.org/10.1007/978-3-030-21657-3_3

healthcare expenses and decreased work productivity. It is estimated that up to 40% of adults in the USA suffer from chronic pain (Von Korff et al. 2016), nearly 20% of patient visits to a healthcare provided are due to pain, and nearly two thirds of chronic pain patients report having the pain for more than 5 years and limited ability to perform activities of daily living (Mäntyselkä et al. 2001; CDC 2018). In addition to the financial burden related to chronic pain, patients and caregivers suffer from many intangible costs related to comorbidities including depression, anxiety, relational stress, and a decreased quality of life (Dahan et al. 2014).

Chronic pain is well-known to be a comorbidity of various diseases and injuries. However, chronic pain has also been identified as a disease in and of itself with its own established definitions and taxonomy (Merskey et al. 1986, 1994; Tracey and Bushnell 2009; Williams and Craig 2016). Thus, it is imperative that chronic pain conditions be diagnosed accurately treated appropriately (Kapur et al. 2014). While the goal of pain management is often to achieve near-total analgesia, this goal is not always realistic for chronic pain due to the ineffectiveness of traditional pharmacological therapies (Curatolo and Bogduk 2001; von Korff et al. 2011). This is largely due to the lack of an identifiable source for chronic pain and the variability in patients' responses to analgesic medications (Kapur et al. 2014). Therefore, it is vitally important to increase our understanding of the pathophysiology, mechanisms, and risk factors of pain so that new therapeutic strategies can be developed to improve chronic pain management and quality of life for patients (Van Hecke et al. 2014; van Hecke et al. 2013). This review provides an introduction to the physiology of pain.

3.1 Pain Defined

When we suffer an injury, it is accompanied by a complex and unpleasant sensory phenomenon that results from synergistic physiological, psychological, and emotional processes (Steeds 2016). While we all recognize this phenomenon as pain, it is a multifaceted, subjective experience that is unique for each individual (Argoff et al. 2009). In addition to the physiological, psychological, and emotional processes that constitute pain, the intensity of the pain experience can be affected by the circumstance of the injury. This phenomenon was described by Beecher (1956) after his study of 150 civilian men and 150 soldiers showed that the soldiers with severe wounds required less morphine to manage their pain than the civilians with similar types of traumatic of injuries. Because of this complexity, the pain experience is very hard to define. The subjectivity of the pain experience led McCaffery to define pain as "whatever the experiencing person says it is, existing whenever the experiencing person says it does" (McCaffery and Beebe 1989). However, this definition does little to incorporate the underlying neurophysiological mechanisms of the pain experience. As pain was gaining a larger footprint in the scientific research arena, the International Association for the Study of Pain (IASP) developed their definition of pain as "an unpleasant sensory and emotional experience associated with actual or potential tissue damage or described in terms of such damage"

(Merskey and Bogduk 1986). This definition goes beyond the subjective "experience" of pain to include the sensory aspects of pain that are related to tissue damage. The IASP, in seeking to improve their definition, recently formed a task force that revised the definition of pain to be "an aversive sensory and emotional experience typically caused by, or resembling that caused by, actual or potential tissue injury" (IASP Definition of Pain Task Force 2019). This new definition captures key aspects of the pan experience; though, it remains fairly subjective. It is currently out for comment from the pain community and may be further revised before the final definition will be published. Thus, despite the efforts of McCaffery, the IASP, and countless others to date, a concrete definition of pain remains elusive.

3.2   Categories of Pain

Though we lack a clear and concrete definition, pain can be classified into by two commonly known and broad categories related to the duration of the experience: acute pain and chronic pain. Acute pain can be thought of as good pain that performs a protective biological function. Acute pain is a warning mechanism that alerts the brain of potential harm and triggers a protective response (Mifflin and Kerr 2014; Crofford 2015). Acute pain is localized to the site of injury and has an identifiable cause (Mifflin and Kerr 2014). In the clinical setting, acute pain is viewed as a symptom of many disease processes in addition to injuries. Acute pain can typically be well-managed with standard analgesic routines and, as the disease or injury resolves, the acute pain will also resolve after a short duration. However, in some cases the pain will persist beyond tissue healing, at which point it becomes a chronic condition that can be very debilitating for the patient (Fayaz et al. 2016).

While acute pain can be thought of a s good pain or protective, chronic pain is a very complex phenomenon that is considered bad pain (Beard and Aldington 2011). According to the International Association for the Study of pain, chronic pain serves no biological function and often has no identifiable cause as it persists beyond the normal tissue healing time (Harstall and Ospina 2003). From a clinical perspective, chronic pain exhibits characteristics that are more like those of a disease rather than being merely a symptom associated with a disease. It does not respond well to traditional therapies; thus, requiring complex treatment strategies for its management and often is associated with a poor prognosis (Argoff et al. 2009).

It is well-established that pain is a subjective experience with characteristics that are unique to each individual. Therefore, a concrete definition of pain may continue to elude us for some time. However, the body of knowledge related to the physiological mechanisms underlying acute pain and the development of chronic pain continues to grow through ongoing research, both in the basic sciences and in the clinical setting. This increase in the understanding of the pathophysiology of pain is leading to the development of new therapeutic strategies and improved clinical management of pain.

Pain is a multidimensional sensory process that involves ascending pathways that transmit information from the peripheral nervous system to the brain and descending pathways that transmit modulatory information from the brain to the spinal cord where it alters the ascending transmission process. This chapter will give an overview of the function of the ascending and descending pathways and the biological processes that occur at key points along those pathways.

3.2.1 Ascending Pain Pathway

The pain process encompasses a complex system that starts with encoding a stimulus in the peripheral tissues and then transmitting information about that stimulus in the form of actions potentials traveling along neurons to the brain where the information is interpreted (Fig. 3.1). Each phase in the ascending transmission process represents an opportunity to identify new therapeutic targets to inhibit the ascending transmission of pain information. These key components of ascending pain transmission are described below.

Nociception The first step in the pain process involves activation of specialized free nerve endings (nociceptors) in the peripheral tissues by a noxious stimulus (Voscopoulos and Lema 2010), which is a stimulus that is potentially tissue-damaging (Sneddon 2018). Receptors in the cell membrane of the nociceptor transduce the noxious stimulus into electrical signals (action potentials) that are transmitted along the axon of the nociceptor to the spinal cord (Steeds 2016). Nearly all body tissues are innervated with nociceptors, which are neurons that have a high

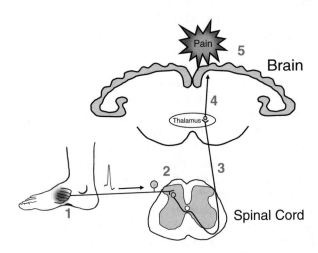

Fig. 3.1 The ascending pain pathway transmits nociceptive information from (1) injured tissues in the periphery as action potentials traveling along primary afferent fibers to (2) the dorsal horn of the spinal cord. The central terminal of the primary afferent fibers synapse with (3) second order neurons that project upward through the spinal cord to higher regions of the CNS in the brainstem and diencephalon. The second order neurons synapse with (4) third order neurons in the higher CNS structures such as the thalamus that transmit the nociceptive information to (5) the cerebral cortex of the brain where the signal is interpreted as pain

threshold of activation and will only respond to tissue-damaging thermal (hot or cold), chemical (endogenous and/or exogenous), or mechanical (force or pressure) stimuli (Fig. 3.2) (Voscopoulos and Lema 2010). The cell damage that occurs with tissue injury results in the release of intracellular contents that trigger the inflammatory process and the attraction of inflammatory cells to the area. Both the damaged cells and the inflammatory cells release chemical mediators that can activate nociceptors (Scholz and Woolf 2002). These chemical mediators include chemokines, cytokines, nerve growth factor, potassium, hydrogen ions, bradykinin, endothelins, interleukins, serotonin, histamine, prostaglandins, among others (Voscopoulos and Lema 2010; Scholz and Woolf 2002; Millan 1999). One area of pain research is focused on investigating ways to inhibit these chemical mediators to reduce the nociceptive input from an area of injury. All tissues in the body are innervated by non-nociceptive mechanoreceptor, thermoreceptor, and chemoreceptor neurons that respond to non-noxious stimuli with an intensity below the tissue-damaging level (Argoff et al. 2009). There are two sensory fiber types in primary afferent nerves that transmit nociceptive information, A-delta fibers and C fibers. A-delta fibers have small diameter, myelinated axons with fast transmission of action potentials. Activation of A-delta fibers produces the sharp first pain that is felt immediately after injury (Millan 1999; Beissner et al. 2010). C fibers have small diameter, unmyelinated axons with slow transmission of action potentials. Activation of C fibers produces the burning second pain that develops after the first pain subsides (Millan 1999; Beissner et al. 2010). Nociceptors terminate centrally in the dorsal horn of the

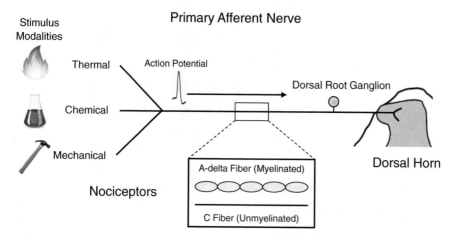

Fig. 3.2 Tissue-damaging stimuli (thermal, chemical, or mechanical) are transduced in peripheral tissues by cell surface receptors in the cell membrane of nociceptors. The transduction of the stimuli generates action potentials that are propagated toward the spinal cord along the axons of nociceptors that are in primary afferent nerves. Nociceptors are one of two fiber types: A-Delta and C fibers (inset). The A-Delta fibers propagate action potentials rapidly along myelinated axons and the C fibers, with unmyelinated axons, propagate action potential more slowly. The dorsal root ganglion is a structure located near the spinal cord that contains the cell bodies of all of the sensory neurons found within a primary afferent nerve

spinal cord where the action potentials trigger the release neurotransmitters from the nociceptor's terminal onto spinal dorsal horn neurons to continue the transmission of the nociceptive signal (Woolf and Ma 2007).

Spinal Dorsal Horn The primary afferent neurons synapse with second order neurons in the dorsal gray matter of the spinal cord. On cross-section, the gray matter (neuronal cell bodies) of the spinal cord is in the shape of a butterfly with two dorsal horns (sensory functions) and two ventral horns (motor functions) (Fig. 3.3). In 1952, Rexed published a detailed analysis of the cytoarchitecture of the spinal cord, which showed that the spinal gray matter can be divided into tens laminae that are identified with the Roman numerals I–IX from dorsal to ventral with lamina X surrounding the central canal (Rexed 1952). Primary afferent neurons terminate predominantly in the superficial laminae I and II and the deeper lamina V (Steeds 2016) and synapse either directly with the second order neuron or indirectly via interneurons. The primary afferent neurons transmit the nociceptive message by releasing neurotransmitters, such as the excitatory amino acid glutamate, substance P, and calcitonin gene related peptide. The neurotransmitters bind with receptors in the cell

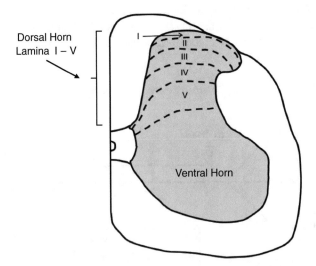

Fig. 3.3 The spinal gray matter is a butterfly shaped area in the center of the cord that contains the cell bodies of second order neurons in the dorsal horn and efferent motor neurons in the ventral horn. Surrounding the gray matter is white matter that contains the myelinated axons of the ascending second order neurons and descending motor neurons. The dorsal horn receives sensory information from the periphery and transmits the sensory information to higher centers in the central nervous system via second order neurons. The ventral horn receives motor information from the brain and transmits the motor information out to the periphery. The gray matter of both the dorsal and ventral horns is divided into ten laminae (layers). The dorsal horn encompasses laminae I-V, starting from the dorsal surface and moving ventrally. Nociceptive sensory input is received primarily in laminae I, II, and V. Whereas, innocuous sensory input is received primarily in laminae III and IV

membrane of the second order neurons and the nociceptive signal is transduced into action potentials that travel along the axon of the second order neurons to higher regions of the central nervous system (CNS).

Ascending Tracts   The second order or projection neurons are arranged in ascending tracts that transmit the nociceptive information from the spinal cord to the brainstem and brain. Each tract terminates in a supraspinal structure in the brainstem and diencephalon, including the medullary reticular formation, periaqueductal gray, and thalamus, among others (Feizerfan and Sheh 2015). The function of the ascending tracts is merely to transmit nociceptive information to higher regions of the CNS. When the second order neurons terminate in the target region or structure, they synapse with third order neurons that transmit the nociceptive signal cortical and limbic structures where the nociceptive signal is decoded as pain (Millan 1999).

The spinothalamic tract is the most well-described of the ascending tracts, which is thought to transmit information regarding sensations of pain, temperature, and touch (Millan 1999; Boadas-Vaello et al. 2016). The spinomesencephalic tract is a second ascending pain transmission pathway that terminates in the periaqueductal gray (PAG) and nucleus cuneiformis, regions of the midbrain among others (Steeds 2016; Wiberg et al. 1987; Keay et al. 1997). A third key ascending tract is the spinoreticular tract, which terminates in the reticular formation of the medulla (Steeds 2016; Millan 1999). In addition to their primary targets, the ascending tracts also send collateral projections to synapse with other regions within the brainstem. These three ascending tracts are critical pathways for the transmission of nociceptive information to the higher regions of the CNS; however, these three pathways are representative of the ascending system and do not constitute a complete list of all ascending sensory pathways. There are other pathways that transmit nociceptive information from the spinal cord to the brainstem as well.

Thalamus   The thalamus is considered the main relay center for processing and integrating somatosensory and nociceptive information for transmission to the cerebral cortex where the signals are interpreted (Boadas-Vaello et al. 2016). In this role, the thalamus receives signal via spinothalamic tract, as well as through collateral projections from the other ascending tracts transmitting nociceptive information. The thalamus processes and integrates the information about the type, temporal pattern intensity, and topographic localization of a stimulus and then encodes it for transmission to cortical and limbic structures for interpretation (Millan 1999; Boadas-Vaello et al. 2016).

Cerebral Cortex   When the nociceptive information is transmitted to the cerebral cortex, it is then that it is processed for cognitive and emotional interpretation as a signal ascending from a noxious stimulus and identified as pain (Millan 1999; Julius and Basbaum 2001; Perini et al. 2013). The nociceptive signal is spread to a number of different cortical sites, including the thalamus and limbic structures, that are linked by a complex neuronal network (Steeds 2016; Perini et al. 2013; Sherman and Guillery 1996). Communication between the structures within this network is necessary for the identification of the sensory-discriminative (perception of the intensity, location, duration, temporal pattern and quality of noxious stimuli) and motivational-affective (relationship between pain and mood, attention, coping, tolerance, and rationalization) aspects of the pain experience (Feizerfan and Sheh 2015).

3.2.2 Descending Pain Modulation

The idea that higher centers in the CNS exert a modulatory influence over pain was first introduced by Head and Holmes (1911). Head and Holmes theory was confirmed several decades later when several studies reported that supraspinal sites exert tonic inhibitory control of spinal dorsal horn neurons to inhibit the transmission of ascending nociceptive input (Hagbarth and Kerr 1954; Carpenter et al. 1965; Wall 1967). Thus, much effort has been put into understanding the mechanisms involved in this modulatory process as potential therapeutic targets to manage acute and chronic pain.

Studies investigating the roles of the various supraspinal structures involved in the modulation of nociception found that the CNS has well-defined descending pathways that convey the brainstem control of nociceptive transmission. The descending pathways exert their modulatory effects in the spinal dorsal horn at the synapse between the primary afferents and second order neurons (Fig. 3.4a). In the absence of noxious input, there is minimal to no descending modulation of sensory input. Thus, there are no changes in the amount of neurotransmitters that are released from the central terminal of the primary afferent, in the sensitivity of the post-synaptic cell surface receptors on the second order neuron, or in the number of actions potential generated (Fig. 3.4b). Early in the pain process, the descending modulatory effect is facilitatory, increasing nociceptive transmission by either increasing the release of neurotransmitters from the primary afferent terminals (Fig. 3.4c) or by increasing the sensitivity of post-synaptic cell surface receptors on the second order neuron (Fig. 3.4e), both of which increase the number of action potentials generated (Ren and Dubner 2002; Vanegas and Schaible 2004; Zhuo 2017). Later in the pain process, the descending modulatory effect becomes inhibitory to decrease nociceptive transmission by either inhibiting neurotransmitter release from the primary afferent terminal (Fig. 3.4d) or by reducing the activation of post-synaptic cell surface receptors on the second order neurons (Fig. 3.4f), both of which reduce or eliminate the generation of action potentials (Ren and Dubner 2002; Millan 2002; Heinricher et al. 2009; Kwon et al. 2014). Several brainstem sites play a key role in the descending modulation of nociception. These include the periaqueductal gray (PAG), the locus coeruleus (LC), the parabrachial nucleus, and the rostral ventromedial medulla (RVM), among others (Millan 2002; Kwon et al. 2014).

Periaqueductal Gray It is well-established that the PAG is a major component of the pain modulatory circuitry. The PAG is a midline gray matter structure comprised of cell bodies surrounding the cerebral aqueduct in the midbrain that exerts a strong descending inhibitory effect (Kwon et al. 2014). The ability of the PAG to inhibit

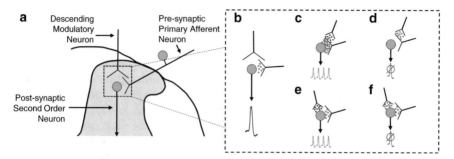

Fig. 3.4 Descending modulatory neurons exert their effects in the dorsal horn on the synapse between a primary afferent neuron and a second order neuron (**a**). In the normal resting state, there is minimal descending modulation exerted with no alteration in the release of neurotransmitters from the primary afferent or function of the second order neuron (**b**). In response to pain, descending modulation can alter the function of the presynaptic terminal of the primary afferent fiber leading to increased neurotransmitter release and action potential propagation during facilitation (**c**) or stop/reduce neurotransmitter release and action potential propagation during inhibition (**d**). Descending modulation can also alter the function cell surface receptors in the membrane of the second order neuron leading to increased receptor activity and action potential propagation during facilitation (**e**) or stop/reduce receptor activity and action potential propagation during inhibition (**f**)

nociception was first reported by Reynolds (1969) after he discovered that electrical stimulation of the ventrolateral PAG in rats produced sufficient analgesia to allow him to perform abdominal surgery in the absence of any other anesthesia (Reynolds 1969). Liebeskind confirmed Reynolds' findings in subsequent studies in both the rat and the cat (Mayer et al. 1971; Oliveras et al. 1974). However, the PAG has very few projection neurons that extend directly to the spinal dorsal horn (Millan 2002). Instead, the PAG exerts its modulatory effect indirectly via neurons projecting to other structures within the brainstem, such as the rostral ventromedial medulla (RVM), parabrachial nucleus, locus coeruleus, and the A5 and A7 noradrenergic cell groups (Steeds 2016; Millan 2002; Kwon et al. 2014).

Rostral Ventromedial Medulla The RVM is a key component in the descending pain modulatory circuitry and serves as the final relay for the descending pain modulatory signals (Zhuo 2017; Ossipov et al. 2014). The RVM has been the focus of an extensive number of studies. In addition to being the final relay of modulatory signals from higher structures in the CNS, the RVM also exerts its own modulatory effects (Heinricher et al. 2009; Ossipov et al. 2014). The RVM occupies a large area of the medulla and can be subdivided into three regions: (1) the midline nucleus raphe magnus (NRM); (2) the nucleus reticularis gigantocellularis pars alpha (GiA); and (3) the nucleus paragigantocellularis lateralis (LPGi) (Ren and Dubner 2002). Projection neurons originating in all three regions of the RVM are found in all levels of the spinal cord and form the majority of neurons projecting to the spinal dorsal horn. All structures included in the descending modulatory circuitry are important; however, because the PAG has very few direct projections to the spinal dorsal horn, the PAG to RVM projection is critical for the PAG to exert its descending modulatory effect (Heinricher et al. 2009; Kwon et al. 2014).

Biphasic Modulation As discussed above, descending pain modulation is both inhibitory and facilitatory. The RVM is the primary site of the opposing modulatory effects and is critical for maintaining the balance inhibition or facilitation. The modulatory effect of activating the RVM depends on the intensity of the stimulus triggering the activation (Vanegas and Schaible 2004). The ability of the RVM to exert both facilitatory and inhibitory effects is the result of two classes of neurons intrinsic to the RVM, on-cells and off-cells (Vanegas and Schaible 2004). Activation of off-cells produces an inhibitory effect and activation of on-cells produces a facilitatory effect (Vanegas and Schaible 2004; Heinricher et al. 2009). While on- and off-cells are distinct populations within the RVM, an anatomical and neurochemical differentiation of the cells has not been identified. They can only be identified using electrophysiological recording of their activity. The balance between descending inhibition and facilitation regulates the net modulatory effect on nociception.

References

Argoff CE, Albrecht P, Irving G, Rice F. Multimodal analgesia for chronic pain: rationale and future directions. Pain Med. 2009;10:S53–66. https://doi.org/10.1111/j.1526-4637.2009.00669.x.

Beard DJ, Aldington D. Chronic pain after trauma. Trauma. 2011;14:57–66.

Beecher HK. Relationship of significance of wound to pain experienced. J Am Med Assoc. 1956;161:1609–13.

Beissner F, Brandau A, Henke C, Felden L, Baumgärtner U, Treede RD, Oertel BG, Lötsch J. Quick discrimination of Adelta and C fiber mediated pain based on three verbal descriptors. PLoS One. 2010;5:e12944. https://doi.org/10.1371/journal.pone.0012944.

Boadas-Vaello P, Castany S, Homs J, Álvarez-Pérez B, Deulofeu M, Verdú E. Neuroplasticity of ascending and descending pathways after somatosensory system injury: reviewing knowledge to identify neuropathic pain therapeutic targets. Spinal Cord. 2016;54:330–40.

Carpenter D, Engberg I, Lundberg A. Differential Supraspinal control of inhibitory and excitatory actions from, the FRA to ascending spinal pathways. Acta Physiol Scand. 1965;63:103–10.

CDC. Prevalence of chronic pain and high-impact chronic pain among adults — United States, 2016. Morb Mortal Wkly Rep. 2018;67:1001–6.

Crofford LJ. Chronic pain: where the body meets the brain. Trans Am Clin Climatol Assoc. 2015;126:167–83.

Curatolo M, Bogduk N. Pharmacologic pain treatment of musculoskeletal disorders: current perspectives and future prospects. Clin J Pain. 2001;17:25–32.

Dahan A, Van Velzen M, Niesters M. Comorbidities and the complexities of chronic pain. Anesthesiology. 2014;121:675–7.

Fayaz A, Croft P, Langford RM, Donaldson LJ, Jones GT. Prevalence of chronic pain in the UK: a systematic review and meta-analysis of population studies. BMJ Open. 2016;6:1–12.

Feizerfan A, Sheh G. Transition from acute to chronic pain. Contin Educ Anaesth Crit Care Pain. 2015;15:98–102.

Hagbarth KE, Kerr DI. Central influences on spinal afferent conduction. J Neurophysiol. 1954;17:295–307. https://doi.org/10.1152/jn.1954.17.3.295.

Harstall C, Ospina M. International Association for the Study of Pain. Pain Clin Update. 2003;11:1–4.

Head H, Holmes G. Sensory disturbances from cerebral lesions. Brain. 1911;34:102–254. https://doi.org/10.1093/brain/34.2-3.102.

Heinricher MM, Tavares I, Leith JL, Lumb BM. Descending control of nociception: specificity, recruitment and plasticity. Brain Res Rev. 2009;60:214–25.

Henschke N, Kamper SJ, Maher CG. The epidemiology and economic consequences of pain. Mayo Clin Proc. 2015;90:139–47.

IASP Definition of Pain Task Force. IASP's proposed new definition of pain released for comment. 2019.

Institute of Medicine. Relieving pain in America : a blueprint for transforming prevention. Care Educ Res. 2011;181:397–9. https://doi.org/10.7205/MILMED-D-16-00012.

James S. Human pain and genetics: some basics. Br J Pain. 2013;7:171–8.

Julius D, Basbaum AI. Molecular mechanisms of nociception. Nature. 2001;413:203–10.

Kapur BM, Lala PK, Shaw JLV. Pharmacogenetics of chronic pain management. Clin Biochem. 2014;47:1169–87.

Keay KA, Feil K, Gordon BD, Herbert H, Bandler R. Spinal afferents to functionally distinct periaqueductal gray columns in the rat: an anterograde and retrograde tracing study. J Comp Neurol. 1997;385:207–29. https://doi.org/10.1002/(SICI)1096-9861(19970825)385:2<207::AID-CNE3>3.0.CO;2-5.

Kroenke K. Patients presenting with somatic complaints - epidemiology. Int J Methods Psychiatr Res. 2003;12:34–43.

Kroenke K, Krebs E, Wu J, et al. Stepped care to optimize pain care effectiveness (SCOPE) trial study design and sample characteristics. Contemp Clin Trials. 2013;34:270–81.

Kwon M, Altin M, Duenas H, Lilly E, Neuroscience EL, Lilly E. The role of descending inhibitory pathways on chronic pain modulation and clinical implications. Pain Pract. 2014;14:656–67.

Mäntyselkä P, Kumpusalo E, Ahonen R, Kumpusalo A, Kauhanen J, Viinamäki H, Halonen P, Takala J. Pain as a reason to visit the doctor: a study in Finnish primary health care. Pain. 2001;89:175–80.

Mayer DJ, Wolfle TL, Akil H, Carder B, Liebeskind JC. Analgesia from electrical stimulation in the brainstem of the rat. Science. 1971;174:1351–4.

McCaffery M, Beebe A. Pain: clinical manual for nursing practice. St. Louis: Mosby; 1989.

Merskey H, Bogduk N. Classification of chronic pain. Descriptions of chronic pain syndromes and definitions of pain terms. Prepared by the international association for the study of pain, subcommittee on taxonomy. Pain Suppl. 1986;3:S1–226.

Merskey H, Bond MR, Bonica JJ, et al. Classification of chronic pain. Pain. 1986;24:S1–S226.

Merskey H, Addison RG, Beric A, et al. Classification of chronic pain: Descriptions of chronic pain syndromes and definitions of pain terms. Second. Classif Chronic Pain. 1994. https://doi.org/10.1002/ana.20394.

Mifflin KA, Kerr BJ. The transition from acute to chronic pain: understanding how different biological systems interact. Can J Anesth. 2014;61:112–22.

Millan MJ. The induction of pain: an integrative review. Prog Neurobiol. 1999;57:1–164.

Millan MJ. Descending control of pain. Prog Neurobiol. 2002;66:355–474.

Nelson R. Decade of pain control and research gets into gear in USA. Lancet. 2003;362:1129. https://doi.org/10.1016/s0140-6736(03)14505-9.

Oliveras JL, Besson J-MM, Guilbaud G, Liebeskind JC. Behavioral and electrophysiological evidence of pain inhibition from midbrain stimulation in the cat. Exp Brain Res. 1974;20:32–44.

Ossipov MH, Morimura K, Porreca F. Descending pain modulation and chronification of pain. Curr Opin Support Palliat Care. 2014;8:143–51.

Perini I, Bergstrand S, Morrison I. Where pain meets action in the human brain. J Neurosci. 2013;33:15930–9. https://doi.org/10.1523/JNEUROSCI.3135-12.2013.

Ren K, Dubner R. Descending modulation in persistent pain: an update. Pain. 2002;100:1–6.

Rexed B. The cytoarchitectonic organization of the spinal cord in the cat. J Comp Neurol. 1952;96:415–95.

Reynolds DV. Surgery in the rat during electrical analgesia induced by focal brain stimulation. Science. 1969;164:444–5.

Scholz J, Woolf CJ. Can we conquer pain? Nat Neurosci. 2002;5(Suppl):1062–7.

Sherman SM, Guillery RW. Functional organization of thalamocortical relays. J Neurophysiol. 1996;76:1367–95. https://doi.org/10.1152/jn.1996.76.3.1367.

Sneddon LU. Comparative physiology of nociception and pain. Physiology. 2018;33:63–73.

Steeds CE. The anatomy and physiology of pain. Surgery. 2016;34:55–9.

Tracey I, Bushnell MC. How neuroimaging studies have challenged us to rethink: is chronic pain a disease? J Pain. 2009;10:1113–20. https://doi.org/10.1016/j.jpain.2009.09.001.

van Hecke O, Torrance N, Smith BH. Chronic pain epidemiology and its clinical relevance. Br J Anaesth. 2013;111:13–8.

Van Hecke O, Austin SK, Khan RA, Smith BH, Torrance N. Neuropathic pain in the general population: a systematic review of epidemiological studies. Pain. 2014;155:654–62.

Vanegas H, Schaible HG. Descending control of persistent pain: inhibitory or facilitatory? Brain Res Rev. 2004;46:295–309.

von Korff M, Kolodny A, Deyo RA, Chou R. Long-term opioid therapy reconsidered. Ann Intern Med. 2011;155:325. https://doi.org/10.7326/0003-4819-155-5-201109060-00011.

Von Korff M, Scher AI, Helmick C, et al. United States national pain strategy for population research: concepts, definitions, and pilot data. J Pain. 2016;17:1068–80.

Voscopoulos C, Lema M. When does acute pain become chronic? Br J Anaesth. 2010;105:i69–85.

Wall PD. The laminar organization of dorsal horn and effects of descending impulses. J Physiol. 1967;188:403–23.

Wiberg M, Westman J, Blomqvist A. Somatosensory projection to the mesencephalon: an anatomical study in the monkey. J Comp Neurol. 1987;264:92–117. https://doi.org/10.1002/cne.902640108.

Williams ACDC, Craig KD. Updating the definition of pain. Pain. 2016;157:2420–3.

Woolf CJ, Ma Q. Nociceptors - noxious stimulus detectors. Neuron. 2007;55:353–64.

Zhuo M. Descending facilitation: from basic science to the treatment of chronic pain. Mol Pain. 2017;13:1744806917699212. https://doi.org/10.1177/1744806917699212.

Pre-Clinical Models of Pain

4

Angela R. Starkweather, Cynthia L. Renn,
and Susan G. Dorsey

Contents

Animal models have been used to study pain responses and mechanisms since the late nineteenth century, providing significant insight into pain variability, modulation, and genetic contributions of pain and analgesia. The benefit of studying animal models is numerous, including the ability to control intrinsic and extrinsic factors that influence pain as well as the application of experimental pain paradigms, behavioral testing and interrogation of mechanisms ranging from individual cells, and neural tissue to multiple systems that contribute to pain and pain outcomes.

Since the advent of using preclinical models to study pain, much has been learned about how the genetic landscape influences pain variability (Mogil 2012) and

A. R. Starkweather
University of Connecticut School of Nursing, Storrs, CT, USA
e-mail: Angela.Starkweather@uconn.edu

C. L. Renn · S. G. Dorsey (✉)
Department of Pain and Translational Symptom Science,
University of Maryland School of Nursing,
Baltimore, MD, USA
e-mail: Renn@umaryland.edu; SDorsey@umaryland.edu

© Springer Nature Switzerland AG 2020
S. G. Dorsey, A. R. Starkweather (eds.), *Genomics of Pain and Co-Morbid Symptoms*, https://doi.org/10.1007/978-3-030-21657-3_4

differences in pain processing across species, for instance, between mice and rats (LaCroix-Fralish et al. 2011). Despite the apparent complexity, well-designed and highly precise mechanistic studies can be conducted using transgenic animal models that allow specific over-production or knockout of specific molecules within a tissue of interest. This approach has been successful for identifying specific therapeutic targets, such as the case for controlled inhibition of peripheral tetrahydrobiopterin (BH4) synthesis to reduce chronic pain (Latremoliere and Costigan 2018). Using a reproducible pain type [inflammatory, neuropathic, visceral, muscle, joint or due to cancer, spinal cord injury, or postoperative incisional pain] that replicates the condition in humans is essential, as well as precise measurement of standardized pain behaviors. Use of multiple pain behavior measures, both reflexive and non-reflexive, as well as measuring avoidance of evoked stimuli to assess the unpleasantness of pain, is recommended to fully understand the multidimensional pain experience (Gregory et al. 2013). Several suggestions have been made to increase rigor and reproducibility in the use of animal models of pain including:

1. Using an animal model of the disease state that replicates the clinical pain context (face validity), including features of pain severity and provocation (spontaneous vs. evoked)
2. Integrating reflexive and non-reflexive pain behaviors with other patient-centered endpoints such as operant learning measures, spontaneous nocifensive behaviors, quality of life, and physical activity measures
3. Providing descriptions of animal environmental conditions, including housing, diet, access to, and use of an enriched environment (running wheel or other equipment)
4. Control of circadian rhythm with lighting and other environmental factors
5. Assessment of the estrous cycle
6. Ensuring adequate representation of each sex
7. Description of littermates, socialization, and breeding
8. Control of the life cycle during the experiment

These components are important for planning out and implementing the study so that there is as little artifact as possible from extraneous or environmental sources. This is particularly important when considering the relationship of stress and pain, as well as inflammatory load and pain. For instance, it has been shown that mice fed with a total western diet have more pronounced and significantly prolonged hypersensitivity after chronic pain induction using complete Freund's adjuvant (Totsch et al. 2016). In another study, rats fed a Standard American Diet for 20 weeks prior to induction of persistent inflammatory pain with Complete Freund's Adjuvant had significantly delayed recovery, significantly increased leptin and pro-inflammatory cytokines in peripheral blood serum, and increased microglial activation in the spinal cord (Totsch et al. 2017). As these physiologic changes can influence pain outcomes, ensuring adequate environmental control is critical to isolating the physiologic events involved in pain processing and direct consequences on behavior, function, and quality of life.

4.1   Reflexive Pain Testing

Reflexive tests mimic evoked pain responses in humans by measuring behavioral responses to noxious stimuli (thermal, mechanical, and electrical stimuli). These tests activate nociceptors to trigger stereotyped motor responses in an intact motor system. Primary hyperalgesia is measured by application of the test at the site of injury where sensitization of nociceptive primary afferents occurs. Secondary hyperalgesia can also be measured by application of the test at a site outside of the area of injury. If present, secondary hyperalgesia manifests as enhanced sensitivity in the absence of peripheral injury and is caused by sensitization of neurons in the spinal cord or higher in the central nervous system (Gregory et al. 2013).

Reflexive pain behavior to thermal, mechanical, or chemical stimuli has provided insight into the mechanisms associated with hyperalgesia and allodynia. Reproducibility is augmented by precise quantification of the stimulus type, evoked response and correlation with electrophysiological and biochemical signaling. Common thermal and mechanical tests are described in Table 4.1.

4.2   Non-reflexive Pain Testing

Paw elevation and paw licking in response to injection of an inflammatory compound, such as dilute formaldehyde (formalin), capsaicin, or mustard oil, can be used to measure spontaneous pain behavior (Cortright et al. 2008; Dubuisson and Dennis 1977). The concentration and volume of the noxious chemical substance injected must be carefully recorded. Limb guarding can also be quantified by assessing weight bearing during walking (Gabriel et al. 2007). Avoidance of evoked noxious stimuli can be used to measure the unpleasant or aversive components of pain with the Thermal Escape Test, Conditioned Place Avoidance (CPP), or Place Escape

Table 4.1 Reflexive tests

Thermal tests	
Tail flick test	Latency to remove the tail upon application of a heat stimulus (D'Amour and Smith 1941)
Hot-plate test	Latency to remove the paw from the hot-plate or lick the paw (Woolfe and MacDonald 1944)
Hargreaves test	Latency to remove the hindpaw from the heat stimuli; control and experimental paw can be tested in the same animal (Hargreaves et al. 1988)
Mechanical tests	
Von Frey filaments	Latency to remove paw from mechanical stimulus or response frequency to repeated application of a single Von Frey filament (Dixon 1980)
Dolorimeter; Calibrated forceps; Randall-Selitto analgesiometer	Animal is restrained and latency to withdraw or vocalization is measured upon application of pressure (Barton et al. 2007)

Avoidance Paradigm (PEAP). The Thermal Escape Test uses a two-chamber box in which the temperature of the floor is manipulated between the two chambers (Mauderli et al. 2000). Place preference, indicated by the chamber that the animal prefers to go to, or latency to withdraw from the platform with the noxious thermal stimulus, is recorded. Conditioned place avoidance uses a pre-conditioning phase in which animals learn to associate a specific chamber with injection of a noxious substance (Johansen et al. 2001). On the following test day, the preference for a specific chamber is assessed without the injection. The PEAP uses a two-chamber box to apply a noxious stimulus to the painful paw in one chamber, or the control paw in the opposite chamber (LaBuda and Fuchs 2000). The chamber preference is assessed, with expectation that the animal will shift to the chamber where the non-painful paw is stimulated.

Another paradigm that can be used is self-administration of analgesics as a behavioral response to inflammatory or nerve injury or preference for analgesics using Conditioned Place Preference (CPP) with a drug-paired chamber (Shippenberg et al. 1988). CPP uses a three-chamber box with a neutral middle compartment and two side chambers that are distinct in terms of the visual, textural, and olfactory cues (Sluka 2002). To condition the animal to a certain side chamber, an analgesic is administered in one of the two chambers so that during testing, the amount of time spent in the drug-paired chamber is measured and indicates a preference if it is greater than the other chambers.

A series of reflexive and non-reflexive tests are used to evaluate behavioral responses to primary and secondary hyperalgesia, spontaneous pain, and higher-order cortical evaluation of noxious stimuli or analgesia. The specific protocol is chosen to study pain characteristics of the target disease state. Various models are used to mimic the target disease state in animals, and some of the more common are described below.

4.3   Disease- or Injury-Specific Pain Models

Inflammation is a common condition associated with pain and can be induced with injection of noxious substances, such as capsaicin, carrageenan, or Complete Freund's Adjuvant (CFA) in the tail, paw, muscle, or joint. These agents, at specific doses and concentrations, cause reproducible patterns of sensitization that are useful for studying inflammatory-induced pain. Capsaicin specifically activates TRPV1-containing nociceptors and produces neurogenic inflammation with discrete zones of primary and secondary hyperalgesia (Sluka 2002). Carrageenan also causes increased sensitivity but results in initial acute inflammation with conversion to chronic inflammation by 2 weeks (Radhakrishnan et al. 2003). Although CFA also causes sensitization with hyperalgesia to thermal and mechanical stimuli, the results are more characteristic of chronic inflammatory pain.

In contrast to inflammatory pain, neuropathic pain is caused by direct nerve damage to the peripheral or central somatosensory system due to disease or injury, direct tissue trauma, metabolic derangements, exposure to neurotoxic substances, and/or

Table 4.2 Neuropathic pain models

Disease or injury	Mode of disease or injury	Expected responses
Cortical or thalamic pain	Microinjection of excitotoxic agents (such as picrotoxin or kainite) into somatosensory cortex or nuclei of the thalamus	Mechanical and thermal hyperalgesia, Spontaneous pain behavior
Spinal cord injury	Direct contusion, surgical lesion, laser irradiation, neurotransmitter-induced excitotoxicity	Heat and cold hyperalgesia Spontaneous pain behavior
Peripheral nerve injury	Ligating, transecting, or lesioning the peripheral nerve or distal branches	Decreased withdrawal thresholds to mechanical and thermal stimuli, spontaneous guarding of affected limb(s)
Peripheral neuropathy	Genetic modification, injection of streptozotocin, or exposure to toxins	Mechanical and thermal hyper- or hypoalgesia

ischemia. The disease or injury under study should cause reproducible results as demonstrated by the measurement of reflexive and non-reflexive behaviors (Table 4.2). As with all animal models of pain, the importance of setting up the experiment to consider elements of rigor and reproducibility is paramount, including sex as a biological variable (Mogil 2009).

Models of fibromyalgia using repeated intramuscular acid injections, hyperalgesic priming, or biogenic amine depletion have been used to study the pathophysiology of pain, effects of various treatments on behavior and functional status, and response to exercise (De Santana et al. 2013). As each of the fibromyalgia models are diverse in terms of induction and potential underlying mechanisms of fibromyalgia, they can be used to ascertain particular pathways involved. An animal model of rheumatoid arthritis uses injection of collagen type II antibodies or serum from K/BxN transgenic mice [which produce autoantibodies that recognize glucose-6-phosphate isomerase (GPI) and cause joint-specific inflammation] (Sluka et al. 2013).

Other pain models consider different treatments such as chemotherapy administration or surgery. Chemotherapy-induced peripheral neuropathy (CIPN) can be modeled by injecting the chemotherapeutic agent systemically to mimic the mechanical and thermal hyperalgesia seen in humans (Cata et al. 2006). To mimic postoperative pain induced by superficial and deep-tissue injury, a model has been used in which a longitudinal incision is made through the plantar aspect of the hindpaw or gastrocnemius (Pogatzki et al. 2002). Two models have been commonly used to mimic visceral pain; the acetic acid writhing and colorectal distension models. In the acetic acid writhing model, acetic acid is administered via intraperitoneal injection and the number of writhing events is counted (Koster et al. 1959). The colorectal distension model entails inserting a balloon into the colon where it is inflated and electromyographic activity of the abdominal muscles is recorded (Ness and Gebhart 1988). Many other protocols of disease or injury-specific pain models have been published, and each of these should be evaluated prior to selecting the best model for a particular experiment.

In six mice models of chronic pain, dorsal root ganglion (DRG) sensory neurons were used to identify dysregulated DRG genes (Bangash et al. 2018). The models included: (1) bone cancer pain using cancer cell injection in the intramedullary space of the femur; (2) neuropathic pain using partial sciatic nerve ligation; (3) osteoarthritis pain using mechanical joint loading; (4) chemotherapy-induced pain with oxaliplatin; (5) chronic muscle pain using hyperalgesic priming; and (6) inflammatory pain using intraplantar Complete Freund's Adjuvant. The microarray analyses using RNA isolated from DRG samples from the six models were compared with sham/vehicle treated controls. Of interest, the transcriptomic profiles from each model exhibited a unique set of altered transcripts, suggesting a distinct cellular response to different pain etiologies. While this study was informative in providing evidence on dysregulated genes across models, there were limitations on finding genes that were common across models because published RNAseq data is derived from pain-free mice, and the neuronal gene expression patterns may dynamically change over time. These are areas that are yet to be explored, but will undoubtedly add to our understanding of the transition from acute to chronic pain.

4.4   Conclusions

Animal pain models provide a means to precisely determine the mechanisms of pain at the behavioral level as well as general physiologic and tissue specific cellular level. These models have been pivotal to understanding pain as a multi-system phenomenon and the complex interactions that take place during the transition from acute to chronic pain. When using animal models, it is important to use an induction procedure that most closely resembles the pathophysiology of the disease or injury in humans, and careful consideration of each element of the experiment to address rigor and reproducibility.

References

Bangash MA, Alles SRA, Santana-Varela S, Millet Q, Sikandar S, de Clauser L, Ter Heegde F, Habib AM, Pereira V, Sexton JE, Emery EC, Li S, Luiz AP, Erdos J, Gossage SJ, Zhao J, Cox JJ, Wood JN. Distinct transcriptional responses of mouse sensory neurons in models of human chronic pain conditions. Wellcome Open Res. 2018;3:78.

Barton NJ, Strickland IT, Bond SM, Brash HM, Bate ST, Wilson AW, Chessell IP, Reeve AJ, Mcqueen DS. Pressure application measurement (PAM): a novel behavioural technique for measuring hypersensitivity in a rat model of joint pain. J Neurosci Methods. 2007;163:67–75.

Cata JP, Weng HR, Lee BN, Reuben JM, Dougherty PM. Clinical and experimental findings in humans and animals with chemotherapy-induced peripheral neuropathy. Minerva Anestesiol. 2006;72:151–69.

Cortright DN, Matson DJ, Broom DC. New Frontiers in assessing pain and analgesia in laboratory rodents. Expert Opin Drug Discovery. 2008;3:1099–108.

D'Amour FE, Smith DL. A method for determining loss of pain sensation. J Pharmacol Exp Ther. 1941;72:74–9.

De Santana JM, da Cruz K, Sluka KA. Animal models of fibromyalgia. Arthritis Res Ther. 2013;15:222.

Dixon WJ. Efficient analysis of experimental observations. Annu Rev Pharmacol Toxicol. 1980;20:441–62.

Dubuisson D, Dennis SG. The formalin test: a quantitative study of the analgesic effects of morphine, meperidine and brainstem stimulation in rats and cats. Pain. 1977;4:161–74.

Gabriel AF, Marcus MA, Honig WM, Walenkamp GH, Joosten EA. The CatWalk method: a detailed analysis of behavioral changes after acute inflammatory pain in the rat. J Neurosci Methods. 2007;163:9–16.

Gregory N, Harris AL, Robinson CR, Dougherty PM, Fuchs PN, Sluka KA. An overview of animal models of pain: disease models and outcome measures. J Pain. 2013;14(11):1–26. https://doi.org/10.1016/j.jpain.2013.06.008.

Hargreaves K, Dubner R, Brown F, Flores C, Joris J. A new and sensitive method for measuring thermal nociception in cutaneous hyperalgesia. Pain. 1988;32:77–88.

Johansen JP, Fields HL, Manning BH. The affective component of pain in rodents: direct evidence for a contribution of the anterior cingulate cortex. Proc Natl Acad Sci U S A. 2001;98:8077–82.

Koster R, Anderson M, DeBeer EJ. Acetic acid for analgesic screening. Fed Proc. 1959;18:412.

LaBuda CJ, Fuchs PN. A behavioral test paradigm to measure the aversive quality of inflammatory and neuropathic pain in rats. Exp Neurol. 2000;163:490–4.

LaCroix-Fralish ML, Austin JS, Zheng FY, Levitin DJ, Mogil JS. Patterns of pain: meta-analysis of microarray studies of pain. Pain. 2011;152:1888–98.

Latremoliere A, Costigan M. Combining human and rodent genetics to identify new analgesics. Neurosci Bull. 2018;34(1):143–55.

Mauderli AP, Acosta-Rua A, Vierck CJ. An operant assay of thermal pain in conscious, unrestrained rats. J Neurosci Methods. 2000;97:19–29.

Mogil JS. Animal models of pain: progress and challenges. Nat Rev Neurosci. 2009;10:283–94.

Mogil JS. Pain genetics: past, present and future. Trends Genet. 2012;28:258–66.

Ness TJ, Gebhart GF. Colorectal distension as a noxious visceral stimulus: physiologic and pharmacologic characterization of pseudaffective reflexes in the rat. Brain Res. 1988;450:153–69.

Pogatzki EM, Niemeier JS, Brennan TJ. Persistent secondary hyperalgesia after gastrocnemius incision in the rat. Eur J Pain. 2002;6:295–305.

Radhakrishnan R, Moore SA, Sluka KA. Unilateral carrageenan injection into muscle or joint induces chronic bilateral hyperalgesia in rats. Pain. 2003;104:567–77.

Shippenberg TS, Stein C, Huber A, Millan MJ, Herz A. Motivational effects of opioids in an animal model of prolonged inflammatory pain: alteration in the effects of kappa- but not of mu-receptor agonists. Pain. 1988;35:179–86.

Sluka KA. Stimulation of deep somatic tissue with capsaicin produces long-lasting mechanical allodynia and heat hypoalgesia that depends on early activation of the cAMP pathway. J Neurosci. 2002;22:5687–93.

Sluka KA, Rasmussen LA, Edgar MM, O'Donnel JM, Walder RY, Kolker SJ, Boyle DL, Firestein GS. ASIC3 deficiency increases inflammation but decreases pain behavior in arthritis. Arthritis Rheum. 2013;65(5):1194–202.

Totsch SK, Waite ME, Tomkovich A, Quinn TL, Gower BA, Sorge RE. Total Western diet alters mechanical and thermal sensitivity and prolongs hypersensitivity following complete Freund's adjuvant in mice. J Pain. 2016;17(1):119–25. https://doi.org/10.1016/j.pain.2015.10.006.

Totsch SK, Quinn TL, Strath LJ, McMeekin LJ, Cowell RM, Gower BA, Sorge RE. The impact of the Standard American diet in rates: effects on behavior, physiology and recovery from inflammatory injury. Scand J Pain. 2017;17:316–24. https://doi.org/10.1016/j.sjpain.2017.08.009.

Woolfe G, MacDonald AD. The evaluation of the analgesic action of pethidine hydrochloride. J Pharmacol Exp Ther. 1944;80:300–7.

Clinical Pain Phenotyping for Omics Studies

5

Shad B. Smith

Contents

5.1 Introduction

When the genetics of pain and pain disorders first began to be explored, it was unclear whether the particular phenotype investigated would produce a set of genes distinctively associated with each trait or disease, or whether there would be a set of "master pain genes" that would be implicated regardless of the condition. Fundamentally, it was an empirical question as to whether different manifestations of pain are influenced by distinct genetic etiologies. The first evidence came from studies of rodent strains with genetically varied backgrounds tested across several pain modalities. It became clear that "pain sensitivity" was not a unitary phenomenon, as knowing that a strain was sensitive or resistant to one type of pain was not predictive of its sensitivity to a different type (Mogil et al. 1996). It is now recognized that while there may be some overlap in the genes responsible for individual

S. B. Smith (✉)
Center for Translational Pain Medicine, Department of Anesthesiology, Duke University Medical Center, Durham, NC, USA
e-mail: shad.smith@duke.edu

© Springer Nature Switzerland AG 2020
S. G. Dorsey, A. R. Starkweather (eds.), *Genomics of Pain and Co-Morbid Symptoms*, https://doi.org/10.1007/978-3-030-21657-3_5

differences in pain, different sets of genes would be implicated depending on the condition or mechanism underlying the pain.

The design of genetic studies of pain should reflect the causal biological mechanisms responsible for producing or modulating the pain. Pain phenotyping necessitates selecting the measurable traits that are most relevant to the underlying biology. Pain is by definition a subjective experience with both sensory and emotional components, and manifests alongside an array of biopsychosocial determinants and consequences (IASP Task Force on Taxonomy 1994). Therefore, a thorough battery of multidimensional measurements is ideal for the characterization of an individual's pain experience. In the context of omics studies, however, the need to recruit potentially thousands or tens of thousands of participants imposes trade-offs between comprehensiveness and convenience in obtaining such measures.

Genetic and other omics studies of pain rely on the assumption that variability at the genetic level is reflected in individual differences in protein function or activity, which in turn interact with environmental factors resulting in more complex phenomena including altered pain sensitivity or susceptibility to pain disorders. As the subjective experience of pain is several steps removed from the gene level, it may be useful to interrogate intermediate phenotypes more closely governed by genetic polymorphism, with a presumably less complex genetic architecture. An intermediate phenotype (also called endophenotype) is a heritable trait that is measurable in both affected and unaffected individuals, and which co-segregates with affected status in the population (Gottesman and Gould 2003). Phenotype domains that may be of particular interest as endophenotypes for the study of chronic pain include pain amplification (increased pain sensitivity) and psychological distress (Diatchenko et al. 2006). A number of quantifiable characteristics can be evaluated to measure these constructs, including quantitative sensory testing (QST) and psychosocial assessments of anxiety, depression, and somatic awareness. The key to rigor and reproducibility in omics studies of pain is the systematic collection of mechanistically relevant phenotypes encompassing the multidimensional components of pain, including sensory, emotional, and cognitive aspects, as well as demographic characteristics, patient history, functional interference, environmental exposures, and comorbidities.

Although rigorous and accurate data collection in omics studies has long been appreciated in other fields, a few examples from the pain field indicate that improvement is needed for the types of large-scale consortia required for truly powerful dissection of the genetic determinants of pain. A meta-analysis of chronic widespread pain (CWP) (Peters et al. 2012) incorporated 1480 cases from nine independent cohorts to replicate findings from a discovery analysis. Heterogeneity across the cohorts was observed, possibly due to four cohorts being specific to joint pain, while the other five cohorts included non-joint pain. When the non-joint specific cohorts were removed, heterogeneity reduced to 0%, resulting in a stronger observed replication effect. Despite the reduced number of subjects in the more homogeneous analysis, power to detect a joint pain-specific effect was increased. The investigators also examined 44 candidate genes at 136 SNPs previously associated with other pain conditions, but were unable to detect associations with CWP in their study for any candidate variants, suggesting the genetic architecture of CWP is distinct from other conditions. More recently, a review of genetic risk factors for neuropathic pain observed large variation in case ascertainment and phenotyping, as well as in selection of pain-free controls, that prevented meta-analysis among the 29 studies reviewed (Veluchamy et al. 2018). The authors identified small sample sizes and differences in case definition as probable reasons for the lack of replication of genome-wide association study (GWAS) findings, and concluded that large-scale GWAS with a consensus neuropathic phenotype in independent cohorts will be necessary to identify true associations.

5.2 Special Considerations for Omics Studies

Quantifying pain and distress is challenging on any scale. The subjectivity of pain experience, and its consequent disconnection from measurable organic injury or tissue damage, results in well-appreciated difficulties in phenotyping pain for research or clinical use. Additionally, there are a number of other considerations for phenotyping in omics studies of pain; this chapter will make recommendations for phenotyping methods highlighting usefulness for large cohort-based data collection.

Statistical Methodology Human studies of pain are most frequently conducted in either case/control cohorts, using a diagnosis or self-report of one or more pain conditions as a basis for comparison with non-disease controls; or else in hospital- or community-based samples in which the distribution of a trait is intended to reflect the population. Due to the enormous size of omics datasets, relatively simple statistical methods are typically employed to identify associated genes, such as logistic (for case/control phenotypes) or linear (for continuous variables) regression. These approaches make critical assumptions about the data that, if violated, can produce spurious results. Notably, assessing quantitative traits in a cohort with non-random sampling (such as a case/control study) can lead to biased estimates of association with traits affected by the ascertainment strategy (Monsees et al. 2009). The selection of phenotypes for analysis should include measures that are appropriate for the recruitment strategy and the planned statistical tests. Demographic and other descriptive variables should be collected in order to appropriately account for potential covariates and other confounding factors (e.g., sex, age, race, and ethnic background).

Sample Size Large cohorts are needed to achieve sufficient power for genetic association and other omics studies, due to a number of factors. Omics studies by definition involve a vast number of individual statistical tests, and stringent multiple testing procedures (e.g., the Bonferroni correction, or a genome-wide significance threshold of $p < 5 \times 10^{-8}$) must be applied to avoid false positives (Sham and Purcell 2014). At the same time, complex traits including pain phenotypes are likely to be influenced by a large number of genetic effects, each with a small contribution to the overall trait variance. Statistical power is also impacted in proportion to the minor allele frequency of associated variants and the prevalence of the trait in the cohort studied. Typically, thousands or tens of thousands of subjects will be needed to be recruited and assessed in such studies. In order to minimize costs and subject burden, data collection for omics studies has an imperative to be as efficient as possible without sacrificing accuracy and precision.

Compatibility with Other Cohorts Because of the large number of subjects required for omics studies, it is becoming increasingly common to combine separate cohorts from different studies before analysis or through meta-analysis to attain sufficient power. It is also expected that genetic associations and other omics results be replicated in independent cohorts before they are accepted as true. Without phenotypes in common, merging data and comparing findings between studies is not possible. As the universe of measurement methods is potentially limitless, working groups within government agencies and scientific societies are developing sets of common data elements (Moore et al. 2016) to harmonize data collection and analysis. Although there is not yet a singular set of endorsed phenotypes for general pain, recommendations for particular classifications of pain are emerging and should be adopted for future studies. For example, the National Institute of Neurological Disorders and Stroke (NINDS) has made available a set of common data elements for headache research (Oshinsky and Tanveer 2018), and a National Institutes of Health (NIH) Pain Consortium Task Force on chronic low back pain has published recommendations for a minimal standard research dataset (Deyo et al. 2014). Additionally, phenotypes collected in existing pain cohorts (and therefore available for replication of novel discoveries) can be reviewed in public data repositories, such as the National Center for Biotechnology Information's Database of Phenotypes and Genotypes (dbGaP) (Mailman et al. 2007) and Phenotype-Genotype Integrator (PheGenI) (Ramos et al. 2014).

5.3 Pain Phenotyping

Phenotyping of human pain should capture the multidimensional character of pain experience. Pain diagnostics have historically relied on the symptoms, signs, and bodily location of the pain, but more recently there has been movement towards an evidence-based taxonomy of chronic pain. The Analgesic, Anesthetic, and Addiction Clinical Trial Translations, Innovations, Opportunities, and Networks (ACTTION) committee has incorporated recommendations for pain phenotyping into the ACTTION-APS Pain Taxonomy (AAPT) (Dworkin et al. 2016). This taxonomy takes into account additional features such as the biopsychosocial mechanisms and risk factors, medical and psychiatric comorbidities, and functional consequences of pain. As our understanding of the mechanisms of chronic pain grows, it is likely that genetic markers will associate more with underlying pathophysiological processes, rather than the present diagnostic categories (Diatchenko et al. 2013; Zorina-Lichtenwalter et al. 2016).

5.3.1   Demographic and Environmental Characteristics

Non-clinical characteristics of subjects are significant sources of variance in large omics cohorts, although they are not explicitly considered in any of the AAPT dimensions. Collecting demographic and environmental data allows for stratification or accounting for confounding factors in downstream analyses. Such steps are critical when these variables associate strongly or subtly with outcome measures, which can bias genetic associations when they are also associated with inherited genetic or ethnic background. A full discussion of population stratification is beyond the scope of this chapter, but phenotyping efforts should collect sufficient data to detect and account for potential bias (Sillanpaa 2011).

Basic demographic information includes innate characteristics that can interact with genetic background, as well as potential sources of environmental exposures and biopsychosocial contributors to variation. Such variables include age, sex, place of residence, race and ethnicity of subjects and one or more previous generations, and marital status and family structure. Socioeconomic stratum and its attendant characteristics of education and occupation can contribute to pain experience, along with employment status, disability and retirement claims, and accessibility of insurance and sick leave. Other characteristics that may be relevant include current and lifetime use of tobacco, alcohol, and other drugs; religion; and immigration status.

5.4   Clinical Pain Phenotyping

The majority of genetic association studies performed to date have examined either clinical pain conditions or syndromes, or chronic pain following surgery. Ideally, case-control categorization in these studies follows established diagnostic criteria and includes a clinical exam by a qualified practitioner, but such efforts are costly in time, labor, and expense. In practice, it is also common in large community-based studies to make use of phone interviews and self-report questionnaires or retrospective medical record reviews to economically gather data, albeit at the expense of precision.

5.4.1   Diagnostic Criteria

The AAPT considers five dimensions in the assessment of chronic pain: (1) core diagnostic criteria; (2) common features; (3) common medical and psychiatric comorbidities; (4) neurobiological, psychosocial, and functional consequences; and (5) putative neurobiological and psychosocial mechanisms, risk factors, and protective factors (Dworkin et al. 2016). The first two dimensions are critical elements of the clinical assessment of pain, but important information pertinent to the other dimensions can also be uncovered in this effort. Genetic studies focused on a specific condition or disease should assess the characteristic symptoms, signs, and test results pertinent to the diagnosis. Both inclusion and exclusion criteria should be considered in the process of differential diagnosis. For chronic pain, duration is typically a defining feature; i.e., pain is considered chronic after 3 months.

Peripheral and Central Nervous System Disorders Painful neuropathies resulting from post-traumatic and surgical injury, diabetic neuropathy, postherpetic neuralgia, HIV infection, and trigeminal neuralgia make up a large proportion of this category, but it also includes pain associated with other disorders such as complex regional pain syndrome, stroke, and multiple sclerosis. Expert recommendations outlined in the Neuropathic Pain Phenotyping by International Consensus (NeuroPPIC) statement provide an "entry-level phenotyping" standard for genetic studies (van Hecke et al. 2015). The primary qualification for neuropathic pain is symptom assessment for neuropathic characteristics. A number of validated screening tools are recommended to assess symptom qualities, including the painDETECT (Freynhagen et al. 2006), Douleur Neuropathique en 4 questions (DN4) interview (Bouhassira et al. 2005), and the Self-report Leeds Assessment of Neuropathic Symptoms and Signs (S-LANSS) (Bennett 2001) instruments. Additional information may be provided using a body chart or checklist to establish a pain distribution consistent with the underlying lesion or disease, and a detailed pain history including the duration and intensity of the pain. Diagnostic certainty can be improved by a clinical evaluation including QST outlined in a later section, particularly measures of gain of function (allodynia, hyperalgesia) and loss of function (thermal or vibratory insensitivity).

Spine Pain The AAPT considers chronic axial musculoskeletal low back pain and chronic lumbosacral radiculopathy (also called sciatica) in the category of spinal pain, while recognizing that back pain may arise from both neuropathic and musculoskeletal mechanisms, and may result in pain referred to other anatomical locations such as legs. Spinal imaging is typically used for anatomical diagnosis, but even a complete physical examination is not always able to determine the source of pain. Genetic studies to date have relied on self-reported functional definitions of chronic low back pain (cLBP), including nonspecific or idiopathic cLBP. A recent NIH Research Task Force created standards for cLBP phenotyping (Deyo et al. 2014) including recommendations for collecting chronicity and functional impact measures. The Oswestry Disability Index (ODI) (Fairbank 1995) and the Roland Morris Disability Questionnaire (RMDQ) (Roland and Morris 1983) are two quick screening tools that have been extensively validated for back pain impact.

Musculoskeletal Pain This category includes arthritic joint pain disorders (osteoarthritis, rheumatoid arthritis, gout, spondyloarthropathies) and muscle pain (chronic myofascial pain), as well as widespread pain and fibromyalgia. Typically, arthritic conditions can be confirmed with blood tests, X-rays, and/or synovial fluid analysis, but diagnosis of fibromyalgia and chronic widespread pain is based on symptom report. The current American College of Rheumatology (ACR) diagnostic criteria for fibromyalgia specifies generalized pain in at least 4 of 5 regions, lasting at least 3 months, and scores above threshold on a widespread pain index and symptom severity scale (Wolfe et al. 2016). Self-report questionnaires for pain and functional interference are condition- and joint-specific. For osteoarthritis, the Western Ontario and McMaster Universities Osteoarthritis Index (WOMAC) (Bellamy et al. 1988) is the most widely used tool for assessing hip and/or knee pain, while the similar Australian/Canadian (AUSCAN)

Osteoarthritis Hand Index (Bellamy et al. 2002) is appropriate for hand pain. Rheumatoid arthritis characteristics can be assessed using the Rheumatoid Arthritis Pain Scale (RAPS) (Anderson 2001). For fibromyalgia, the Revised Fibromyalgia Impact Questionnaire (FIQR) (Bennett et al. 2009) provides an assessment of symptoms, function, and overall impact.

Orofacial and Head Pain When only considering the genetics underlying pain itself (rather than disease processes), migraine is to date the only pain disorder that has been rigorously examined by well-powered GWAS (Nyholt et al. 2017). This is likely due to the relative ease of migraine diagnosis, including categorization into subtypes, as well as the high prevalence of headaches of various types. However, these large genetic studies have generally relied on ad hoc self-reported headache questions, rather than validated instruments or clinical examinations. Phenotyping of head pain has been standardized by the International Classification of Headache Disorders (ICHD-3) (Society IH 2018). As the third edition of the ICHD has only recently been finalized, available screening instruments may not precisely follow updated guidelines. For migraine, the ID Migraine (Lipton et al. 2003) screener assesses three items: disability, nausea, and sensitivity to light; while the Visual Aura Rating Scale (VARS) (Eriksen et al. 2005) is useful for the diagnosis of migraine with aura. Temporomandibular disorders (TMD), a collection of conditions producing pain in the jaw joint or musculature, also falls under this AAPT category. Diagnosis of TMD follows the Research Diagnostic Criteria (TMD-RDC) (Dworkin and LeResche 1992), while the Jaw Function Limitation Scale (JFLS) (Ohrbach et al. 2008) is a useful tool to assess functional status of the region.

Abdominal, Pelvic, and Urogenital Pain Functional pain disorders of these regions include urologic chronic pelvic pain syndromes such as interstitial cystitis (IC) and chronic prostatitis (CP), irritable bowel syndrome (IBS), and vulvodynia (VVD). IBS is diagnosed and subcategorized using the consensus Rome criteria, the latest of which is the Rome IV update (Mearin et al. 2016). Widely accepted diagnostic criteria for other disorders have not yet been adopted, so care is advised in formulating entry criteria for research studies (Gewandter et al. 2018). A number of recommendations for diagnosing IC or bladder pain syndrome have been provided by groups such as the European Association of Urology (van de Merwe et al. 2008) and the American Urological Association (Hanno et al. 2011). VVD can be diagnosed using Friedrich's criteria (Friedrich Jr 1987), and the Vulvar Pain Assessment Questionnaire (VPAQ) (Dargie et al. 2016) is a useful survey instrument.

5.4.2 Self-report Measures

The second dimension of the ACTTION-APS Pain Taxonomy describes common features of the clinical pain experience, including symptoms and signs. This additional information may include sensory and affective qualities of the pain, its duration and temporal characteristics, functional consequences, and non-pain features such as numbness and fatigue. As a number of short, validated scales are widely used for patient self-report of pain, these instruments are suitable for omics studies that require large numbers of subjects similarly phenotyped in independent cohorts.

Pain Intensity As a subjective phenomenon, the gold standard for measuring pain is self-report by the patient. Rating scales such as visual analogue scales (VAS), numerical rating scales (NRS), and verbal rating scales (VRS) are the most frequently used measures of the intensity of pain, but are also used to assess other qualities of pain such as unpleasantness or interference. The Initiative on Methods, Measurement, and Pain Assessment in Clinical Trials (IMMPACT)-II consensus recommendation for the measurement of pain intensity is an 11-point NRS with 0 described as "no pain" and 10 as "pain as bad as you can imagine" (Dworkin et al. 2005). The 3-item PEG is a very brief screening tool assessing pain intensity (P), interference with enjoyment of life (E), and interference with general activity (G) using such a scale (Krebs et al. 2009). The Gracely Pain Scale (Gracely et al. 1978) uses a 0–20 scale for intensity and unpleasantness, but it benchmarks the numerical ratings with verbal descriptors for patients that have difficulty with arbitrary numbers. Other instruments include assessments of the functional impact of pain. The Brief Pain Inventory Short Form (BPI-SF) is a 9-item instrument that incorporates an 11-point NRS for current, average, worst, and least pain with ratings of interference with normal daily activities. The Multidimensional Pain Inventory (MPI; also known as the West Haven-Yale Multidimensional Pain Inventory, or WHYMPI) (Kerns et al. 1985) uses 7-point Likert-type scales to assess pain level as well as aspects of function, interference, and emotional state. Condition-specific instruments, such as the WOMAC mentioned previously as well as the Disability of Arm, Shoulder and Hand (DASH) (Hudak et al. 1996) and Neck Disability Index (NDI) (Vernon and Mior 1991) questionnaires, frequently include self-report ratings of pain intensity and interference similar to the generalized surveys, but assess more relevant functional aspects.

Pain Qualities Patient descriptions of pain qualities may differ depending on the underlying nature of the pain. Assessing qualitative characteristics of pain to subdivide into pain types may therefore allow for genetic dissection of pain mechanisms. The Short-Form McGill Pain Questionnaire (MPQ-SF) (Melzack 1987) uses a VAS for intensity, and generates scores for sensory and affective dimensions using a list of pain descriptors. The Pain Quality Assessment Scale (PQAS) (Jensen et al. 2006) assesses pain intensity overall as well as for both deep and surface pain, and includes separate ratings of 16 descriptors such as "sharp," "tender," and "radiating."

5.4.3 Controls

The selection and characterization of appropriate controls for case-control studies of pain disorders is an underappreciated aspect of study design. In addition to being free of the condition of interest, controls should be representative of the same risk population as cases, meaning they should have the same opportunity for exposure. As large groups of cases and controls are needed for omics studies, researchers have frequently used cohorts conveniently available to them. For cases, selecting patients from hospital or specialty clinic registries is an efficient method of identifying people with the disease or condition in question. Lacking a similar stream of potential recruits for controls, omics researchers may adopt strategies that introduce bias. One pitfall arises from selection bias, or using different criteria to select cases and controls. For example, cases of patients with post-operative pain should not be contrasted with community-sampled individuals who have not been exposed to the surgical procedure. Similarly, it is inappropriate to apply differing exclusion criteria, such as screening only controls for the presence of any chronic pain disorder, when no similar exclusion was made for comorbid pain in cases. Omics studies are also prone to information bias, when different information is available for cases and controls about exposure and risk factors. This situation is common when genotyped biobank participants are used as controls without detailed knowledge of their pain status.

To minimize the potential for bias, controls should be drawn from the same risk exposed population. Recruitment and phenotyping protocols should apply the same inclusion/exclusion criteria for cases and controls, and all participants should ideally be evaluated identically and concurrently.

5.5   Psychosocial Phenotyping

While the first two dimensions of the AAPT framework deal primarily with the diagnostic signs, symptoms, and characteristics of the pain itself, the other three dimensions capture a holistic view of the experience of pain (Dworkin et al. 2016). Dimension 3 incorporates assessment of common comorbidities of chronic pain, including both medical and psychiatric disorders. Dimension 4 assesses the neuro-biological, psychosocial, and functional consequences that follow the development of chronic pain. The pathophysiological mechanisms potentially underlying the chronic pain are explored along with factors impacting risk and protection in Dimension 5. It should be noted that important domains, such as psychosocial distress and heightened sensitivity to stimuli, overlap between these dimensions as both contributors to and consequences of the development of chronic pain.

Poor psychological health is known to be a strong risk factor for pain, likely in a bidirectional manner that confounds the relationship between antecedent and consequence. Numerous studies have also found psychiatric disorders to be commonly comorbid with chronic pain. Comprehensive phenotyping of pain therefore should include rigorous psychosocial assessment, including affective traits such as anxiety, depression, and mood, as well as indicators of distress related to environmental stressors and sleep disturbance. Pain-related behaviors and attitudes, such as catastrophizing, self-efficacy, and coping, may also be assessed. A number of validated self-report instruments useful for research are presented in Table 5.1. These surveys are not intended to diagnose psychiatric disorders, but to quantify psychological characteristics in chronic pain patients and pain-free individuals. This review will focus, where possible, on instruments that have been developed or validated for use with pain populations, or that have normative data from such groups allowing appropriate interpretation.

Table 5.1 Psychosocial survey instruments

	Measure	Subject burden	Notes
Anxiety	Beck Anxiety Inventory (BAI)	21 items (5–10 min)	Focuses on somatic symptoms associated with anxiety
	Hospital Anxiety and Depression Scale-Anxiety (HADS-A)	7 items (<5 min)	Developed to screen for clinically significant anxiety symptoms in medically ill patients
	Pain Anxiety Symptoms Scale (PASS) and Short Form (PASS-20)	40 items (10 min); PASS-20: 20 items (5–10 min)	Developed to evaluate pain-related anxiety and fear in individuals with chronic pain disorders
	State-Trait Anxiety Inventory (STAI)	40 items, 20 per scale (10 min)	Two subscales: State anxiety (current symptoms) and trait anxiety (propensity to be anxious)
Depression	Beck Depression Inventory-II (BDI-II)	21 items (5–10 min)	A revision of the original BDI that omits items related to weight loss, body image, and work disability to correspond more closely with the DSM-IV criteria for major depressive disorder
	Center for Epidemiological Studies-Depression (CES-D) and Screening version	20 items (5–10 min); Screening version: 5 items (<5 min)	For use in the general population to assess perceived mood and level of functioning
	Hospital Anxiety and Depression Scale-Depression (HADS-D)	7 items (<5 min)	Developed to screen for clinically significant depressive symptoms in medically ill patients
	Patient Health Questionnaire (PHQ-9)	9 items (<5 min)	Assesses DSM-IV criteria for major depressive disorder as well as psychosocial impairment
Stress and Mood	Perceived Stress Scale (PSS)	10 items (<5 min)	Assesses the degree to which the subject appraises life situations as stressful
	Profile of mood states second edition (POMS 2) and short form (POMS 2-SF)	65 items (10 min); POMS 2-SF 35 items (5 min)	Measures six dimensions of transient and persistent mood states: Tension or Anxiety, Anger or Hostility, Vigor or Activity, Fatigue or Inertia, Depression or Dejection, Confusion or Bewilderment
	Scale of Positive and Negative Experiences (SPANE)	12 items (<5 min)	Assesses a broad range of negative and positive experiences and feelings, including a balance score

Table 5.1 (continued)

	Measure	Subject burden	Notes
Symptoms and Somatization	Medical outcome study short form 36 (SF-36) and short form (SF-12)	36 items (10–12 min); SF-12: 12 items (<5 min)	Multi-item scale that assesses eight health concepts including functional limitation, pain, and mental health. The SF-12 is limited to physical and mental summary scores
	Pennebaker Inventory of Limbic Languidness (PILL)	54 items (10–15 min)	Measures frequency of noticing and endorsing common physical symptoms and sensations
	Symptom Checklist-90-R (SCL-90R)	90 items (12–15 min)	Assesses 9 primary symptom dimensions of somatization and mental distress; shorter subscale-specific versions exist, e.g., Brief Symptom Inventory-18 (BSI-18) which only includes anxiety, depression, and somatization
Beliefs/ Catastrophizing	Pain Beliefs Questionnaire (PBQ)	12 items (<5 min)	Assesses organic and psychological beliefs about pain
	Pain Catastrophizing Scale (PCS)	13 items (5 min)	Includes subscale scores for rumination, magnification, and helplessness
	Pain Locus of Control (PLOC)	36 items (10–12 min)	Assesses perception of internal and external control over pain
Coping and Self-efficacy	Chronic Pain Self-efficacy Scale (CPSS)	22 items (5–10 min)	Developed to measure chronic pain patients' perceived self-efficacy to cope with the consequences of chronic pain
	Coping Strategies Questionnaire (CSQ)	44 items (10 min)	Measures frequency of specific cognitive and behavioral coping strategies
	Pain Self-efficacy Questionnaire (PSEQ)	10 items (<5 min)	Assesses the confidence people with ongoing pain have in performing activities while in pain
Fatigue and Sleep	Chalder Fatigue Questionnaire (CFQ)	11 items (5 min)	Developed to assess physical and mental fatigue in clinical and community populations
	Pittsburgh Sleep Quality Index (PSQI)	19 items + 5 items for bed partner (5–10 min)	Measures quality and quantity of sleep and sleep disruption
	Sleep Problems Scale (SPS)	4 items (<5 min)	Assesses difficulty in achieving restful sleep

5.5.1   Anxiety

Anxiety disorders are more prevalent among chronic pain sufferers, and high levels of anxiety may contribute to pain-related disability through a fear-avoidance cycle.

The Hospital Anxiety and Depression Scale-Anxiety (HADS-A) (Zigmond and Snaith 1983) is commonly administered in clinical settings to screen for significant anxiety symptoms, and is considered to be relatively unaffected by the physical illness of patients. The Beck Anxiety Inventory (BAI) (Beck et al. 1988), conversely, emphasizes somatic symptoms associated with anxiety, rather than cognitive components, in an effort to distinguish between anxiety and depression. Such symptoms may better represent biological intermediate phenotypes, more closely governed by genetic factors. The Spielberger State-Trait Anxiety Inventory (STAI) (Spielberger et al. 1983) is a lengthier instrument (although shortened subscale forms exist), with 20 items that assess situational anxiety levels "at this moment," and a separate set of 20 items that distinguish more stable qualities of proneness to anxiety. The Pain Anxiety Symptoms Scale (PASS) (McCracken et al. 1992) was developed to assess cognitive, behavioral, and physiological aspects of pain-related anxiety that may be important in the persistence of pain behaviors.

5.5.2   Depression

Clinically significant depression is also common among chronic pain sufferers. Assessments of depression and suicidality are frequently used to screen pain patients for emotional distress, and they may be useful to omics research as indicators of a major psychological risk factor for chronic pain. As with its counterpart for anxiety, the Hospital Anxiety and Depression Scale-Depression (HADS-D) (Zigmond and Snaith 1983) is brief and widely adopted for use in clinical populations. The Patient Health Questionnaire-9 (PHQ-9) (Kroenke et al. 2001) is an alternative concise screening instrument that scores the 9 DSM-IV criteria for depression. The widely used Beck Depression Index (revised to the BDI-II version) (Beck et al. 1996) is longer but assesses intensity of a number of cognitive, affective, somatic, and vegetative symptoms of depression. The Center for Epidemiological Studies-Depression (CES-D) (Radloff 1977) scale was developed for research purposes in the general population and may be more sensitive to differences in subclinical populations.

5.5.3   Stress and Mood

Predisposition or resilience to stress and negative effect is a related component of pain experience that can greatly impact cognition, behaviors, and function. The social and environmental context of pain can be difficult to quantify, but self-reported impressions of stress and mood may be as or more important than the stressors themselves, and may represent innate characteristics governed by etiological pathways. The Perceived Stress Scale (PSS) (Cohen et al. 1983) is a short instrument that measures the degree to which life situations are appraised as stressful, and has been shown to predict the effects of stress on outcomes better than life-event scores themselves. Mood may be encapsulated in a single dimension of general well-being, or it may be represented by multiple distinct facets. The Scale of Positive and Negative Experience (SPANE) (Diener et al. 2010) broadly assesses the frequency of a range of emotions, which are grouped into either subjective well-being or ill-being scores. The Profile of Mood States is a longer instrument that measures six domains: Tension or Anxiety, Anger or Hostility, Vigor or Activity, Fatigue or Inertia, Depression or Dejection, Confusion or Bewilderment. A short version (POMS-SF) (Shacham 1983) that reduces the length from 65 down to 37 questions may be more useful for omics research to balance the breadth of subscales with a more expedient form.

5.5.4 General Symptoms and Somatization

Overall mental and physical health encompasses more domains than can typically be addressed in a single instrument or even survey battery. However, as risk of pain chronicity increases with diminished general health, concise assessments of physical and emotional functioning are desired. The general tendency to attend to or linger on unpleasant sensations, referred to as somatization or somatic awareness, is also a strong risk factor for chronic pain. The Medical Outcomes Study 36-Item Short-Form Health Survey (SF-36) (Ware Jr. and Sherbourne 1992) is a brief instrument that assesses bodily pain and general health perceptions, as well as limitations in physical, emotional, and social functioning, general mental health, and vitality. A 12-item version, the SF-12 (Ware Jr. et al. 1996) contracts these elements to a physical and a mental summary score. The Symptom Checklist 90-Revised (SCL-90R) (Derogatis 1994) measures multiple aspects of psychological distress: somatization, obsessive-compulsive, interpersonal sensitivity, depression, anxiety, hostility, phobic anxiety, paranoid ideation and psychoticism. A short version of this scale, the Brief Symptom Inventory-18 (BSI-18) (Derogatis and Melisaratos 1983) is available as well, consisting of only the somatization, anxiety, and depression subscales. Focusing specifically on common physical symptoms, the Pennebaker Inventory of Limbic Languidness (PILL) (Pennebaker 1982) is a useful measure of somatization.

5.5.5 Catastrophizing and Coping

Cognitions and emotions around pain influence perception, and may interact with neural mechanisms impacted by omic regulation such as ascending and descending pain modulatory systems. The Pain Beliefs Questionnaire (PBQ) (Edwards et al. 1992) measures attributional beliefs about the organic and psychological components underlying causes and consequences of pain. The Pain Locus of Control scale (PLOC) (Toomey et al. 1993) also assesses perceived attribution of pain control, as either internal or external to the individual. The Pain Catastrophizing Scale (PCS) (Sullivan et al. 1995) assesses attitudes, emotions, and behaviors contributing to the exaggeration of negative experiences, including subscales for rumination, magnification, and helplessness.

Behavioral and cognitive strategies used by patients to deal with pain can affect their experience of pain and attendant emotional and functional difficulties. The Coping Strategies Questionnaire-Revised (CSQ-R) (Robinson et al. 1997) is a widely used instrument that assesses cognitions and behaviors used to reduce pain, such as diverting attention, reinterpreting pain, coping self-statements, ignoring pain, praying or hoping, catastrophizing, and increasing activity. Two shorter instruments that focus on patients' perceived confidence in their ability to perform daily activities include the Chronic Pain Self-Efficacy Scale (CPSS) (Anderson et al. 1995) and the Pain Self-Efficacy Questionnaire (PSEQ) (Nicholas 2007).

5.5.6   Fatigue and Sleep

Chronic pain has been long known to result in difficulties with sleep and daytime fatigue, but it is recently becoming more generally accepted that poor quality sleep can disrupt nociception and emotional distress pathways leading to pain chronicity. Objective measurements of sleep, such as polysomnography and actigraphy, are ideal for studies exploring the biological basis of disturbed sleep. These measures require overnight observation with specialized instruments, however, limiting their utility for large omics studies. For the pain researcher interested in including sleep quality in a biopsychosocial profile, there are a number of questionnaires that can assess subjective fatigue and sleep problems more conveniently. The Pittsburgh Sleep Quality Index (PSQI) (Buysse et al. 1989) collects information on sleep quality and duration from the subject and the subject's bed partner. The Sleep Problems Scale (SPS) (Jenkins et al. 1988) is a short survey that captures the frequency of common sleep symptoms. The Chalder Fatigue Questionnaire (CFQ) (Chalder et al. 1993) is useful to assess both physical and mental sensations of fatigue in clinical and community populations.

5.6   Experimental Pain Phenotyping

Sensitivity to painful stimuli is believed to be partially under genetic control, and has been moderately associated with risk of chronic pain. The assessment of threshold and tolerance to various sensory modalities is a common component of phenotyping in pain cohorts in order to characterize the function of somatosensory and nociceptive afferents. Quantitative sensory testing (QST) is the use of standardized protocols to assess psychophysical responses to calibrated sensory stimuli. Unlike assessments of disease-linked pain, these measures may be used to phenotype both patients with pain or sensory deficits as well as healthy individuals. A number of methods have been proposed to test sensitivity across a range of somatosensory dimensions:

- stimulus modality (e.g., thermal, mechanical, vibrotactile, etc.);
- cutaneous versus deep pain;

- nerve fiber populations (e.g., Aβ fibers, which usually transmit innocuous touch sensations, versus C fibers which conduct signals from polymodal nociceptors);
- threshold detection versus supra-threshold ratings;
- naïve versus sensitized tissue.

A large number of QST procedures have been devised to assess somatosensory function, many of which require extensive examinations with specialized equipment. A comprehensive battery of measures would therefore strain the time and cost constraints of most omics pain research. Such studies should include a judiciously selected series of tests that address the specific biological questions under investigation. A number of representative QST batteries have been previously published for omics studies (see, for example, (Rolke et al. 2006; Greenspan et al. 2011; Schmid et al. 2019)); consulting such protocols may be useful for identifying widely adopted measures likely to be available in existing cohorts. This section outlines a number of common measures, most of which can be performed using simple hand-held devices, although some require specialized laboratory equipment and software.

5.6.1 Best Practices for QST Experiments

While many QST procedures are feasible for omics research, the conduct of such studies requires attention to a number of considerations. QST is usually performed as a series of measures in rapid succession within a short time frame (less than 1 h). With few exceptions, these measures should be non-invasive and non-injurious, both for ethical reasons as well as to avoid sensitization in peripheral tissue or central nociceptive processing, confounding subsequent measures. Tests that initiate or are intended to evaluate sensitizing or inhibitory responses should be performed after less severe measures. The order of tests and anatomical placement of stimuli should be standardized across all subjects. Within subjects, tests may be performed bilaterally or at multiple body sites to avoid sensitizing a single area. It may also be useful to use one side or location as a test site to familiarize the participant with the protocol, reserving the other side for experimental measurements. The test side may be chosen randomly, but handedness (right or left dominant) should be recorded. Although right and left sides of the body do not generally show differences, it has been observed that QST parameters vary across anatomical site, with lower thresholds noted in the face, followed by hands and feet, compared with other sites (Rolke et al. 2006). For most tests, replicate measurements should be taken and averaged to minimize variability; it is common to exclude first and/or last measurements which tend to be more inconsistent in some assays. As anxiety and emotional state can influence pain perception during QST, an appropriate psychological instrument (such as those described previously) may be administered to detect and control for any confounding effects. Variation can also arise from differences between examiners in large or lengthy studies, so careful calibration of examiners and equipment is necessary prior to enrollment and periodically over the course of the study.

5.6.2   Tests of Thermal Sensitivity

Heat Pain Threshold and Tolerance   Sensitivity to heat is assessed using a Peltier thermode, a contact thermal stimulator, usually applied to the ventral forearm but potentially to the dorsal skin of the hand or foot. Beginning at a non-noxious temperature (32 °C), the temperature is ramped up 0.5–1.0 °C/s until the participant signals the perception of pain and the probe is removed. The mean temperature of 3–5 consecutive measurements, moving the contact site after each trial, is taken as the heat pain threshold. Heat pain tolerance is assessed using a similar paradigm, except that the participant signals to stop when the pain is perceived as intolerable. A cutoff temperature at which the trial is terminated (50 °C) should be enforced in order to avoid tissue damage.

Cold Pressor Test   Devised to test vascular responses, the cold pressor test has found common use in population pain research due to its simplicity and ease of administration. The subject immerses one hand up to the wrist into a circulating cold water bath held at 3–5 °C. The subject is allowed to remove his or her hand when the pain is considered unbearable, or at a cutoff time (2–5 min). Pain ratings, using a 0–10 NRS, are reported by the participant shortly after submersion (4 s) and at regular intervals over the test period; the overall pain intensity is calculated as the mean of these measurements. Pain threshold is recorded as the time at which pain is first reported, and pain tolerance is the time at which the hand is removed from the water.

Other Thermal Sensitivity   The Peltier thermal stimulator can be used to assess cold sensations, beginning at a neutral temperature (32 °C) and ramping downward at 0.5–1.0 °C/s until pain is perceived or becomes intolerable. A cutoff value of 0 °C should be used. Non-painful thermal sensitivity can also be measured using the thermal stimulator. Participants are told to signal when the thermal probe temperature is observed to change from neutral to cold or warmth.

5.6.3   Tests of Mechanical Sensitivity

Mechanical Pain   Mechanical detection threshold is measured with a standardized set of von Frey hairs, nylon filaments calibrated to exert differing forces (between 10 and 3000 mN). These fibers are applied in an ascending series until the participant first detects the stimulus. A second series is applied, starting from the strongest fiber and descending until the stimulus is no longer detected. The threshold is calculated by the "method of limits," as the geometric mean of five series of ascending and descending stimuli. The mechanical pain threshold can be similarly assessed by noting the first stimulus perceived as sharp or pricking when ascending, or until the stimulus is first experienced as blunt in a descending series.

An alternate method for assessing mechanical cutaneous pain threshold is to use weighted pinprick stimulators, each exerting a different force (typically between 8 and 512 mN) to a flat contact area of 0.2 mm diameter. The stimuli may be applied to the dorsum of the hand or foot, or to the digits. The pinprick threshold is the lightest weighted needle consistently perceived as sharp, as determined by the "method of limits."

Pressure Pain   To assess pain sensitivity of deep tissue, the pressure pain threshold may be obtained using an algometer device, a mechanical force gauge with a probe area of 1 cm^2 that indicates pressure up to 2000 kPa. The algometer is applied by hand to the test site with slowly increasing pressure of 30–50 kPa/s until the participant signals that the pressure is painful. For general use, the test site may be the center of the temporalis muscle, the center of the trapezius muscle, or overlying the lateral epicondyle; other sites relevant to the pain condition under investigation may also be chosen. A cutoff threshold of 600 kPa should be recorded if the participant has not signaled pain before this value is reached. The mean of 3–5 trials is calculated to arrive at a final estimate of pressure pain threshold.

5.6.4   Dynamic QST Measures

Temporal Summation   Chronic pain is believed to be associated with increased sensitization at the level of the spinal cord dorsal horn, which can be observed during QST as enhanced temporal summation and allodynia. Temporal summation (TS), or wind-up, occurs when closely repeated painful stimuli result in an increased perception of pain. Protocols to assess TS may use either the heat pain thermode or pinprick stimulators described above, for thermal or mechanical wind-up, respectively. For thermal TS, the temperature of the thermode is raised from a base of 38 °C to a super-threshold stimulus temperature (46–50 °C for general use) in a series of 10 pulses at 2.5 s interstimulus intervals. After each pulse, the participant verbally reports pain intensity using a 0–100 NRS, where "0" represents "no pain" and "100" represents "the most intense pain imaginable." The series is stopped if the participant rates any pulse as 100. Additional trials using alternate temperatures or body sites may be performed. Mechanical TS is performed similarly, applying a series of 10 pinprick stimuli of the same force (128–256 mN, depending on the body site) at a rate of 1/s. The wind-up ratio is calculated by dividing the mean rating of the trains of stimuli by the mean rating of a single stimulus. Alternately, TS may be expressed as the area under the response curve, the highest rating minus the first rating, or the slope of the response curve for the first 3 ratings.

Conditioned Pain Modulation   QST can also be used to assess conditioned pain modulation using a Diffuse Noxious Inhibitory Controls (DNIC)-like paradigm. Pain at a particular site can be diminished by a separate, spatially distant noxious stimulus via inhibition arising from endogenous pain modulatory systems. Chronic pain and central sensitization are associated with inefficient DNIC response. The assessment of DNIC function may be performed using heat pain as a primary stimulus, or "test" pain, and a cold stimulus as the "conditioning" pain. Participants first rate a super-threshold heat stimulus (47 °C for 4 s) from a Peltier thermode applied to the forearm on a 0–100 NRS, as a baseline measure. The conditioning stimulus is delivered by immersing the other arm in a cold water bath (12 °C). After 15 s in the cold water, the participant rates a second identical heat stimulus to the test arm. The participant withdraws his or her hand from the water bath at 30 s, and then two more heat pain stimuli are given and rated 15 and 30 s after removal. Magnitude of DNIC at each test rating is calculated as the percentage decrease from the baseline heat pain rating.

5.7   Electronic Phenotyping

The current era of portable and wearable computers, constant wireless internet, and interconnected electronic medical record (EMR) databases heralds a potential revolution in pain phenotyping. Electronic tools are already becoming more useful for genomic studies as patients and hospitals increase their use of computerized data collection and warehousing. These methods offer expedited participant screening and recruitment, improved diagnostic accuracy, greater temporal resolution, and more efficient and comprehensive assessments. In a conceivable future in which genotyping all patients is integrated into routine clinical practice, the bottleneck for pain genetics research may be effectively harnessing vast amounts of existing medical and lifestyle data.

Medical Records Many of the successes of large omics consortia to date have relied on hospital and national health system data collected in standard practice. Administrative datasets of potential use to pain research include diagnostic (e.g., International Statistical Classification of Diseases and Related Health Problems, or ICD) and procedural (Current Procedural Terminology, CPT) codes relating to chronic pain disorders. Phenotypes can be computationally derived from these datasets, permitting efficient screening of large numbers of patients for genome-wide association and other omics studies. To facilitate this effort, the NIH created the Electronic Medical Records and Genomics (eMERGE) Network to combine biorepositories with EMR systems for genomic discovery (Gottesman et al. 2013). While economical, this practice has proven challenging in the pain field due to the large degree of variance between clinicians in the usage of codes, introducing heterogeneity in case-control categorization. Future iterations of medical coding schemes, including the imminent ICD-11 system (Treede et al. 2019), should improve the fidelity of disease classification algorithms. Medication records, especially post-operative analgesic dosing, represent a more precise measure, and have been successfully used in exploring the genetics of opioid consumption (Nishizawa et al. 2014).

Computer Adaptive Testing Traditional phenotyping methods, including clinical examinations and psychometric assessments and surveys, are often too burdensome and time-consuming for practical use in omics research. More recent survey approaches have used question banks with items presented according to participants' previous responses. Algorithms are used to select the most efficient sequence of questions in a method known as Computer Adaptive Testing (CAT), resulting in a streamlined assessment of numerous domains with a relatively small number of targeted questions. The NIH Patient Reported Outcomes Measurement Information System (PROMIS) (Cella et al. 2010) is one such algorithm-based platform useful for pain research, as it includes item banks for pain quality, interference, and behaviors appropriate for both adult and pediatric populations, as well as a variety of psychosocial domains.

Mobile Apps To counteract recollection biases in retrospective self-report data, researchers have used daily diary methods that allow subjects to track pain levels over time in their home environment. With the widespread adoption of mobile devices and health monitoring accessories, a growing number of phone-based apps are available for scheduled or prompted pain data collection. Such data may be more accurate and detailed with respect to temporal variation and other factors that affect pain. Omics researchers may be eager to take advantage of the wealth of data emerging from such apps, but to date few publicly available tools have been evaluated scientifically (de la Vega and Miro 2014). The Painometer (de la Vega et al. 2014) app is one such example of a validated tool, presaging convenient and systematic electronic pain monitoring.

5.8 Summary

After years of largely disappointing and contradictory findings from candidate gene studies, the molecular determinants of pain have begun to be elucidated through well-designed GWAS and other omic methods. It is now recognized that the key to reliable and replicable "pain-omics" discoveries is to examine stable and heritable phenotypes, relevant to the underlying biological pathology, in large, well-characterized cohorts. Attaining the statistical power necessary for these studies is going to rely increasingly on collaborative consortia with consistent protocols. The application of accepted diagnostic criteria and other phenotyping recommendations (e.g., the ACTTION-APS Pain Taxonomy) should engender harmonization of data between studies, while broadening the scope of data collection to include clinical, environmental, affective, and sensory characteristics. Innovative phenotyping methods can be utilized and improved to capture the pain experience with greater precision and resolution without burdening patients.

References

Anderson DL. Development of an instrument to measure pain in rheumatoid arthritis: rheumatoid arthritis pain scale (RAPS). Arthritis Rheum. 2001;45(4):317–23.
Anderson KO, Dowds BN, Pelletz RE, Edwards WT, Peeters-Asdourian C. Development and initial validation of a scale to measure self-efficacy beliefs in patients with chronic pain. Pain. 1995;63(1):77–84.

Beck AT, Epstein N, Brown G, Steer RA. An inventory for measuring clinical anxiety: psychometric properties. J Consult Clin Psychol. 1988;56(6):893–7.

Beck AT, Steer RA, Brown GK. Manual for the Beck depression inventory–second edition. San Antonio: Psychological Corporation; 1996.

Bellamy N, Buchanan WW, Goldsmith CH, Campbell J, Stitt LW. Validation study of WOMAC: a health status instrument for measuring clinically important patient relevant outcomes to antirheumatic drug therapy in patients with osteoarthritis of the hip or knee. J Rheumatol. 1988;15(12):1833–40.

Bellamy N, Campbell J, Haraoui B, Buchbinder R, Hobby K, Roth JH, et al. Dimensionality and clinical importance of pain and disability in hand osteoarthritis: development of the Australian/Canadian (AUSCAN) osteoarthritis hand index. Osteoarthr Cartil. 2002;10(11):855–62.

Bennett M. The LANSS pain scale: the Leeds assessment of neuropathic symptoms and signs. Pain. 2001;92(1–2):147–57.

Bennett RM, Friend R, Jones KD, Ward R, Han BK, Ross RL. The revised fibromyalgia impact questionnaire (FIQR): validation and psychometric properties. Arthritis Res Ther. 2009;11(4):R120.

Bouhassira D, Attal N, Alchaar H, Boureau F, Brochet B, Bruxelle J, et al. Comparison of pain syndromes associated with nervous or somatic lesions and development of a new neuropathic pain diagnostic questionnaire (DN4). Pain. 2005;114(1–2):29–36.

Buysse DJ, Reynolds CF 3rd, Monk TH, Berman SR, Kupfer DJ. The Pittsburgh sleep quality index: a new instrument for psychiatric practice and research. Psychiatry Res. 1989;28(2): 193–213.

Cella D, Riley W, Stone A, Rothrock N, Reeve B, Yount S, et al. The patient-reported outcomes measurement information system (PROMIS) developed and tested its first wave of adult self-reported health outcome item banks: 2005–2008. J Clin Epidemiol. 2010;63(11):1179–94.

Chalder T, Berelowitz G, Pawlikowska T, Watts L, Wessely S, Wright D, et al. Development of a fatigue scale. J Psychosom Res. 1993;37(2):147–53.

Cohen S, Kamarck T, Mermelstein R. A global measure of perceived stress. J Health Soc Behav. 1983;24:385–96.

Dargie E, Holden RR, Pukall CF. The vulvar pain assessment questionnaire inventory. Pain. 2016;157(12):2672–86.

de la Vega R, Miro J. mHealth: a strategic field without a solid scientific soul. A systematic review of pain-related apps. PLoS One. 2014;9(7):e101312.

de la Vega R, Roset R, Castarlenas E, Sanchez-Rodriguez E, Sole E, Miro J. Development and testing of painometer: a smartphone app to assess pain intensity. J Pain. 2014;15(10):1001–7.

Derogatis LR. The SCL-90-R: administration, scoring and procedures manual. Minneapolis: National Computer Systems, Inc.; 1994.

Derogatis LR, Melisaratos N. The brief symptom inventory: an introductory report. Psychol Med. 1983;13:595–605.

Deyo RA, Dworkin SF, Amtmann D, Andersson G, Borenstein D, Carragee E, et al. Report of the NIH task force on research standards for chronic low back pain. Pain Med. 2014;15(8):1249–67.

Diatchenko L, Nackley A, Slade G, Fillingim R, Maixner W. Idiopathic pain disorders--pathways of vulnerability. Pain. 2006;123(3):226–30.

Diatchenko L, Fillingim RB, Smith SB, Maixner W. The phenotypic and genetic signatures of common musculoskeletal pain conditions. Nat Rev Rheumatol. 2013;9(6):340–50.

Diener E, Wirtz D, Tov W, Kim-Prieto C, Choi D-W, Oishi S, et al. New Well-being measures: short scales to assess flourishing and positive and negative feelings. Soc Indic Res. 2010;97(2):143–56.

Dworkin S, LeResche L. Research diagnostic criteria for temporomandibular disorders: review, criteria, examinations and specifications, critique. J Craniomandib Disord. 1992;6(4):301–55.

Dworkin RH, Turk DC, Farrar JT, Haythornthwaite JA, Jensen MP, Katz NP, et al. Core outcome measures for chronic pain clinical trials: IMMPACT recommendations. Pain. 2005;113(1–2):9–19.

Dworkin RH, Bruehl S, Fillingim RB, Loeser JD, Terman GW, Turk DC. Multidimensional diagnostic criteria for chronic pain: introduction to the ACTTION-American Pain Society Pain Taxonomy (AAPT). J Pain. 2016;17(9 Suppl):T1–9.

Edwards LC, Pearce SA, Turner-Stokes L, Jones A. The pain beliefs questionnaire: an investigation of beliefs in the causes and consequences of pain. Pain. 1992;51(3):267–72.

Eriksen MK, Thomsen LL, Olesen J. The Visual Aura Rating Scale (VARS) for migraine aura diagnosis. Cephalalgia. 2005;25(10):801–10.

Fairbank J. Use of Oswestry Disability Index (ODI). Spine (Phila Pa 1976). 1995;20(13):1535–7.

Freynhagen R, Baron R, Gockel U, Tolle TR. painDETECT: a new screening questionnaire to identify neuropathic components in patients with back pain. Curr Med Res Opin. 2006;22(10):1911–20.

Friedrich EG Jr. Vulvar vestibulitis syndrome. J Reprod Med Obstetrician Gynecol. 1987;32(2):110–4.

Gewandter JS, Chaudari J, Iwan KB, Kitt R, As-Sanie S, Bachmann G, et al. Research design characteristics of published pharmacologic randomized clinical trials for irritable bowel syndrome and chronic pelvic pain conditions: an ACTTION systematic review. J Pain. 2018;19(7):717–26.

Gottesman II, Gould TD. The endophenotypic concept in psychiatry: etymology and strategic intentions. Am J Psychiatry. 2003;160:636–45.

Gottesman O, Kuivaniemi H, Tromp G, Faucett WA, Li R, Manolio TA, et al. The electronic medical records and genomics (eMERGE) network: past, present, and future. Genet Med. 2013;15(10):761–71.

Gracely RH, McGrath F, Dubner R. Ratio scales of sensory and affective verbal pain descriptors. Pain. 1978;5(1):5–18.

Greenspan JD, Slade GD, Bair E, Dubner R, Fillingim RB, Ohrbach R, et al. Pain sensitivity risk factors for chronic TMD: descriptive data and empirically identified domains from the OPPERA case control study. J Pain. 2011;12(11, Supplement):T61–74.

Hanno PM, Burks DA, Clemens JQ, Dmochowski RR, Erickson D, Fitzgerald MP, et al. AUA guideline for the diagnosis and treatment of interstitial cystitis/bladder pain syndrome. J Urol. 2011;185(6):2162–70.

Hudak PL, Amadio PC, Bombardier C. Development of an upper extremity outcome measure: the DASH (disabilities of the arm, shoulder and hand) [corrected]. The upper extremity collaborative group (UECG). Am J Ind Med. 1996;29(6):602–8.

IASP Task Force on Taxonomy. Part III: pain terms, a current list with definitions and notes on usage. Recommended by the IASP Subcommittee on taxonomy. In: Merskey H, Bogduk N, editors. Classification of chronic pain. 2nd ed. Seattle: IASP Press; 1994. p. 209–14.

Jenkins CD, Stanton BA, Niemcryk SJ, Rose RM. A scale for the estimation of sleep problems in clinical research. J Clin Epidemiol. 1988;41(4):313–21.

Jensen MP, Gammaitoni AR, Olaleye DO, Oleka N, Nalamachu SR, Galer BS. The pain quality assessment scale: assessment of pain quality in carpal tunnel syndrome. J Pain. 2006;7(11):823–32.

Kerns RD, Turk DC, Rudy TE. The west haven-yale multidimensional pain inventory (WHYMPI). Pain. 1985;23(4):345–56.

Krebs EE, Lorenz KA, Bair MJ, Damush TM, Wu J, Sutherland JM, et al. Development and initial validation of the PEG, a three-item scale assessing pain intensity and interference. J Gen Intern Med. 2009;24(6):733–8.

Kroenke K, Spitzer RL, Williams JB. The PHQ-9: validity of a brief depression severity measure. J Gen Intern Med. 2001;16(9):606–13.

Lipton RB, Dodick D, Sadovsky R, Kolodner K, Endicott J, Hettiarachchi J, et al. A self-administered screener for migraine in primary care: the ID migraine validation study. Neurology. 2003;61(3):375–82.

Mailman MD, Feolo M, Jin Y, Kimura M, Tryka K, Bagoutdinov R, et al. The NCBI dbGaP database of genotypes and phenotypes. Nat Genet. 2007;39(10):1181–6.

McCracken LM, Zayfert C, Gross RT. The pain anxiety symptoms scale: development and validation of a scale to measure fear of pain. Pain. 1992;50(1):67–73.

Mearin F, Lacy BE, Chang L, Chey WD, Lembo AJ, Simren M, et al. Bowel disorders. Gastroenterology. 2016;150(6):1393–407.

Melzack R. The short-form McGill pain questionnaire. Pain. 1987;30(2):191–7.

Mogil JS, Sternberg WF, Marek P, Sadowski B, Belknap JK, Liebeskind JC. The genetics of pain and pain inhibition. Proc Natl Acad Sci U S A. 1996;93(7):3048–55.

Monsees GM, Tamimi RM, Kraft P. Genome-wide association scans for secondary traits using case-control samples. Genet Epidemiol. 2009;33(8):717–28.

Moore SM, Schiffman R, Waldrop-Valverde D, Redeker NS, McCloskey DJ, Kim MT, et al. Recommendations of common data elements to advance the science of self-management of chronic conditions. J Nurs Scholarsh. 2016;48(5):437–47.

Nicholas MK. The pain self-efficacy questionnaire: taking pain into account. Eur J Pain. 2007;11(2):153–63.

Nishizawa D, Fukuda K, Kasai S, Hasegawa J, Aoki Y, Nishi A, et al. Genome-wide association study identifies a potent locus associated with human opioid sensitivity. Mol Psychiatry. 2014;19:55–62.

Nyholt DR, Borsook D, Griffiths LR. Migrainomics – identifying brain and genetic markers of migraine. Nat Rev Neurol. 2017;13(12):725–41.

Ohrbach R, Larsson P, List T. The jaw functional limitation scale: development, reliability, and validity of 8-item and 20-item versions. J Orofac Pain. 2008;22(3):219–30.

Oshinsky ML, Tanveer S. Improving research through NINDS headache common data elements. Cephalalgia. 2018;38(14):2083–4.

Pennebaker JW. The psychology of physical symptoms (appendix B). New York: Raven Press; 1982.

Peters MJ, Broer L, Willemen HLDM, Eiriksdottir G, Hocking LJ, Holliday KL, et al. Genome-wide association study meta-analysis of chronic widespread pain: evidence for involvement of the 5p15.2 region. Ann Rheum Dis. 2012;72(3):427–36.

Radloff LS. The CES-D scale: a self-report depression scale for research in the general population. Appl Psychol Meas. 1977;1(3):385–401.

Ramos EM, Hoffman D, Junkins HA, Maglott D, Phan L, Sherry ST, et al. Phenotype-genotype integrator (PheGenI): synthesizing genome-wide association study (GWAS) data with existing genomic resources. Eur J Hum Genet. 2014;22(1):144–7.

Robinson ME, Riley JL 3rd, Myers CD, Sadler IJ, Kvaal SA, Geisser ME, et al. The coping strategies questionnaire: a large sample, item level factor analysis. Clin J Pain. 1997;13(1):43–9.

Roland M, Morris R. A study of the natural history of back pain. Part I: development of a reliable and sensitive measure of disability in low-back pain. Spine (Phila Pa 1976). 1983;8(2):141–4.

Rolke R, Magerl W, Campbell KA, Schalber C, Caspari S, Birklein F, et al. Quantitative sensory testing: a comprehensive protocol for clinical trials. Eur J Pain. 2006;10(1):77–88.

Schmid AB, Adhikari K, Ramirez-Aristeguieta LM, Chacon-Duque JC, Poletti G, Gallo C, et al. Genetic components of human pain sensitivity: a protocol for a genome-wide association study of experimental pain in healthy volunteers. BMJ Open. 2019;9(4):e025530.

Shacham S. A shortened version of the profile of mood states. J Pers Assess. 1983;47(3):305–6.

Sham PC, Purcell SM. Statistical power and significance testing in large-scale genetic studies. Nat Rev Genet. 2014;15(5):335–46.

Sillanpaa MJ. Overview of techniques to account for confounding due to population stratification and cryptic relatedness in genomic data association analyses. Heredity. 2011;106(4):511–9.

Society IH. Headache classification Committee of the International Headache Society (IHS) the international classification of headache disorders, 3rd edition. Cephalalgia. 2018;38(1):1–211.

Spielberger CD, Gorsuch RL, Lushene R, Vagg PR, Jacobs GA. Manual for the state-trait anxiety inventory (form Y1). Palo Alto: Consulting Psychologists Press; 1983.

Sullivan MJL, Bishop SR, Pivik J. The pain catastrophizing scale: development and validation. Psychol Assess. 1995;7(4):524–32.

Toomey TC, Mann JD, Abashian SW, Carnrike CL Jr, Hernandez JT. Pain locus of control scores in chronic pain patients and medical clinic patients with and without pain. Clin J Pain. 1993;9(4):242–7.

Treede RD, Rief W, Barke A, Aziz Q, Bennett MI, Benoliel R, et al. Chronic pain as a symptom or a disease: the IASP classification of chronic pain for the international classification of diseases (ICD-11). Pain. 2019;160(1):19–27.

van de Merwe JP, Nordling J, Bouchelouche P, Bouchelouche K, Cervigni M, Daha LK, et al. Diagnostic criteria, classification, and nomenclature for painful bladder syndrome/interstitial cystitis: an ESSIC proposal. Eur Urol. 2008;53(1):60–7.

van Hecke O, Kamerman PR, Attal N, Baron R, Bjornsdottir G, Bennett DL, et al. Neuropathic pain phenotyping by international consensus (NeuroPPIC) for genetic studies: a NeuPSIG systematic review, Delphi survey, and expert panel recommendations. Pain. 2015;156(11):2337–53.

Veluchamy A, Hebert HL, Meng W, Palmer CNA, Smith BH. Systematic review and meta-analysis of genetic risk factors for neuropathic pain. Pain. 2018;159(5):825–48.

Vernon H, Mior S. The neck disability index: a study of reliability and validity. J Manip Physiol Ther. 1991;14(7):409–15.

Ware JE Jr, Sherbourne CD. The MOS 36-item short-form health survey (SF-36). I. Conceptual framework and item selection. Med Care. 1992;30(6):473–83.

Ware J Jr, Kosinski M, Keller SD. A 12-item short-form health survey: construction of scales and preliminary tests of reliability and validity. Med Care. 1996;34(3):220–33.

Wolfe F, Clauw DJ, Fitzcharles MA, Goldenberg DL, Hauser W, Katz RL, et al. 2016 revisions to the 2010/2011 fibromyalgia diagnostic criteria. Semin Arthritis Rheum. 2016;46(3):319–29.

Zigmond AS, Snaith RP. The hospital anxiety and depression scale. Acta Psychiatr Scand. 1983;67(6):361–70.

Zorina-Lichtenwalter K, Meloto CB, Khoury S, Diatchenko L. Genetic predictors of human chronic pain conditions. Neuroscience. 2016;338:36–62.

Genomics of Breast Cancer and Treatment-Related Pain and Comorbid Symptoms

6

Angela R. Starkweather, Gee Su Yang, Debra Lynch Kelly, and Debra E. Lyon

Contents

6.1 Introduction

Symptoms are a perception or sensation that can alter normal function, reflecting a change in health status, onset of illness or disease. The term symptom has also been referred to as a subjective indicator of disease or physical disturbance (National Institutes of Health, National Institute of Nursing Research 2013) and

A. R. Starkweather (✉) · G. S. Yang
University of Connecticut School of Nursing, Storrs, CT, USA
e-mail: Angela.Starkweather@uconn.edu; geesu.yang@uconn.edu

D. L. Kelly · D. E. Lyon
College of Nursing, University of Florida, Gainesville, FL, USA
e-mail: dlynchkelly@ufl.edu; delyon@ufl.edu

© Springer Nature Switzerland AG 2020
S. G. Dorsey, A. R. Starkweather (eds.), *Genomics of Pain and Co-Morbid Symptoms*, https://doi.org/10.1007/978-3-030-21657-3_6

more recently as a sensation or perception of change related to health function that an individual experiences (Danzer et al. 2012). Although symptoms and symptom patterns are used to help prioritize the clinician's differential diagnosis and/or identify etiological mechanisms, patients often report non-specific symptoms that commonly co-occur over the course of chronic disease, including cancer. The recognition of common co-occurring symptoms dates back to observations in animal models in response to inflammatory activation, which were then coined as "sickness behaviors" (Hart 1988). This set of behaviors, which includes fever, fatigue, loss of appetite, hypersomnia, and depression, has been compared to symptoms observed in patients with cancer (Cleeland et al. 2003). In translating this phenomenon to humans, the term symptom clusters arose. Dodd et al. (Dodd et al. 2001) defined a symptom cluster as three or more concurrent and related symptoms that may or may not have a common etiology while Miaskowski et al. (2004) suggested that symptoms may be related through a common mechanism, by sharing common variance or producing different outcomes than individual symptoms. Subsequently, Kim et al. (2005) defined a symptom cluster as a stable group of two or more concurrent symptoms that are related to one another and independent of other symptom clusters.

To date, research on symptom clusters has provided a deeper understanding of the variations of the symptom experience during different disease states and evidence of the negative effects of multiple, co-occurring symptoms on quality of life. Collectively, this body of research has shown that symptoms can be more or less prevalent across different stages of disease and during treatment (Dodd et al. 2004; Bender et al. 2005; Roscoe et al. 2002). Individuals who report more symptoms at diagnosis appear to have a greater symptom burden during treatment (Kim et al. 2009a; Palesh et al. 2007; Miaskowski et al. 2006) and significantly lower quality of life (Byar et al. 2006; Kim et al. 2009b). Besides the number of symptoms reported, symptom severity has also shown an inverse relationship with quality of life (So et al. 2009; Liu et al. 2009; Dodd et al. 2010; Fu et al. 2009). For some individuals, however, the symptoms that begin with the onset of disease persist over time, even after adequate treatment and recovery/relapse is attained (Gwede et al. 2008).

Understanding the symptom experience and symptoms clusters is an important component of patient-centered care. By gaining insight on how symptoms impact the individual, their functional status, and quality of life, as well as investigating plausible mechanisms underlying the frequency, severity, and duration of symptoms, nurses and other healthcare professionals can: (1) predict the symptom trajectory of individual patients based on their personal factors, such as psychosocial and genomic data; (2) develop more precise and effective symptom management strategies that target common mechanisms; and, (3) devise risk assessment tools of the symptom experience to inform patients of the potential options to prevent persistent symptoms. Several symptom-based theoretical models exist to help guide understanding of the factors involved in the symptom experience and how nurses can use this information to improve the delivery of symptom interventions.

6.2   Theoretical Models Used in Cancer Symptom Research

The San Francisco Symptom Management Theory (SMT) developed at the University of California (Larson et al. 1994) posits a relationship among key components of the symptom experience, symptom management strategies, and patient outcomes. The symptom assessment by the individual is paramount in this model including the individuals' perception of the frequency, intensity, distress, and meaning of symptoms as they occur. Symptoms are evaluated within the context of three dimensions of nursing science (person, health and illness, environment). The person dimension includes demographic, psychological, sociological, physiological, and developmental factors. The health and illness dimension includes the patient's current health status, degree of disease or injury, and associated risk factors for these conditions. The environment dimension includes physical, social, and cultural factors. Symptom management is a dynamic process that evolves and changes over time with critical factors including who delivers the strategies, strategies delivered, and when and how they are delivered. Individual outcomes include symptom status, emotional and function status, quality of life, cost, mortality, and morbidity. Each component of the model can influence and be influenced by the other components.

Lenz's theory of unpleasant symptoms (Lenz et al. 1997) asserts relationships among physiological, psychological, and situational antecedents of symptoms as well as functional outcomes of symptoms. The antecedents (physiological, psychological, and/or situational) influence the symptom experience. Symptom attributes include intensity, distress, quality, and timing. The presence of multiple symptoms have a catalytic effect. The consequence of unpleasant symptoms results in changes in functional and/or cognitive activities, activities of daily living, social activities and interactions and role performance.

The Symptom Interpretation Model (SIM) (Teel et al. 1997) posits that symptoms are understood in terms of: (1) *input*, or process of recognizing a physiological disturbance and distinguishing it as an experience that is different from normal; (2) *interpretation*, including conceptual identification of a symptom through a cognitive process, knowledge structure activation in which symptoms are compared to expectations and past experiences, and reasoning in which judgment heuristic and affective responses serve as filters for the symptom experience; and (3) *outcome*, the response to the symptom experience through care-seeking or self-management behaviors. The model incorporates important variables such as the presence and severity of disease, patient experience, beliefs, expectations and preferences, as well as duration of time. All of these factors work together to create and influence the symptom experience. In addition, both cognitive and emotional processes can affect patients' symptom interpretations as well as the outcome stage, which involves symptom management behaviors (Leventhal et al. 1992).

Due to the importance of the temporal aspects of the symptom experience, Brant and colleagues proposed a revision to the symptom management model (Brant et al. 2009), which incorporated previously published symptom management theories/models, including the theory of symptom management (Larson et al. 1994), the theory of unpleasant symptoms (Lenz et al. 1997), the symptoms experience model (Armstrong 2003), and the symptoms experience in time model (Henly et al. 2003).

The main components of the model include concepts that are operationalized for research and used to assess relationships among them. These include antecedents of symptom experience, the symptom trajectory, consequences, and symptom management intervention(s).

Each of these theoretical models offers a structure for understanding the critical cofactors of the symptom experience and posit relationships among the dimensions (such as among antecedents, symptom trajectory, and consequences) that can be useful for testing hypotheses and validating research findings. By far, the largest body of work focused on symptom clusters has been conducted in women with breast cancer. However, many other chronic health conditions share common symptoms such as pain, anxiety/depression, fatigue, sleep disturbances, and cognitive impairment. Due to the breadth of studies integrating symptoms and genomics, this chapter will focus on summarizing the ways in which symptoms can be studied in cancer populations and the most recent findings among breast cancer survivors.

6.3   Incorporation of Salient Factors in Cancer Symptoms Research

When designing research studies to elucidate genomic mechanisms of the symptom experience in cancer survivors, it is important to capture the factors that may influence the study endpoint, such as a symptom (pain) or symptom cluster (Fig. 6.1).

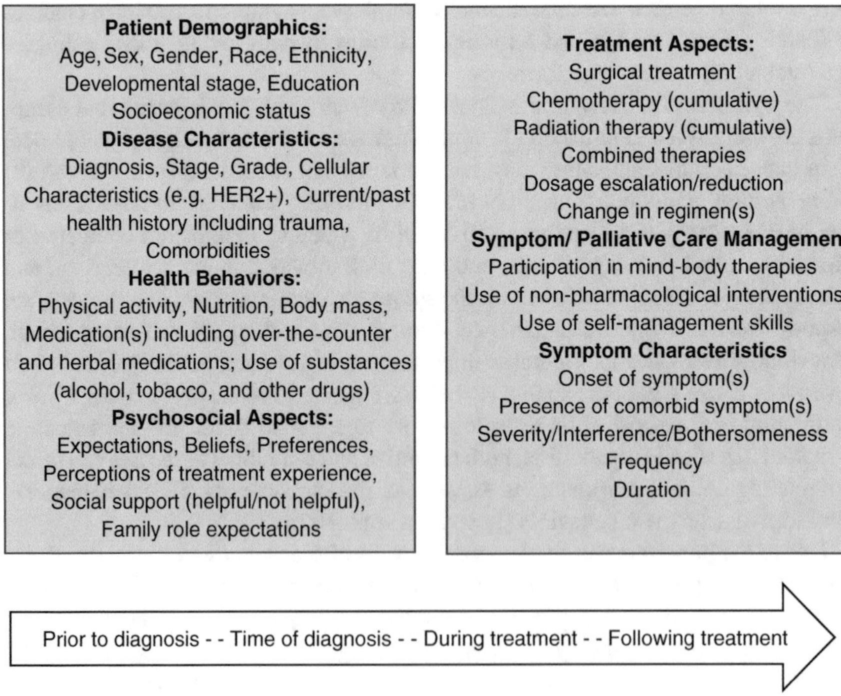

Patient Demographics:
Age, Sex, Gender, Race, Ethnicity,
Developmental stage, Education
Socioeconomic status
Disease Characteristics:
Diagnosis, Stage, Grade, Cellular
Characteristics (e.g. HER2+), Current/past
health history including trauma,
Comorbidities
Health Behaviors:
Physical activity, Nutrition, Body mass,
Medication(s) including over-the-counter
and herbal medications; Use of substances
(alcohol, tobacco, and other drugs)
Psychosocial Aspects:
Expectations, Beliefs, Preferences,
Perceptions of treatment experience,
Social support (helpful/not helpful),
Family role expectations

Treatment Aspects:
Surgical treatment
Chemotherapy (cumulative)
Radiation therapy (cumulative)
Combined therapies
Dosage escalation/reduction
Change in regimen(s)
Symptom/ Palliative Care Management
Participation in mind-body therapies
Use of non-pharmacological interventions
Use of self-management skills
Symptom Characteristics
Onset of symptom(s)
Presence of comorbid symptom(s)
Severity/Interference/Bothersomeness
Frequency
Duration

Prior to diagnosis - - Time of diagnosis - - During treatment - - Following treatment

Fig. 6.1 Critical factors for cancer symptoms research

Major challenges for cancer symptom researchers include the temporal nature of symptoms, treatments and patient exposure to a variety of pharmacological and non-pharmacological therapies. Advancements in providing individualized care to patients with cancer also produce wide variation in treatment regimens, which is difficult to control in mechanistic research. In addition, the integration of genomic, epigenomic, and metabolomic data to explicate the measurable cellular changes associated with the symptom(s) is a major milestone that will require further investigation.

6.4 Genotypic Studies in Breast Cancer Populations: Focus on Pain

It is well known that genetic variability plays an important role in modulating pain signaling transmission in the central and peripheral nervous system in individuals (Mogil 2012; Zorina-Lichtenwalter et al. 2016; Dorsey et al. 2018). Single nucleotide polymorphisms (SNPs), which are considered as a stable biomarker, may result in an increase or decrease in individuals' pain susceptibility and response to analgesics (Mogil 2012; Zorina-Lichtenwalter et al. 2016; Dorsey et al. 2018). Genotypic studies can help strategize effective precision medicine approaches in aspects of prediction/screening and therapeutics development in patients suffering from cancer-related symptoms (Low et al. 2018).

A recent systematic review of genetic variants associated with cancer pain (i.e. pain resulting from primary cancer or metastases), post-cancer treatment pain (i.e. pain resulting from surgery, chemotherapy, or radiation therapy), and response to opioid analgesics was conducted, demonstrating the contribution of multiple genes involved in each phenomenon (Yang et al. 2019a). In breast cancer, it was found that genetic polymorphisms of interleukin(*IL*)-*1R1* [rs2110726], *IL-13* [rs1295686], fatty acid amide hydrolase, *FAAH* [rs324430, rs1571138, rs3766246, and rs4660928], catechol-O-methyltransferase, *COMT* [rs4680] and opioid receptor mu 1, *OPRM1* [rs1799971] influence variability in preoperative breast pain. Whereas persistent postsurgical pain in breast cancer patients was associated with genetic polymorphisms of *IL1R2* [rs11674595], *IL-10* [rs3024505], *IL-6* [rs20069840], *CXCL8* [rs4073], *TNF* [rs1800610], *COMT* [rs4680, rs165774, rs887200, and rs4818], *CACNG2* [rs4820242], *KCNA1* [rs4766311], *KCND2* [rs1072198], *KCNJ3* [rs12995382, rs17641121], *KCNJ6* [rs858003], and *KCNK9* [rs2542424 and rs2545457]. Other studies that followed women with breast cancer who were treated with aromatase inhibitors (AIs) found that polymorphisms in *CYP17A1* [rs4919686, rs4919683, rs4919687, rs3781287, rs10786712, rs6163, and rs743572], *CYP19A1* [rs4775936], *CYP27B1* [rs4646536], *VDR* [rs11568820], and *OPG* [rs2073618] contributed to arthralgia. In addition, *OPRM1* [rs1799971] was associated with response to opioids in patients with postsurgical breast pain, with carriers of at least one minor allele in *OPRM1* experiencing more opioid requirement (e.g., oxycodone) to relieve pain compared to those of two major alleles.

6.5   Focus on the Psychoneurological Symptom Cluster

A systematic review of the genetic variants found to be associated with the psycho-neurological symptom cluster, which includes anxiety, cognitive impairment, depressive symptoms, fatigue, sleep disturbances, and pain (Kim et al. 2012; Lengacher et al. 2012; Starkweather et al. 2013), was recently published (Yang et al. 2019b). The review found that most polymorphisms associated with psychoneuro-logical symptoms were involved in immune regulation, metabolism, signal trans-duction, and neuronal system pathways. Of note, pain was associated with the largest number of polymorphisms. In the findings, specific pathways were found to contribute to key symptoms including:

- Fatigue: signal transduction pathway genes (including *BDNF, ADRB, GALR1, NPY1R*, and *NTRK2*)
- Fatigue and Pain: metabolism pathway genes (including *COMT, GCH1, FAAH, VCR, CYP3A4, CYP17A1, CYP19A1*, and *CYP27B1*)
- Anxiety and Pain: neuronal system pathway genes (including *SERT, SLC6A4, KCNA1, KCND2, KCNJ3*, and *KCNJ6*)

The complexity of identifying associations among symptoms and genetic poly-morphisms provides a way to possibly recognize common pathways. For instance, variants of the *COMT* gene were associated with the symptom cluster of pain, fatigue, and cognitive impairment. According to this review, sleep disturbances were associated with polymorphisms in *IL13, IL1R2, NFKB2*, and *NTRK2*; depres-sive symptoms were associated with polymorphisms in *BDNF, IL6, INFGR1, TNFA*, and *IL1B*; and cognitive impairment may be attributed to polymorphisms in *COMT, APOE, IL1R1*, and *BDNF*. Very recently, Chan et al. (2019) evaluated a potential association between SNPs in DNA methyltransferase 1 (*DNMT1*) and cognitive impairment in a cohort of breast cancer patients ($n = 351$). In this study, patients carrying the A allele in *DNMT1* [rs2162560] had lower odds of cognitive impairment in functional interference and concentration cognitive domains, impli-cating a neuroprotective effect.

Another emerging area of research will be investigating responses to various treat-ments based on genetic variation. In assessing various treatments for arthralgia due to AIs in women with breast cancer, studies have found large effect sizes for pharmaco-logical approaches, acupuncture and relaxation techniques (Yang et al. 2017). The next stage of this research will be to examine whether there were differences in response among individuals based on their genomic background or other factors.

6.6   Additional Considerations of Sex, Race, Ethnicity, and Social Determinants of Health

The research on cancer-related symptoms has become more complex as investiga-tors have sought to explain long-standing differences in treatment outcomes among various populations. Social determinants, such as age, race/ethnicity, income/

insurance, education, comorbidity, employment, beliefs/communication on pre-scribed medication, may affect symptoms and overall treatment experiences in women with breast cancer (Feinberg and Longo 2018). Miaskowski et al. (2014) found that oncology patients (breast, lung, gynecological, or gastrointestinal can-cer) receiving chemotherapy included in a high symptom cluster were significantly younger; more likely to be female and nonwhite; had a higher comorbid condition and poorer functional status; and, reported lower socioeconomic status, suggesting that social determinants of health are contributory factors to cancer-related symp-toms. Noting that black women die from breast cancer at a rate 42% higher than white women (Richardson et al. 2016), Rosenzweig's team developed a novel explanatory model that incorporates race and social determinants of health, symp-tom experience, management, and omic methods (genomics, epigenomics, and metabolomics) for the disparity in breast cancer symptoms and treatment (McCall et al. 2019). Current evidence is used to support the inclusion of each concept in the model and relationships among concepts and understand applicable targets in envi-ronmental and social contexts for translation into clinical practice. Although the only symptom reviewed in the article was taxane-induced peripheral neuropathy, the authors make the case for integration social genomics within symptom science research.

6.7　Capturing Changes Over Time: Gene Expression and Epigenomics

In contrast with genetic variants, gene expression profiles and epigenetic alterations can be used to investigate changes in the function of genes over time. Post-translational regulations involving chromatin organization, DNA methylation, and histones can cause epigenetic alterations in humans (Feinberg and Longo 2018). Miaskowski's team (Kober et al. 2018) recently studied gene expression profiles using RNA-sequencing to identify differentially expressed genes in women with breast cancer who received paclitaxel and either develop paclitaxel-induced periph-eral neuropathy ($n = 25$) or did not ($n = 25$). The study identified five differentially expressed genes (e.g., *PRDX5*, *GLRX5*, *FIS1*, *UBB*) and nine pathways (e.g., mitophagy, mTOR signaling, PI3K-AKT signaling, p53 signaling, cellular senes-cence) associated with mitochondrial dysfunction related to oxidative stress, homeostasis, mitochondrial fission, apoptosis, and autophagy. This team also exam-ined potential epigenetic mechanisms of evening fatigue in breast cancer women receiving chemotherapy (Flowers et al. 2019). In this study, 23 significant differen-tially methylated positions in genes related to inflammation (e.g., *HOXA9*, *LRRCI5*, *TFPI*), skeletal muscle energy (e.g., *PAX7*), neurotransmission (e.g., *CNTN2*), renal function (e.g., *UTS2*), transcription (e.g., *CCDC93*, *PAPOLG*), and cell-cycle regu-lation (e.g., *CEP41*) were found between patients in the moderate evening fatigue and very high fatigue classes. Lyon's team (Lyon et al. 2014) has been examining the psychoneurological symptom cluster in women with breast cancer and found relationships among fatigue and objective cognitive performance, suggesting that these symptoms are functionally related but may differ in underlying mechanisms and severity over time (Gullett et al. 2019). More recently, the team found acquired epigenetic changes, such as differential DNA methylation ratios, were associated with changes in cognitive function over 1-year as an adverse effect of cancer and its treatment in breast cancer survivors (Yang et al. 2019c). Differential methylation ratio changes associated with lack of memory improvement were localized to genes involved in neural function (*ECE2, PPFIBP2*) or signaling processes (*USP6NL, RIPOR2, KLF5, UBE2V1, DGKA, RPS6KA1*), and differences in memory domain scores were inversely correlated with differences in methylation patterns, suggesting that increases in memory domain scores were associated with decreases in DNA methylation. Another study found that chemotherapy significantly changes approximately 4.2% of CpG sites in comparison to stable DNA methylome in healthy controls (without cancer or chemotherapy administration) (Yao et al. 2019). Importantly, leukocyte composition was controlled for in the analysis as well as adjustments for the chemotherapeutic regimen, cumulative infusions, growth factors, and steroids. The team found four CpGs that were significantly altered following chemotherapy, with cg16936953 in *VMP1/MIR21*, cg01252023 in *CORO1B*, cg11859398 in *SDK1*, and cg19956914 in *SUMF2* associated with cognitive decline in breast cancer patients.

6.8   Conclusions

Symptom science research in cancer populations is still in its infancy. Although many important findings have been described for predicting the symptom experience in women with breast cancer, no studies to date have prospectively studied and validated the predictive models that have been developed. Several important milestones must be reached in order to translate current knowledge into the clinical settings. These include validation of findings across datasets, validation of findings in prospective studies, and integration of omic measures for improved prediction of symptoms and/or response to treatment(s). Advancements in this area of science hold promise for improving the symptom experience for individuals diagnosed and receiving treatment, and possibly for reduction of persistent symptoms after treatment has commenced.

References

Armstrong TS. Symptoms experience: a concept analysis. Oncol Nurs Forum. 2003;30(4):601–6. https://doi.org/10.1188/03.ONF.601-606.

Bender CM, Senuzen Ergun F, Rosenzweig MQ, Cohen SM, Sereika SM. Symptom clusters in breast cancer across 3 phases of the disease. Cancer Nurs. 2005;28(3):219–25. https://doi.org/10.1097/00002820-200505000-00011.

Brant JM, Beck S, Miaskowski C. Building dynamic models and theories to advance the science of symptom management research. J Adv Nurs. 2009;66(1):228–40. https://doi.org/10.1111/j.1365-2648.2009.05179.x.

Byar KL, Berger AM, Bakken SL, Cetak MA. Impact of adjuvant breast cancer chemotherapy on fatigue, other symptoms, and quality of life. Oncol Nurs Forum. 2006;33(1):E18–26. https://doi.org/10.1188/06.ONF.E18-E26.

Chan A, Yeo A, Shwe M, Tan CJ, Foo KM, Chu P, Khor CC, Ho HK. An evaluation of DNA methyltransferase 1 (DNMT1) single nucleotide polymorphisms and chemotherapy-associated cognitive impairment: a prospective, longitudinal study. Sci Rep. 2019;9(1):14570. https://doi.org/10.1038/s41598-019-51203-y.

Cleeland CS, Bennett GJ, Dantzer R, Dougherty PM, Dunn AJ, Meyers CA, Miller AH, Payner R, Reuben JM, Wang XS, Lee BN. Are the symptoms of cancer and cancer treatment due to a shared biologic mechanism? A cytokine-immunologic model of cancer symptoms. Cancer. 2003;97(11):2919–25. https://doi.org/10.1002/cncr.11382.

Danzer R, Meagher MW, Cleeland CS. Translational approaches to treatment-induced symptoms in cancer patients. Nat Rev Clin Oncol. 2012;9(7):414–26. https://doi.org/10.1038/nrclinonc.2012.88.

Dodd M, Janson S, Facione N, Fauceet J, Froelicher ES, Humphreys J, Lee K, Miaskowski C, Puntillo K, Rankin S, Taylor D. Advancing the science of symptom management. J Adv Nurs. 2001;33:668–76. https://doi.org/10.1046/j.1365-2648.2001.01697.x.

Dodd MJ, Miaskowski C, Lee KA. Occurrence of symptom clusters. J Natl Cancer Inst Monogr. 2004;32:76–8. https://doi.org/10.1093/jncimonographs/lgh008.

Dodd MJ, Cho MH, Cooper BA, Miaskowski C. The effect of symptom clusters on functional status and quality of life in women with breast cancer. Eur J Oncol Nurs. 2010;14(1):101–10. https://doi.org/10.1016/j.ejon.2009.09.005.

Dorsey SG, Resnick BM, Renn CL. Precision health: use of omics to optimize self-management of chronic pain in aging. Res Gerontol Nurs. 2018;11:7–13. https://doi.org/10.3928/19404921-20171128-01.

Feinberg AP, Longo JD. The key role of epigenetics in human disease prevention and mitigation. N Engl J Med. 2018;378(14):1323–34. https://doi.org/10.1056/NEJMra1402513.

Flowers E, Flentje A, Levine J, Olshen A, Hammer M, Paul S, Conley Y, Miaskowski C, Kober K. A pilot study using a multistaged integrated analysis of gene expression and methylation to evaluate mechanisms for evening fatigue in women who received chemotherapy for breast cancer. Biol Res Nurs. 2019;21(2):142–56.

Fu OS, Crew KD, Jacobson JS, Greenlee H, Yu G, Campbell J, Ortiz Y, Hershman DL. Ethnicity and persistent symptom burden in breast cancer survivors. J Cancer Surviv. 2009;3(1):241–50. https://doi.org/10.1007/s11764-009-0100-7.

Gullett JM, Cogen RA, Yang GS, Menzies VS, Fieo RA, Kelly DL, Starkweather AR, Jackson-Cook CK, Lyon DE. Relationship of fatigue with cognitive performance in women with early-stage breast cancer over 2 years. Psychooncology. 2019;28(5):997–1003. https://doi.org/10.1002/pon.5028.

Gwede CK, Small BJ, Munster PN, Andrykowski MA, Jacobsen PB. Exploring the differential experience of breast cancer treatment-related symptoms: a cluster analytic approach. Support Care Cancer. 2008;16(8):925–33. https://doi.org/10.1007/s00520-007-0364-2.

Hart BL. Biological basis of the behavior of sick animals. Neurosci Biobehav Rev. 1988;12(2):123–37. https://doi.org/10.1016/s0149-7634(88)80004-6.

Henly SJ, Kallas KD, Klatt CM. The notion of time in symptom experiences. Nurs Res. 2003;52(6):410–7. https://doi.org/10.1097/00006199-200311000-00009.

Kim H, McGuire DB, Tulman L, Barsevick AM. Symptom clusters: concept analysis and clinical implications for cancer nursing. Cancer Nurs. 2005;28:270–82. https://doi.org/10.1097/00002820-200507000-00005.

Kim HJ, Barsevick AM, Tulman L. Predictors of the intensity of symptoms in a cluster in patients with breast cancer. J Nurs Scholar. 2009a;41(2):158–65. https://doi.org/10.1111/j.1547-5069.2009.01267.x.

Kim E, Jahan T, Aouizerat BE, Dodd MJ, Cooper BA, Paul SM, West C, Lee K, Swift PS, Wara W, Miaskowski C. Differences in symptom clusters identified using occurrence rates versus symptom severity rating in patients at the end of radiation therapy. Cancer Nurs. 2009b;32(6):429–36. https://doi.org/10.1097/NCC.0b013e3181b046ad.

Kim HJ, Barsevick AM, Fang CY, Miaskowski C. Common biological pathways underlying the psychoneurological symptom cluster in cancer patients. Cancer Nurs. 2012;35:E1–E20. https://doi.org/10.1097/NCC.0b013e318233a811.

Kober KM, Olshen A, Conley YP, Schumacher M, Topp K, Smoot B, Mazor M, Chesney M, Hammer M, Paul SM, Levine JD, Miaskowsk C. Expression of mitochondrial dysfunction-related genes and pathways in paclitaxel-induced peripheral neuropathy in breast cancer survivors. Mol Pain. 2018;14:1744806918816462. https://doi.org/10.1177/1744806918816462.

Larson P, Carrieri-Kohlman V, Dodd M, et al. A model for symptom management. The University of California, San Francisco School of Nursing Symptom Management Faculty Group. Image J Nurs Sch. 1994;26(4):272–6.

Lengacher CA, Reich RR, Post-White J, Moscoso M, Shelton MM, Barta M, Le N, Budhrani P. Mindfulness based stress reduction in post-treatment breast cancer patients: an examination of symptoms and symptom clusters. J Behav Med. 2012;35:86–94. https://doi.org/10.1007/s10865-011-9346-4.

Lenz ER, Pugh LC, Milligan RA, Gift A, Suppe F. The middle-range theory of unpleasant symptoms: an update. ANS Adv Nurs Sci. 1997;19(3):14–27. https://doi.org/10.1097/00012272-199703000-00003.

Leventhal H, Diefenbach M, Leventhal EA. Illness cognition: using common sense to understand treatment adherence and affect cognition interactions. Cogn Ther Res. 1992;16:143–63. https://doi.org/10.1007/BF01173486.

Liu L, Fiorentino L, Natarajan L, Parker BA, Mills PJ, Robins Sadler G, Dimsdale JE, Rissling M, He F, Ancoli-Israel S. Pre-treatment symptom cluster in breast cancer patients is associated with worse sleep, fatigue and depression during chemotherapy. Psychooncology. 2009;18(2):187–94. https://doi.org/10.1002/pon.1412.

Low SK, Zembutsu H, Nakamura Y. Breast cancer: the translation of big genomic data to cancer precision medicine. Cancer Sci. 2018;109:497–506. https://doi.org/10.1111/cas.13463.

Lyon D, Elmore L, Aboalela N, Merrill-Schools J, McCain N, Starkweather A, Elswick RK Jr, Jackson-Cook C. Potential epigenetic mechanism(s) associated with the persistence of psychoneurological symptoms in women receiving chemotherapy for breast cancer: a hypothesis. Biol Res Nurs. 2014;16(2):160–74. https://doi.org/10.1177/1099800413483545.

McCall MK, Connolly M, Nugent B, Conley YP, Bender CM, Rosenzweig MQ. Symptom experience, management, and outcomes according to race and social determinants including genomics, epigenomics, and metabolomics (SEMOARS + GEM): an explanatory model for breast cancer treatment disparity. J Cancer Educ, 2019. https://doi.org/10.1007/s13187-019-01571-w.

Miaskowski C, Dodd M, Lee K. Symptom clusters: the new frontier in symptom management research. J Natl Cancer Inst Monogr. 2004;32:17–21. https://doi.org/10.1093/jncimonographs/lgh023.

Miaskowski C, Cooper BA, Paul SM, Dodd M, Lee K, Aouizerat BE, West C, Cho M, Bank A. Subgroups of patients with cancer with different symptom experiences and quality-of-life outcomes: a cluster analysis. Oncol Nurs Forum. 2006;33(5):E79–89. https://doi.org/10.1188/06.ONF.E79-E89.

Miaskowski C, Cooper BA, Melisko M, Chen LM, Mastick J, West C, Paul SM, Dunn LB, Schmidt BL, Hammer M, Cartwright F, Wright F, Langford DJ, Lee K, Aouizerat BE. Disease and treatment characteristics do not predict symptom occurrence profiles in oncology outpatients receiving chemotherapy. Cancer. 2014;120:2371–8. https://doi.org/10.1002/cncr.28699.

Mogil JS. Pain genetics: past, present and future. Trends Genet. 2012;28:258–66. https://doi.org/10.1016/j.tig.2012.02.004.

National Institutes of Health, National Institute of Nursing Research. Symptom science. Bethesda: National Institutes of Health; 2013. Retrieved from https://www.ninr.nih.gov/newsandinformation/iq/symptom-science-workshop.

Palesh OG, Collie K, Batiuchok D, Tilston J, Koopman C, Perlis ML, Butler LD, Carlson R, Spiegel D. A longitudinal study of depression, pain, and stress as predictors of sleep disturbance among women with metastatic breast cancer. Biol Psychol. 2007;75(1):37–44. https://doi.org/10.1016/j.biopsycho.2006.11.002.

Richardson LC, Henley SJ, Miller JW, Massetti G, Thomas CC. Patterns and trends in age-specific black-white differences in breast cancer incidence and mortality – United States, 1999-2014. MMWR Morb Mortal Wkly Rep. 2016;65:1093–8. https://doi.org/10.15585/mmwr.mm6540a1.

Roscoe JA, Morrow GR, Hickok JT, Bushunow P, Matteson S, Rakita D, Andrews PL. Temporal interrelationships among fatigue, circadian rhythm and depression in breast cancer patients undergoing chemotherapy treatment. Support Care Cancer. 2002;10(4):329–36. https://doi.org/10.1007/s00520-001-0317-0.

So WK, Marsh G, Ling WM, Leung FY, Lo JC, Yeung M, Li GK. The symptom cluster of fatigue, pain, anxiety, and depression and the effect on the quality of life of women receiving treatment for breast cancer: a multicenter study. Oncol Nurs Forum. 2009;36(4):E205–14. https://doi.org/10.1188/09.ONF.E205-E214.

Starkweather AR, Lyon DE, Elswick RK Jr, Montpetit AJ, Conley Y, McCain NL. A conceptual model of psychoneurological symptom cluster variation in women with breast cancer: bringing nursing research to personalized medicine. Curr Pharmacogenomics Pers Med. 2013;11:224–30. https://doi.org/10.2174/1875692111311319990004.

Teel CS, Meek P, McNamara AM, Watson L. Perspectives unifying symptom interpretation. Image J Nurs Sch. 1997;29(2):175–81. https://doi.org/10.1111/j.1547-5069.1997.tb01553.x.

Yang GS, Kim HJ, Griffith KA, Zhu S, Dorsey SG, Renn CL. Interventions for the treatment of aromatase inhibitor-associated arthralgia in breast cancer survivors: a systematic review and meta-analysis. Cancer Nurs. 2017;40(4):E26–41. https://doi.org/10.1097/NCC.0000000000000409.

Yang GS, Barnes NM, Lyon DE, Dorsey SG. Genetic variants associated with cancer pain and response to opioid analgesics: implications for precision pain management. Semin Oncol Nurs. 2019a;35(1):291–9. https://doi.org/10.1016/j.soncn.2019.04.011.

Yang GS, Kumar S, Dorsey SG, Starkweather AR, Kelly DL, Lyon DE. Systematic review of genetic polymorphisms associated with psychoneurological symptoms in breast cancer survivors. Support Care Cancer. 2019b;27(2):351–71. https://doi.org/10.1007/s00520-018-4508-3.

Yang GS, Mi X, Jackson-Cook CK, Starkweather AR, Lynch Kelley D, Archer KJ, Zou F, Lyon DE. Differential DNA methylation following chemotherapy for breast cancer is associated with lack of memory improvement at one year. Epigenetics. 2019c;18:1–12. https://doi.org/10.1080/15592294.2019.1699695.

Yao S, Hu Q, Kerns S, Yan L, Onitilo AA, Misleh J, Young K, Lei L, Bautista J, Mohamed M, Mohile SG, Ambrosone CB, Liu S, Janelsins MC. Impact of chemotherapy for breast cancer on leukocyte DNA methylation landscape and cognitive function: a prospective study. Clin Epigenetics. 2019;11(1):45. https://doi.org/10.1186/s13148-019-0641-1.

Zorina-Lichtenwalter K, Meloto CB, Khoury S, Diatchenko L. Genetic predictors of human chronic pain conditions. Neuroscience. 2016;338:36–62. https://doi.org/10.1016/j.neuroscience.2016.04.041.

Low Back Pain

7

Angela R. Starkweather and Susan G. Dorsey

Contents

7.1   Introduction

Low back pain is a heterogenous condition that can arise from a variety of sources, including direct nerve injury, spinal cord compression, pulled muscles or injured ligaments, inflammation or infection. The location and characterization of the pain can often provide clues to the etiology, however, addressing the source of pain in clinical practice currently requires diligence, problem-solving, and a significant amount of trial and error to ascertain optimal treatment. Multi-omics based research is focused on identifying the specific mechanisms contributing to musculoskeletal pain so that more precise, targeted methods can be used to identify the source of pain and deliver therapeutic treatment to reduce pain, disability and prevent pain chronicity.

A. R. Starkweather (✉)
University of Connecticut School of Nursing, Storrs, CT, USA
e-mail: Angela.Starkweather@uconn.edu

S. G. Dorsey
Department of Pain and Translational Symptom Science, University of Maryland School of Nursing, Baltimore, MD, USA
e-mail: SDorsey@umaryland.edu

© Springer Nature Switzerland AG 2020
S. G. Dorsey, A. R. Starkweather (eds.), *Genomics of Pain and Co-Morbid Symptoms*, https://doi.org/10.1007/978-3-030-21657-3_7

Heritability of LBP is estimated to range from 30% to 60% with higher familial incidence associated with increased severity of LBP (Livshits et al. 2011; MacGregor et al. 2004). However, in approximately 90% of individuals who seek healthcare for low back pain there is no identifiable etiology on imaging and the pain is presumed to be of mechanical or non-specific origin (Hansen et al. 2016). With each different potential source of pain (nerve impingement from a herniated disc versus diffuse low back pain from muscle strain) the chemical response initiated results in nociceptor activation with impulses generated that travel through the nerves to the spinal cord and brain. Multi-omic measures can be used to define the specific chemical response initiated at the peripheral and/or central level to produce a "biological pain profile." Once a common profile is identified and verified in a specific population, it can then be used to detect individuals at risk for an adverse outcome (severe, disabling pain or chronic pain) or evaluate response to different treatments.

Theories about pain variability and acute to chronic pain transition hold certain assumptions that must be considered in the study of musculoskeletal pain:

- Each individual has a certain level of vulnerability to severe, disabling pain or chronic pain and that level of vulnerability can be identified through their biopsychosocial profile.
- Vulnerability to severe, disabling pain or chronic pain may be present at the time of pain onset or acquired over time.
- Pain sensitization, which occurs through lowering the threshold of producing an action potential and pain signaling in primary afferent nerves, or through centrally mediated pain pathways, increases vulnerability to severe, disabling pain or chronic pain.
- Alterations in descending pain-modulatory systems (pain facilitatory and/or inhibitory circuits) interact with high centers of the brain and influence the risk of severe, disabling pain or chronic pain.
- Omic measures, including those from circulating blood cells, can be used to differentiate levels of vulnerability by providing a biochemical pain profile.

To address each of these assumptions, the study design, methods of phenotyping pain, and the integration of omic measures need to be thoroughly examined and accurately implemented. Each of these issues will be reviewed and summarized in the context of low back pain.

7.2 Muscles of the Lower Back

The musculature of the lower spine is composed of multiple layers designed to provide 360 degree support of spinal structures during movement and at rest (Fig. 7.1). The iliocostalis muscles run along the ribs and spine to facilitate trunk and neck rotation. Longissimus and multifidus muscles run down the sides of the spine and extend the neck and spine. The spinalis muscle runs down the middle of the spine and extends the head and neck.

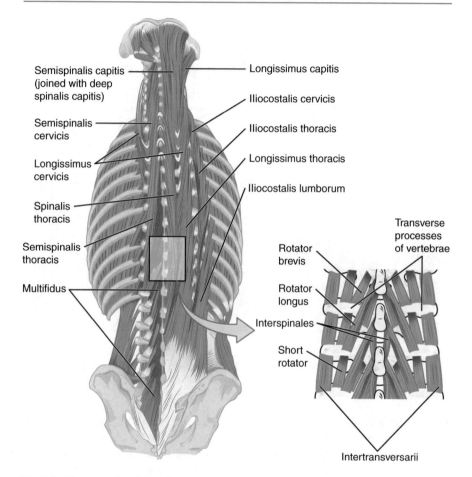

Fig. 7.1 Major muscles of the spine. Anatomy and Physiology by Rice University, licensed under a Creative Commons Attribution 4.0 International License, Chap. 11.3 Axial Muscles of the Head, Neck, and Back. Download for free at http://cnx.org/contents/14fb4ad7-39a1-4eee-ab6e-3ef248 2e3e22@8.24

The semispinalis muscle transverses down the middle section of the spine and sides, extending and rotating the spine. The erector spinae lie on top of deeper muscles running down the sides of the spine and its distributions are categorized as the long head, lateral head, and medial head. Each head acts on specific regions of the lumbar, thoracic, and cervical spine. The latissimus dorsi extends from the lower back to shoulders and controls shoulder and arm abduction and adduction.

The abdominals, gluteus muscles, and hamstrings also support the lumbar muscles, and when weak can lead to a forward tipped pelvis (Fig. 7.2). Weak thoracic erector spinae, hip flexors, and quadriceps can lead to a backward tipped pelvis, which is another potential source of low back pain. The extensor, flexor, and oblique muscles provide another source for supporting the lumbar vertebrae. Extensor

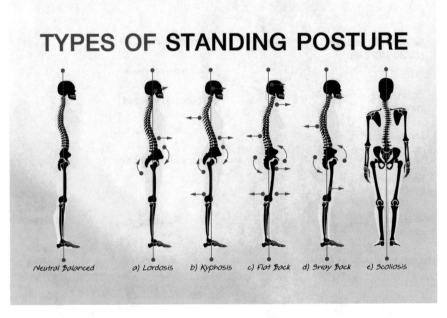

Fig. 7.2 Standing posture. Is Your Posture Hurting You? by Body Fit Solutions, licensed under a Creative Commons Attribution 4.0 International License, Types of Standing Posture. Download for free at https://bodyfitsolutions.co.uk/is-your-posture-hurting-you/

muscles include the erector spinae that attach to the posterior spine and gluteal muscles, and enable standing and lifting objects. The flexor muscles attach to the anterior spine, and enable flexing, bending forward, lifting and arching the lower back. The oblique muscles attach to the sides of the spine, rotating the spine and maintaining posture.

Damage, injury, and/or inflammation of the muscles surrounding the lower back or structures supporting posture and movement can cause pain. For instance, pronated foot posture and function have been found to significantly influence low back pain in women [odds ratio (OR) = 1.51, 95% CI 1.1, 2.07, $p = 0.011$] even after adjusting for age, weight, smoking and depressive symptoms [OR = 1.48, 95% CI 1.07, 20.05, $p = 0.018$] (Menz et al. 2013). In addition, occupational postures of nurses, physiotherapists, dentists, and dieticians have been studied with findings that trunk and head posture distribution of individuals with and without low back pain are different ($p < 0.05$) (Cinar-Medeni et al. 2017). This infers that healthcare professionals with low back pain could modify their posture of the trunk and neck to potentially alleviate pain.

The load of the spine by posture, movement, and muscle activation is assumed to contribute to low back pain onset, persistence, and recovery. Modification of motor control, the motor, sensory, and central processes for control of posture and movement, has been used to address low back pain (Hides et al. 2019). However, when participants are categorized by the type of movement eliciting their pain and either instructed to perform stretching and strengthening exercises to correct posture and movement, or given general instructions, pain outcomes are equivalent (Van Dillen et al. 2016). To date, synthesis of the research demonstrates that exercise of any kind is beneficial for improving low back pain and reducing disability compared to no exercise, and this evidence has been incorporated into current treatment guidelines with exercise remaining as first line therapy (Foster et al. 2018).

At the cellular level, injured cells due to overload, compression, overstretch, and anoxia release hydrogen ions, adenosine triphosphate, and glutamate, which lower the stimulus required to activate action-potential generation in afferent nerves thereby sensitizing the nerves to further stimuli (Amaya et al. 2013; Barbe and Barr 2006; Pongs 1999). Injured mast cells degranulate and release histamine, bradykinin, inflammatory cytokines, and proteases, which further sensitize afferent terminals and increase vascular permeability, while macrophages release proinflammatory cytokines that are chemotactic, facilitating immune cell infiltration (Barr and Barbe 2004; Barbe et al. 2013). Although these coordinated events are part of an acute inflammatory response, a repetitive or chronic injury may lead to prolonged inflammation, and release of cytotoxic free radicals and inflammatory cytokines that further sensitize primary afferents causing cold and mechanical allodynia and hyperalgesia, and promote tissue fibrosis (Bove et al. 2009; Fisher et al. 2015). Localized inflammation and tissue fibrosis within the fascia surrounding the muscles may generate myofascial pain (Pavan et al. 2014). The increased production and release of inflammatory cytokines in local tissues can also enter circulation and stimulate a systemic inflammatory response (Amaya et al. 2013; Elliott et al. 2009). Immune cell activation, including M2-type macrophages that regulate tissue repair, releases proinflammatory and profibrogenic cytokines, which promote fibrosis through activation of fibroblasts and collagen matrix deposition in and around muscle, tendons, and peripheral nerve tissue (Ji et al. 2016; Laskin 2009). Substance P produced from neurons, macrophages, mast cells, endothelial cells, fibroblasts, and tenocytes further contributes to fibroblast proliferation, collagen production and remodeling (Fong et al. 2013). The process of tissue fibrosis can produce prolonged pain due to chronic nerve compression and the tethering of muscles, tendons, and the nerves within these structures, which also causes alterations in tissue mechanics (Fisher et al. 2015).

7.3 Vertebrae, Tendons, Ligaments, and Intervertebral Discs

Supporting tissues, bones, and intervertebral discs of the spine can also contribute to low back pain through similar mechanisms—release of proinflammatory mediators and substance P with infiltration of immune cells, tissue fibrosis, and remodeling with extension to surrounding muscle tissue (James et al. 2018a).

The intrasegmental and intersegmental ligament systems provide structural stability of the spine. The intrasegmental system, including the ligamentum flavum,

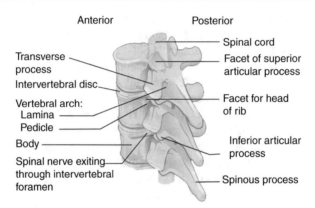

Fig. 7.3 Vertebrae of the lower back. Wikimedia Commons. Source: Jmarchn. This file is licensed under the Creative Commons Attribution-Share Alike 3.0 Unported license. https://commons.wikimedia.org/wiki/File:Vertebra_Posterolateral-en.svg

interspinous and intertransverse ligaments, holds individual vertebrae together, while the intersegmental system, including the anterior and posterior longitudinal ligaments and supraspinous ligaments, holds many vertebrae together (Fig. 7.3). Tendons of the spine connect muscle to vertebrae of the spine.

The bones and intervertebral discs of the spine provide motion and are subject to wear and tear with aging. Intervertebral discs are fibrous structures that absorb pressure and provide cushioning between the vertebrae. The cartilaginous endplate of the intervertebral disc controls the flow of oxygen and nutrients into the disc space. The exterior of the disc is composed of concentric rings of collagen, named the annulus fibrosis, that bend and twist with motion of the spine. Nucleus pulposus is at the interior of each disc and the mixture of water and proteins act as a shock absorber between the vertebrae. With aging, the nucleus pulposus can decrease in hydration making it less flexible, smaller, and prone to tearing. The loss of disc height can cause structural and mechanical alterations. Degenerative disc disease (DDD) is often cited as a cause of low back pain although disc degeneration is ubiquitous among adults and characterized by increased desiccation of the spinal disk matrix, proteoglycan degradation, and disc tissue levels of proinflammatory cytokines (Binch et al. 2014). Given that the degree of DDD does not correlate with pain severity or duration, surgical options are typically reserved only after conservative measures have failed to improve function and quality of life.

7.4 Phenotyping Low Back Pain

Many characteristics of low back pain are important for understanding potential sources of pain including the location of pain, radiation, whether pain is intermittent or constant, and whether it occurs or worsens with motion or rest (Table 7.1). Duration of pain is often categorized as acute (<6 weeks), subacute (6–12 weeks), or chronic (>12 weeks).

Table 7.1 Low back pain terminology

Term	Definition
Axial low back pain	Refers to pain isolated to the low back region without radiation to the lower extremities
Inflammatory low back pain	Includes ankylosing spondylitis and other spondyloarthritides
Low back pain with radiculopathy	Pain located in the lower back region with radiation to the hip and leg that follows the distribution of the sciatic nerve
Lumbar degenerative disc disease	General term used to describe erosion or damage of the vertebral discs
Mechanical low back pain	Term used when the pain is triggered by movements of the spine
Myofascial pain	Can refer to any part of the body; pressure on sensitive points (trigger points) cause pain in the muscle and may include referred pain in an unrelated part of the body
Sacroiliac joint dysfunction (SI joint pain)	Pain caused by the sacroiliac joint that connects the sacrum and pelvis
Sciatica	Pain that radiates down the distribution of the sciatic nerve

Inflammatory causes of low back pain, such as ankylosing spondylitis and other spondyloarthritides have unique characteristics that include an onset that typically occurs prior to age 35 and persists for greater than 3 months. It worsens with immobility particularly at night and early morning, tends to ease with physical activity and exercise, and is typically relieved effectively with non-steroidal anti-inflammatory drugs (NSAIDs).

A classification strategy has been proposed for neuropathic, nociceptive, or nociplastic/central pain (Smart et al. 2010). This classification scheme is based on verification of a neurological lesion or disease and neuroanatomically plausible pain distribution for neuropathic pain, localized pain in the areas of mechanical load and replicable stimulus-response profile for nociceptive pain and variable/diffuse characteristics with allodynia and/or hyperalgesia found outside the segmental area of primary nociception for nociplastic/central sensitization. The problem with this classification system is that many individuals have mixed or overlapping pain characteristics and the categories do not account for the psychological features that are known to play a significant role in pain perception (Hodges et al. 2019).

To address this issue, the NIH Task Force on Research Standards for Chronic Low Back Pain developed a minimal dataset that incorporates classification of chronic low back by its impact including pain intensity, interference, and physical function, as well as psychological and behavioral characteristics (Deyo et al. 2014). The dataset includes items on depression, catastrophizing, and other dimensions of low back pain that have a wealth of evidence for affecting low back pain outcomes and the tool's responsiveness has been shown to be adequate to change (Dutmer et al. 2019).

Quantitative Sensory Testing
Other phenotyping methods may be used to characterize low back pain, such as diagnostic testing, however, clinically these tests are not indicated for non-specific

low back pain and only add to medical care costs (Maher et al. 2017). Quantitative sensory testing (QST) can be used to evaluate peripheral and central pain sensitization by assessing responses to mechanical, thermal, and electrical stimuli in a standardized format (Starkweather et al. 2016a). Our team has successfully used QST to identify unique phenotypes with peripheral and/or central pain sensitization that have differential RNA and lipidomic profiles (Starkweather et al. 2017). Many important characteristics about the individual's pain can be derived from QST:

- Presence of peripheral and/or central sensitization
- Function of the descending inhibitory control system
- Specific pain receptors that are sensitized
  - A-delta and C fibers respond to heat and touch
  - Pacinian corpuscles respond to pressure
- Ion channels in the skin that may be affected
  - TRPV4—warm 27–34 °C
  - TRPV3—warmer 34–39 °C
  - TRPV1—hot >43 °C/also activated by capsaicin
  - TRPV2—painfully hot >52 °C
  - TRPM8—cold 10–25 °C/also activated by menthol.

Imaging Another tool used for phenotyping pain is the use of brain functional imaging. Data from brain imaging can provide information on the areas that respond to affective or sensory domains of pain, the resting state of the brain and in response to specific stimuli, and the connectivity among regions of the brain. Many regions of brain are activated in pain processing, interpretation, and modulation. The periaqueductal gray (PAG) receives inputs from higher brain centers and can activate analgesia through descending opioid-mediated inhibition of nociceptive inputs. Reciprocal connections with the rostroventromedial medulla (RVM) allow the PAG to influence descending pain modulation (Henderson and Keay 2018). The RVM includes major serotonergic nucleus raphe magnus (NRM), nucleus reticularis gigantocellularis-pars alpha, and nucleus paragigantocellularis lateralis (NpGC) as well as GABAergic and glycinergic neuronal populations, which all have descending projections to the spinal cord. The RVM receives inputs from the PAG, thalamus, parabrachial region, and noradrenergic locus coeruleus, and is considered as the final common relay in descending modulation of pain. Assessing the strength of connections between these regions and patterns in response to stimuli can provide valuable information. In addition, central sensitization is demonstrated on functional imaging by increased brain activity in regions known to be involved in acute pain sensations (insula, anterior cingulate cortex, and prefrontal cortex or PFC) and regions not normally involved in pain processing.

There is some evidence that functional imaging can differentiate states of pain versus pain-free as well as changes over time. Using resting-state functional magnetic resonance imaging, participants with chronic neuropathic orofacial pain show increased functional connectivity between the ventromedial medulla (RVM) and the ventrolateral periaqueductal gray (vlPAG) and locus ceruleus, which are crucial brainstem pain-modulatory regions (Mills et al. 2018; Ossipov et al. 2014). To examine changes over time in response to pain, a longitudinal study was carried out in adult rats exposed to peripheral nerve injury and underwent brain imaging prior to and over 20 weeks (Seminowicz et al. 2009). This study found that compared to noninjured, sham-operated controls, the frontal cortex of neuropathic rats was smaller at the end of the study. However, a trial in humans with chronic low back pain who had a positive response to conservative treatment did not change brain gray matter morphological features up to 12-months follow-up (Malfliet et al. 2018). In another study, however, patients with chronic low back pain who had structural pain-related pathology and underwent spine surgery or facet injections showed a normalization of previous pain-related reduction in the thickness of the PFC and altered functional activity in those reporting improvement in pain and disability (Čeko et al. 2015; Seminowicz et al. 2011).

Recent work in animal models has shown that a proinflammatory response due to degenerative disc is generated in the multifidus muscle, including dysregulation in interleukin (IL)-1beta, IL6, IL10, transforming growth factor-beta, tumor necrosis factor, adiponectin, and leptin (James et al. 2018b). Specifically the degree of disc degeneration was associated with IL-1beta in the multifidus muscle. However, physical activity reduced the proinflammatory response to disc degeneration and prevented fibrotic changes in the multifidus compared to non-active mice (James et al. 2019). Such investigations performed in coordination with imaging could provide new insights on the coordination of peripheral and central sensitization and response to various treatments.

7.5   Genetic Studies on Low Back Pain

Given the wide variability in low back pain outcomes, there has been a significant interest in identifying genetic variants that may contribute to the risk of chronic LBP.

Several previous studies have identified associations between genetic polymorphisms that influence the development and progression of DDD or specific low back pain conditions such as herniated disc, lumbar stenosis or spondylolisthesis and increased risk of LBP.

The GG genotype of *Caspase-9* (*CASP-9*), an initiator caspase protease for apoptosis, was significantly associated with discogenic LBP in two studies (Guo et al. 2011; Mu et al. 2013). Matrix metalloproteinases (MMPs) are involved in the degradation of the extracellular matrix of the intervertebral disc and polymorphisms of the *MMP1* gene have been shown to increase expression of *MMP1*. Two studies have examined a SNP for guanine insertion/deletion (G/D), the -1607 promoter polymorphism of the *MMP1* gene (rs1799750). One study reported that the D allelic variant was significantly associated with DDD (Song et al. 2008), while another found that the 2G allele SNP insertion was associated with more pain and reduced function compared with patients carrying the 1G allele (Jacobsen et al. 2013).

The expression of *SPARC* (secreted protein, acidic, rich in cysteine; aka osteo-nectin or BM-40) has been associated with DDD. *SPARC* encodes a cysteine-rich acidic matrix-associated protein required for the collagen in bone to become calci-fied but is also involved in extracellular matrix synthesis and promotion of changes to cell shape. Decreased expression of *SPARC* and increased methylation of the *SPARC* promoter from lumbar intervertebral discs were associated with LBP in patients undergoing lumbar fusion surgery (Tajerian et al. 2011).

It has been suggested that local expression of proinflammatory cytokines in disc tissue may sensitize surrounding nerves and increase pain. In particular, interleukin (IL)-1 activity has been implicated in progression of DDD and LBP and investiga-tors have examined several SNPs on *IL-1α* (C^{889}-T), *IL-1B* (C^{3954}-T), and *IL-1 recep-tor antagonist* (*IL-1RA* or *IL-1RN*; G^{1812}-A, G^{1887}-C, and T^{11100}-C). Carriers of the *IL-1RN* A1812 allele had an increased risk of LBP and carriers of this allele in combination with the *IL1α* T889 or *IL-1B* T3954 allele had a higher risk of and more days with LBP (Solovieva et al. 2004). In contrast, no association was found between *IL-B* (C^{3954}-T) and LBP at both the genotypic ($p = 0.104$) and the allelic ($p = 0.098$) level in Chinese soldiers (Mu et al. 2013). Another study analyzed the association between single nucleotide polymorphism (SNP) in *IL-1B* (rs1143634) and *IL-1RN* (rs2234677) with chronic low back pain (LBP) in chronic post-traumatic stress disorder (PTSD). The study found no association of rs1143634 in *IL-1B* with LBP. Permutation test showed significant association of rs1143634 in *IL-1RN* with LBP group and presence of wild type allele A was protective in LBP group. The same SNP (rs1143634) in *IL-1B* was associated with the intensity of pain (Loncar et al. 2013).

The association of SNPs in *IL-1α* gene and lumbar radicular pain due to disc herniation was examined by Schistad et al. (2014). The *IL-1α* C > T polymorphism (rs1800587) was associated with the visual analogue scale and pressure pain thresh-old scores in patients with symptomatic disc herniation. Patients with the CT/TT genotype reported a higher VAS leg pain intensity and lower PPT in the gluteal muscles compared with patients with the CC genotype during the 1 year of follow-up.

In patients with chronic LBP and DDD undergoing fusion surgery or active reha-bilitation, association analysis of individual SNPs revealed association of *IL18RAP* polymorphism (rs1420100) with severe degeneration ($p = 0.05$) and more than one degenerated disc ($p = 0.02$) (Omair et al. 2013). From the same gene two SNPs, rs917997 and rs1420106, were found to be in strong linkage disequilibrium (LD) and were associated with posttreatment improvement in disability ($p = 0.02$). Haplotype association analysis of 5 SNPs spanning across *IL18RAP, IL18R1,* and *IL1A* genes revealed significant associations with improvement in disability ($p = 0.02$) and reduction in pain ($p = 0.04$). An association was found between *MMP3* polymorphism (rs72520913) and improvement in pain ($p = 0.03$) and with severe degeneration ($p = 0.006$).

Given the mixed findings from studies evaluating pathologic mechanisms of DDD and LBP, other studies have investigated genetic polymorphisms associated with   variability   in   pain   sensitivity.   These   include   variants   of catechol-O-methyltransferase (*COMT*), beta-2 adrenergic receptor (*ADRB2*), µ-opioid receptor 1 (*OPRM1*), and GTP cyclohydrolase (*GCH1*).

The COMT enzyme metabolizes catecholamines and thus modulates adrenergic, noradrenergic, and dopaminergic signaling. Rut et al. (2014) evaluated the relationship between polymorphisms of *COMT* (rs4680:A > G—Val158Met, rs6269:A > G, rs4633:C > T, rs4818:C > G) and pain sensitivity after lumbar discectomy. All patients had one-level symptomatic disc herniation from L3 to S1. Carriers of rs6269 *AA*, rs4633 *TT*, rs4818 *CC*, and rs4680 *AA* genotypes were characterized by the lowest preoperative scores related to pain intensity and lower pain intensity at 1 year after the surgery. The rs4633 *CC*, rs4680 *GG* genotypes demonstrated significant clinical improvement in pain intensity score at 1 year after the surgery. Patients with *COMT* haplotype associated with low metabolic activity of enzyme (A_C_C_G) showed better scores on disability and pain intensity. There was no significant correlation between leg pain and single-nucleotide polymorphisms in the *COMT* gene. Another study examined *COMT* polymorphisms in patients with chronic LBP for duration of >1 year and DDD who were treated with lumbar fusion or cognitive therapy and exercises (Omair et al. 2012). SNPs adjusted for covariates revealed association of rs4633 and rs4680 with posttreatment improvement in pain intensity, with a tendency for greater improvement among heterozygous patients compared to the homozygous. No association was observed for the analysis of the common haplotypes that the SNPs were situated on.

In contrast, other studies have not shown associations among *COMT* polymorphisms with pain sensitivity, risk of LBP or poor outcomes. A study to test associations among 14 single nucleotide polymorphisms (SNPs) in two candidate genes—beta-2 adrenergic receptor (*ADRB2*) and catecholamine-O-methyltransferase (*COMT*) with risk for chronic disabling comorbid neck and low back pain was performed in a population cohort study (Skouen et al. 2012). Only SNP rs2053044 (*ADRB2*, recessive model) displayed an association with neck and back pain and pain in three to four areas in the last month. The *COMT* Val158Met SNP was not a risk factor for disc herniation in another study, however, patients with Met/Met had more pain and slower recovery than those with Val/Met, which in turn also had more pain and slower recovery than those with Val/Val suggesting the SNP contributes to the progression of the symptoms of disc herniation (Jacobsen et al. 2012). In an 11-year follow-up study of patients who underwent lumbar fusion or non-operative treatment for painful DDD, five single nucleotide polymorphisms (SNPs) in the *COMT* gene, and one SNP in the *OPRM1* gene were examined for associations with outcomes (Omair et al. 2015). Disability at baseline was significantly associated with *COMT* SNPs rs4818 ($p = 0.02$), rs6269 ($p = 0.007$), rs4633 ($p = 0.04$) rs2075507 ($p = 0.009$), two haplotypes ($p < 0.002$), age, gender, and smoking ($p \leq 0.002$). No significant associations with clinical variables were observed for *OPRM1* or *COMT* at long-term follow-up.

The association between *OPRM1* A118G (rs1799971) genotype and subjective health complaints in patients with radicular pain and disc herniation was also evaluated (Hasvik et al. 2013). All patients except female carriers of the G-allele reported a decrease in pain from baseline to 1 year. Female carriers of the G-allele reported significantly higher subjective health complaints score during the study time span than male carriers of the G-allele when controlling for pain and pain duration. Roh and colleagues (Roh et al. 2013) examined the association between estrogen receptor α (*ERα*) polymorphisms and pain intensity in symptomatic female degenerative spondylolisthesis patients. They reported a significant association between *Xba*I polymorphism and pain intensity of the lower back. The pain intensity in patients with a GG genotype was significantly higher than in patients with the AG or the AA genotypes. In addition, the presence of the CG haplotype was found to be associated with back pain intensity in the haplotype analysis of the *Pvu*II and *Xba*I polymorphisms of *ERα*.

The rate-limiting enzyme for tetrahydrobiopterin (BH4) synthesis, *GTP cyclohydrolase (GCH1)*, is a key modulator of peripheral neuropathic and inflammatory pain (Tegeder et al. 2006). BH4 is an essential cofactor for catecholamine, serotonin, and nitric oxide production. A haplotype of the *GCH1* gene (population frequency 15.4%) was significantly associated with less pain following diskectomy for persistent radicular low back pain (Binch et al. 2014). Healthy individuals homozygous for this haplotype exhibited reduced experimental pain sensitivity, and forskolin-stimulated immortalized leukocytes from haplotype carriers upregulated *GCH1* less than did controls. The authors concluded that BH4 is an intrinsic regulator of pain sensitivity and chronicity, and the *GTP cyclohydrolase* haplotype is a marker for these traits.

In a genome-wide association meta-analysis of 5 cohorts, consisting of 4683 patients of European ancestry with lumbar DDD, identified four SNPs with $p < 5 \times 10^{-8}$ the threshold set for genome-wide significance (Williams et al. 2012). Of the 4 identified SNPs, 3 were on chromosome 6 (rs926849; rs2187689; rs7767277) and one was an intergenic marker on chromosome 3 (rs17034687). The analysis adjusted for age and sex revealed the strongest signal for SNP rs926849, which lies on an intronic region of the Parkinson protein 2, E3 ubiquitin protein ligase (*PARK2*) gene on chromosome 6. The minor (C allele) of rs926849 was associated with lower level of lumbar DDD suggesting a protecting role of the SNP. For verification, the team tested the association between lumbar DDD and DNA methylation variants at 3 CpG sites in the PARK2 promoter and found a significant association between DNA methylation at CpG site cg15832436 and lumbar DDD ($\beta = 8.74 \times 10^{-4}$, SE $= 2.49 \times 10^{-4}$, $p = 0.006$).

In a genome-wide association study of 32,642 subjects consisting of 4043 lumbar DDD cases and 28,599 control subjects, the carbohydrate sulfotranferase 3 (*CHST3*) variant was associated with lumbar DDD (Song et al. 2013). The *CHST3* gene codes for an enzyme that catalyzes proteoglycan sulfation and was identified as a susceptibility gene for lumbar DDD. In a meta-analysis using multiethnic populations an association was observed at rs4148941, which lies within the microRNA-513a-5p (miR-513-5p) binding site. In vitro, the interaction between miR-513a-5p and mRNA transcribed from the susceptibility allele (A allele) of rs4148941 was enhanced compared with transcripts from other alleles. In the intervertebral disc cells of subjects carrying the A allele of rs4148941, expression of CHST3 mRNA was significantly reduced.

A follow-up study analyzed twin from TwinsUK registry to assess whether persons reporting episodes of low back pain from lumbar DDD had detectable levels of altered IgG glycosylation (Freidin et al. 2016). Using weighted correlation network analysis seven modules of correlated glycans were found and an association was identified between low back pain and glycan modules featured by glycans either promoting or blocking antibody-dependent cell-mediated cytotoxicity (ADCC). The levels of four glycan traits representing two modules were statistically significantly different in monozygotic twins who were discordant for low back pain. Core fucosylation of IgG serves as the safety switch for reducing ADCC. Twins with low level of fucosylated glycans and high level of non-fucosylated glycans had a higher prevalence of systemic inflammatory disorders suggesting the involvement of ADCC and inflammation in low back pain.

Alternative alleles can produce coding changes and affect gene function through their effect on mRNA expression level or splicing. SNPs associated with the expression level of a gene or exon are referred to as expression quantitative trait loci (eQTLs), with SNPs that produce local expression effects (*cis*-eQTLs) differentiated from SNPs that affect expression levels of distant genes (*trans*-eQTLs). In a study that analyzed 214 human dorsal root ganglia (DRG) specimens on DNA genotyping and RNA expression, the investigators compared DRG to other tissues on eQTL and transcriptome levels with ten different regions of the brain and from whole blood (Parisien et al. 2017). In this investigation, the human HLA region expressing major histocompatibility complex class II (MHCII) class genes was significantly associated with gene expression levels in DRG and low back pain phenotypes (as well as temporomandibular joint disorder). Using a mouse model, it was also demonstrated that MHCII expression influences the duration of an inflammatory pain episode.

An investigation by our team supports that MHCII expression is an important factor in the transition from acute to chronic low back pain. A comparison of individuals who had resolution of low back pain within 6 weeks (acute) compared to individuals who developed chronic low back pain cLBP over 6 months found that baseline neurophysiological factors were different in acute vs. cLBP patients at both the pain site (low back) and non-pain site (remote) (Starkweather et al. 2016b). Specifically, we found significant differences in cold pain thresholds at the pain site and warm detection threshold at both sites. We also found that the wind-up ratio (a measure of temporal summation, which is a perceived increase in pain intensity over time given a repeated stimulus) was significantly higher in chronic low back pain (cLBP) vs. acute LBP patients at a remote site. These findings suggest that neurophysiological differences at baseline may help predict those who transition from acute to cLBP. Following these findings, we examined gene expression profiles in $N = 64$ participants ($n = 21$ samples from healthy controls; $n = 11$ samples at LBP onset from patients whose pain resolved within 6 weeks (acute); $n = 13$ samples at LBP onset from patients whose pain lasted 6 months (chronic time 1 [T1]), and $n = 19$ samples from cLBP patients at 6 months (chronic time 5 [T5]) (Dorsey et al. 2019). Our goal was twofold: (*a*) characterize gene expression differences between the healthy controls and cLBP patients at 6 months to define the cLBP transcriptome; and (*b*) characterize gene expression differences at baseline in acute vs. cLBP to gain insight into potential gene expression biomarkers of the risk to transition from acute to cLBP. All participants had LBP of musculoskeletal origin. Subjects were excluded if the LBP was due to trauma or structural issues (e.g., disc degeneration). After completing gene expression profiling using RNA-sequencing (RNA-seq), we first examined the transcriptome data using Euclidean clustering methods to determine whether the participants in this study had unique transcriptional profiles by group assignment (healthy controls [normal], acute, cLBP at baseline (T1), and cLBP at six-months (T5). We found that normal and acute LBP patients clustered together, whereas at baseline (T1) and the 6-month time point (T5), those who progressed to cLBP clustered together. These results suggest that the transcriptome of healthy controls and acute LBP patients is more alike than different, compared with cLBP patients.

To demonstrate the cLBP pain transcriptome is significantly different than healthy controls, we conducted RNA-seq to examine differential gene expression in cLBP patients at the 6-month time point compared with healthy controls to identify the cLBP transcriptome. Total RNA was extracted from PAXgene tubes, run on Bioanalyzer gels to obtain the RNA integrity number (RIN; those used were at least 8/10 on the quality score), and sequenced on the Illumina HiSeq instrument. We obtained at least 80 million read pairs per sample, for an average of 32X coverage of the genome and aligned 150 bp paired end reads to the latest build of the Ensembl human genome reference with the short read aligner TopHat (Trapnell et al. 2009). We conducted differential gene expression using the count method with the HTSeq program, then normalized and analyzed the data in the R package DESeq (Anders and Huber 2010). Genes were considered differentially expressed between cases (chronic pain) and matched healthy controls if the False Discovery Rate (FDR; used to account for multiple testing) corrected p-value was ≤ 0.05 and ± 1.5 fold-change. We found $N = 5632$ differentially expressed genes in the cLBP patients compared with healthy controls that include many genes that have been implicated in acute and chronic pain as well as novel genes. The majority were protein-coding genes, as would be expected. Next, we conducted unbiased pathway analysis to determine whether the differentially expressed genes in cLBP patients were enriched for known gene ontologies (GO) and/or canonical signaling pathways that could provide mechanistic insight into the cLBP transcriptome. The top significant pathway was extracellular matrix receptor (ECM) interaction. This is interesting because genetic variants in genes encoding proteins that impact on intervertebral disc stability (e.g., collagen genes *COL9A3*, *COL11A1*, *COL11A2*, *COL1A1*) have been associated with chronic back pain (Tegeder and Lötsch 2009; Manek and MacGregor 2005).

Next, we directly compared acute LBP gene expression changes at baseline with those that progressed to cLBP at 6 months. Using RNA-seq, as described above, we found $N = 3288$ differentially expressed genes in the cLBP baseline (T1) patients compared with the acute LBP patients that include many of the genes that have been implicated in acute and chronic pain as well as novel genes. As with the cLBP versus healthy controls, the majority were protein coding. We next conducted unbiased pathway analysis and found that the top differentially expressed pathway was the antigen processing and presentation pathway, which includes many of the major histocompatibility complex (MHC) I and MHC II genes including the human leukocyte antigen (HLA) genes. Given this novel finding, we constructed a database of all genes on the extended MHC locus (Horton et al. 2004) and computed a one-tailed Fisher's Exact test to obtain a hypergeometric p-value, the results of which ($p = 1.43E-02$) demonstrate a significant enrichment of MHC locus genes in our dataset. The MHC Fisher's exact test was non-significant for cLBP patients at 6 months versus healthy controls, suggesting that this is uniquely characteristic of a baseline gene expression profile that might be predictive of the transition from acute to chronic pain. This data supports the hypothesis that neurophysiological and gene expression differences can be used to define the cLBP phenotype and transcriptome, as well as predict those at risk for transitioning from acute to cLBP. Moreover, the genes and pathways that define the chronic pain transcriptome (known pain genes and the ECM-receptor interaction pathway) are unique, and distinct, from those that predict risk for transition from acute to chronic pain (MHC locus genes, antigen processing and presentation pathway). This suggests important changes in gene expression profiles from baseline over time may be altered, in addition to inherent differences that are present at baseline, contributing to chronic pain phenotypes.

It will be very important for future studies to precisely identify the phenotype of low back pain (isolated non-specific low back pain vs. low back pain from herniated disc vs. low back pain from herniated disc and radicular leg pain) and examine differential gene expression or other omic measures over time. Through such inquiry along with matched study populations by age, sex, and psychosocial factors, the precise mechanisms of chronic low back pain and other forms of musculoskeletal pain can be targeted and refined to improve individual outcomes.

7.6   Conclusions

It is well established that the experience of pain is influenced by biological, psychological, and social factors. Low back pain is a heterogenous condition that can involve the muscles, tendons, ligaments, and intervertebral discs. A large number of genetic polymorphisms have been shown to influence the severity and duration of low back pain; however, most studies to date have included participants with overlapping pain characteristics, etiologies, and psychosocial factors. In order to make progress in this area of science, research studies with careful phenotyping, genome-wide analyses of DNA heritability and changes over time such as RNA expression, DNA methylation, and proteomics of local tissues and circulating blood cells will generate the biopsychosocial profile necessary for accurate prediction and targeted treatment to prevent and reduce the transition to chronic low back pain.

References

Amaya F, Izumi Y, Matsuda M, Sasaki M. Tissue injury and related mediators of pain exacerbation. Curr Neuropharmacol. 2013;11:592–7. https://doi.org/10.2174/1570159X11311060003.

Anders S, Huber W. Differential expression analysis for sequence count data. Genome Biol. 2010;11(10):R106. https://doi.org/10.1186/gb-2010-11-10-r106.

Barbe MF, Barr AE. Inflammation and the pathophysiology of work-related musculoskeletal disorders. Brain Behav Immun. 2006;20:423–9. https://doi.org/10.1016/j.bbi.2006.03.001.

Barbe MF, Gallagher S, Popoff SN. Serum biomarkers as predictors of stage of work-related musculoskeletal disorders. J Am Acad Orthop Surg. 2013;21:644–6. https://doi.org/10.5435/JAAOS2110-644.

Barr AE, Barbe MF. Inflammation reduces physiological tissue tolerance in the development of work-related musculoskeletal disorders. J Electromyogr Kinesiol. 2004;14:77–85. https://doi.org/10.1016/j.jelekin.2003.09.008.

Binch AL, Cole AA, Breakwell LM, Michael AL, Chiverton N, Cross AK, Le Maitre CL. Expression and regulation of neurotrophic and angiogenic factors during human intervertebral disc degeneration. Arthritis Res Ther. 2014;16(5):416.

Bove GM, Weissner W, Barbe MF. Long lasting recruitment of immune cells and altered epiperineurial thickness in focal nerve inflammation induced by complete Freund's adjuvant. J Neuroimmunol. 2009;213:26–30. https://doi.org/10.1016/j.jneuroim.2009.06.005.

Čeko M, Shir Y, Ouellet JA, Ware MA, Stone LS, Seminowicz DA. Partial recovery of abnormal insula and dorsolateral prefrontal connectivity to cognitive networks in chronic low back pain after treatment. Hum Brain Mapp. 2015;36:2075–92. https://doi.org/10.1002/hbm.22757.

Cinar-Medeni O, Elbasan B, Duzgun I. Low back pain prevalence in healthcare professionals and identification of factors affecting low back pain. J Back Musculoskelet Rehabil. 2017;30(3):451–9. https://doi.org/10.3233/BMR-160571.

Deyo RA, Dworkin SF, Amtmann D, Andersson G, Borenstein D, Carragee E, Carrino J, Chou R, Cook K, DeLitto A, Goertz C, Khalsa P, Loeser J, Mackey S, Panagis J, Rainville J, Tosteson T, Turk D, Von Korff M, Weiner DK. Report of the NIH task force on research standards for chronic low back pain. Pain Med. 2014;15(8):1249–67. https://doi.org/10.1111/pme.12538.

Dorsey SG, Renn CL, Griffioen M, Lassiter CB, Zhu S, Huot-Creasy H, McCracken C, Mahurkar A, Shetty AC, Jackson-Cook CK, Kim H, Henderson WA, Saligan L, Gilll J, Colloca L, Lyon DE, Starkweather AR. Whole blood transcriptomic profiles can differentiate vulnerability to chronic low back pain. PLoS One. 2019;14(5):e0216539. https://doi.org/10.1371/journal.pone.0216539.

Dutmer AL, Reneman MF, Schiphorst Preuper HR, Wolff AP, Speijer BL, Soer R. The NIH minimal dataset for chronic low back pain: responsiveness and minimal clinically important change. Spine. 2019;44(20):E1211. https://doi.org/10.1097/BRS.0000000000003107.

Elliott MB, Barr AE, Clark BD, Amin M, Amin S, Barbe MF. High force reaching task induces widespread inflammation, increased spinal cord neurochemicals and neuropathic pain. Neuroscience. 2009;158:922–31. https://doi.org/10.1016/j.neuroscience.2008.10.050.

Fisher PW, Zhao Y, Rico MC, et al. Increased CCN2, substance P and tissue fibrosis are associated with sensorimotor declines in a rat model of repetitive overuse injury. J Cell Commun Signal. 2015;9:37–54. https://doi.org/10.1007/s12079-015-0263-0.

Fong G, Backman LJ, Hart DA, Danielson P, McCormack B, Scott A. Substance P enhances collagen remodeling and MMP-3 expression by human tenocytes. J Orthop Res. 2013;31:91–8. https://doi.org/10.1002/jor.22191.

Foster NE, Anema JR, Cherkin D, Chou R, Cohen SP, Gross DP, Ferreira PH, Fritz JM, Koes BW, Peul W, Turner JA, Maher CG, Lancet Low Back Pain Series Working Group. Prevention and treatment of low back pain: evidence, challenges and promising directions. Lancet. 2018;391:2368–83. https://doi.org/10.1016/S0140-6736(18)30489-6.

Freidin MB, Keser T, Gudelj I, Stambuk J, Vucenovic D, Allegri M, Pavic T, Simurina M, Fabiane SM, Lauc G, Williams FMK. The association between low back pain and composition of IgG glycome. Sci Rep. 2016;6:26815.

Guo T-M, Liu M, Zhang Y-G, Guo W-T, Wu S-X. Association between Caspase-9 promoter region polymorphisms and discogenic low back pain. Connect Tissue Res. 2011;52(2):133–8.

Hansen BB, Hansen P, Carrino JA, Fournier G, Rasti Z, Boesen M. Imaging in mechanical back pain: anything new? Best Pract Res Clin Rheumatol. 2016;30(40):766–85.

Hasvik E, Schistad EI, Grvle G, Haugen AJ, Re C, Gjerstad J. Subjective health complaints in patients with lumbar radicular pain and disc herniation are associated with a sex-OPRM1 A118G polymorphism interaction: a prospective 1-year observational study. BMC Musculoskelet Disord. 2013;15(1):161–7.

Henderson LA, Keay KA. Imaging acute and chronic pain in the human brainstem and spinal cord. Neuroscientist. 2018;24(1):84–96. https://doi.org/10.1177/1073858417703911.

Hides JA, Donelson R, Lee D, Prather H, Sahrmann SA, Hodges PW. Convergence and divergence of exercise-based approaches that incorporate motor control for the management of low back pain. J Orthop Sports Phys Ther. 2019;49(6):437–52. https://doi.org/10.2519/jospt.2019.8451.

Hodges PW, Barbe MF, Loggia ML, Nijs J, Stone LS. Diverse role of biological plasticity in low back pain and its impact on sensorimotor control of the spine. J Orthop Sports Phys Ther. 2019;49(6):389–401. https://doi.org/10.2519/jospt.2019.8716.

Horton R, Wilming L, Rand V, et al. Gene map of the extended human MHC. Nat Rev Genet. 2004;5(12):889–99. https://doi.org/10.1038/nrg1489.

Jacobsen L, Schistad E, Storesund A, Pedersen L, Rygh L, Re C, Gjerstad J. The COMT rs4680 met allele contributes to long-lasting low back pain, sciatica and disability after lumbar disc herniation. Eur J Pain. 2012;16(7):1064–9.

Jacobsen LM, Schistad EI, Storesund A, Pedersen LM, Espeland A, Rygh LJ, Re C, Gjerstad J. The MMP1 rs1799750 2G allele is associated with increased low back pain, sciatica, and disability after lumbar disk herniation. Clin J Pain. 2013;29(11):967–71.

James G, Sluka KA, Blomster L, et al. Macrophage polarization contributes to local inflammation and structural change in the multifidus muscle after intervertebral disc injury. Eur Spine J. 2018a;27:1744–56. https://doi.org/10.1007/s00586-018-5652-7.

James G, Millecamps M, Stone LS, Hodges PW. Dysregulation of the inflammatory mediators in the multifidus muscle after spontaneous intervertebral disc degeneration [in] SPARC-null mice is ameliorated by physical activity. Spine. 2018b;43:E1184–94. https://doi.org/10.1097/BRS.0000000000002656.

James G, Klyne DM, Millecamps M, Stone LS, Hodges PW. ISSLS Prize in basic science 2019: physical activity attenuates fibrotic alterations to the multifidus muscle associated with intervertebral disc degeneration. Eur Spine J. 2019;28(5):893–904. https://doi.org/10.1007/s00586-019-05902-9.

Ji RR, Chamessian A, Zhang YQ. Pain regulation by non-neuronal cells and inflammation. Science. 2016;354:572–7. https://doi.org/10.1126/science.aaf8924.

Laskin DL. Macrophages and inflammatory mediators in chemical toxicity: a battle of forces. Chem Res Toxicol. 2009;22:1376–85. https://doi.org/10.1021/tx900086v.

Livshits G, Popham M, Malkin I, Sambrook PN, Macgregor AJ, Spector T, Williams FM. Lumbar disc degeneration and genetic factors are the main risk factors for low back pain in women: the UK twin spine study. Ann Rheum Dis. 2011;70:1740–5. https://doi.org/10.1136/ard.2010.137836.

Loncar Z, Curic G, Havelka Mestrovic A, Mickovic V, Bilic M. Do IL-1B and IL-1RN modulate chronic low back pain in patients with post-traumatic post-traumatic stress disorder? Coll Antropol. 2013;37(4):1237–44.

MacGregor AJ, Andrew T, Sambrook PN, Spector TD. Structural, psychological, and genetic influences on low back and neck pain: a study of adult female twins. Arthritis Rheum. 2004;51:160–7. https://doi.org/10.1002/art.20236.

Maher C, Underwood M, Buchbinder R. Non-specific low back pain. Lancet. 2017;389:736747.

Malfliet A, Kregel J, Coppieters I, De Pauw R, Meeus M, Roussel N, Cagnie B, Danneels L, Nijs J. Effect of pain neuroscience education combined with cognition targeted motor control training on chronic spinal pain: a randomized clinical trial. JAMA Neurol. 2018;75:808–17. https://doi.org/10.1001/jamaneurol.2018.0492.

Manek NJ, MacGregor AJ. Epidemiology of back disorders: prevalence, risk factors, and prognosis. Curr Opin Rheumatol. 2005;17(2):134–40. https://doi.org/10.1097/01.bor.0000154215.08986.06.

Menz HB, Dufour AB, Riskowski JL, Hillstrom HJ, Hannan MT. Foot posture, foot function and low back pain: the Framingham study. Rheumatology. 2013;52(12):2275–82. https://doi.org/10.1093/rheumatology/ket298.

Mills EP, Di Pietro F, Alshelh Z, Peck CC, Murray GM, Vickers ER, Henderson LA. Brainstem pain-control circuitry connectivity in chronic neuropathic pain. J Neurosci. 2018;38(2):465–573. https://doi.org/10.1523/JNEUROSCI.1647-17.2017.

Mu J, Ge W, Zuo X, Chen Y, Huang C. Analysis of association between IL-1b, CASP-9, and GDF5 variants and low-back pain in Chinese male soldiers: clinical article. J Neurosurg Spine. 2013;19(2):243–7.

Omair A, Lie BA, Reikeras O, Holden M, Brox JI. Genetic contribution of catechol-O-methyltransferase variants in treatment outcome of low back pain: a prospective genetic association study. BMC Musculoskeletal Dis. 2012;13(1):76–85.

Omair A, Holden M, Lie BA, Reikeras O, Brox JI. Treatment outcome of chronic low back pain and radiographic lumbar disc degeneration are associated with inflammatory and matrix degrading gene variants: a prospective genetic association study. BMC Musculoskelet Disord. 2013;14(1):105–14.

Omair A, Mannion AF, Holden M, Fairbank J, Lie BA, Hagg O, Fritzell P, Brox JI. Catechol-O-methyltransferase (COMT) gene polymorphisms are associated with baseline disability but not long-term treatment outcome in patients with chronic low back pain. Eur Spine J. 2015;24:2425–31.

Ossipov MH, Morimura K, Porreca F. Descending pain modulation and chronification of pain. Curr Opin Support Palliat Care. 2014;8(2):143–51. https://doi.org/10.1097/SPC.0000000000000055.

Parisien M, Khoury S, Chabot-Doré A-J, et al. Effect of human genetic variability on gene expression in dorsal root ganglia and association with pain phenotypes. Cell Rep. 2017;19(9):1940–52. https://doi.org/10.1016/j.celrep.2017.05.018.

Pavan PG, Stecco A, Stern R, Stecco C. Painful connections: densification versus fibrosis of fascia. Curr Pain Headache Rep. 2014;18:441. https://doi.org/10.1007/s11916-014-0441-4.

Pongs O. Voltage-gated potassium channels: from hyperexcitability to excitement. FEBS Lett. 1999;452:31–5. https://doi.org/10.1016/S0014-5793(99)00535-9.

Roh HL, Lee JS, Suh KT, Kim JI, Lee HS, Goh TS, Park SH. Association between estrogen receptor gene polymorphism and back pain intensity in female patients with degenerative lumbar spondylolisthesis. J Spinal Disord Tech. 2013;26(2):E53–7.

Rut M, Machoy Mokrzynska A, Rezcawowicz D, Soniewski P, Kurzawski M, Drozdzik M, Safranow K, Morawska M, Biaecka M. Influence of variation in the catechol-O-methyltransferase gene on the clinical outcome after lumbar spine surgery for one-level symptomatic disc disease: a report on 176 cases. Acta Neurochir. 2014;156(2):245–52.

Schistad EI, Jacobsen LM, Re C, Gjerstad J. The interleukin-1a gene C T polymorphism rs1800587 is associated with increased pain intensity and decreased pressure pain thresholds in patients with lumbar radicular pain. Clin J Pain. 2014;30(10):869–74.

Seminowicz DA, Laferriere AL, Millecamps M, Yu JS, Coderre TJ, Bushnell MC. MRI structural brain changes associated with sensory and emotional function in a rat model of long-term neuropathic pain. NeuroImage. 2009;47:1007–14. https://doi.org/10.1016/j.neuroimage.2009.05.068.

Seminowicz DA, Wideman TH, Naso L, Hatami-Khouroushahi Z, Fallatah S, Ware MA, Jarzem P, Bushnell MC, Shir Y, Ouellet JA, Stone LS. Effective treatment of chronic low back pain in humans reverses abnormal brain anatomy and function. J Neurosci. 2011;31:7540–50. https://doi.org/10.1523/JNEUROSCI.5280-10.2011.

Skouen J, Smith A, Warrington N, O'Sullivan P, McKenzie L, Pennell C, Straker L. Genetic variation in the beta-2 adrenergic receptor is associated with chronic musculoskeletal complaints in adolescents. Eur J Pain. 2012;16(9):1232–42.

Smart KM, Blake C, Staines A, Doody C. Clinical indicators of 'nociceptive', 'peripheral neuropathic' and 'central' mechanisms of musculoskeletal pain. A Delphi survey of expert clinicians. Man Ther. 2010;15(1):80–7. https://doi.org/10.1016/j.math.2009.07.005.

Soloviva S, Leino-Arjas P, Saarela J, Luoma K, Raininko R, Riihimaki H. Possible association of interleukin 1 gene locus polymorphisms with low back pain. Pain. 2004;109(1):8–19.

Song Y-Q, Ho DW, Karppinen J, Kao PY, Fan B-J, Luk KD, Yip S-P, Leong JC, Cheah KS, Sham P, et al. Association between promoter-1607 polymorphism of MMP1 and lumbar disc disease in southern Chinese. BMC Med Genet. 2008;9(1):38–44.

Song YQ, Karasugi T, Cheung KM, Chiba K, Ho DW, Miyake A, et al. Lumbar disc degeneration is linked to a carbohydrate sulfotransferase 3 variant. J Clin Investig. 2013;123(11):4909–17.

Starkweather AR, Heineman A, Storey S, Rubia G, Lyon D, Greenspan J, Dorsey SG. Methods to measure peripheral and central sensitization using quantitative sensory testing: a focus on individuals with low back pain. Appl Nurs Res. 2016a;29:237–41. PMID: 26856520.

Starkweather AR, Ramesh D, Lyon DE, Siangphoe U, Deng X, Sturgill J, Heineman A, Elswick RK Jr, Dorsey SG. Acute low back pain: differential somatosensory function and gene expression compared with healthy no-pain controls. Clin J Pain. 2016b;32(11):933–9. https://doi.org/10.1097/AJP.0000000000000347.

Starkweather A, Julian T, Ramesh D, Heineman A, Sturgill J, Dorsey S, Lyon DE, Wijesinghe DS. Circulating lipids and acute pain sensitization: an exploratory analysis. Nurs Res. 2017;66(6):454–61. PMID: 29095376.

Tajerian M, Alvarado S, Millecamps M, Dashwood T, Anderson KM, Haglund L, Ouellet J, Szyf M, Stone LS. DNA methylation of SPARC and chronic low back pain. Mol Pain. 2011;7:65.

Tegeder I, Lötsch J. Current evidence for a modulation of low back pain by human genetic variants. J Cell Mol Med. 2009;13(8b):1605–19. https://doi.org/10.1111/j.1582-4934.2009.00703.x.

Tegeder I, Costigan M, Griffin RS, Abele A, Belfer I, Schmidt H, Ehnert C, Nejim J, Marian C, Scholz J, et al. GTP cyclohydrolase and tetrahydrobiopterin regulate pain sensitivity and persistence. Nat Med. 2006;12(11):1269–77.

Trapnell C, Pachter L, Salzberg SL. TopHat: discovering splice junctions with RNA-Seq. Bioinformatics. 2009;25(9):1105–11. https://doi.org/10.1093/bioinformatics/btp120.

Van Dillen LR, Norton BJ, Sahrmann SA, Evanoff BA, Harris-Hayes M, Holtzman GW, Early J, Chou I, Strube MJ. Efficacy of classification-specific treatment and adherence on outcomes in people with chronic low back pain. A one-year follow-up, prospective, randomized, controlled clinical trial. Man Ther. 2016;24(1):52–64. https://doi.org/10.1016/j.math.2016.04.003.

Williams FM, Bansal AT, van Meurs JB, Bell JT, Meulenbelt I, Suri P, Rivadeneira F, Sambrook PN, Hofman A, Bierma-Zeinstra S, Menni C, Kloppenburg M, Slagboom PE, Hunter DJ, Macgregor AJ, Uitterlinden AG, Spector TD. Novel genetic variants associated with lumbar disc degeneration in northern Europeans: a meta-analysis of 4600 subjects. Ann Rheum Dis. 2012;72(7):1141–8. https://doi.org/10.1136/annrheumdis-2012-201551.

Genomics of Chronic Orofacial Pain

8

Raymond A. Dionne and Sharon M. Gordon

Contents

Chronic orofacial pain presents as a spectrum of disorders with multi-factorial etiology manifesting clinically in the muscles, joints, and innervation of the craniofacial region. The complex anatomy and mechanistic factors resulted in a tradition of "eminence-based" hypothesis formulation leading to non-validated treatment strategies contributing to iatrogenic injury and potentiation of the condition. It was generally accepted, for example, that dental malocclusion resulted in masticatory muscle spasms that caused pain (Berry 1967) and that mechanical correction of occlusal discrepancies was curative (Guichet 1969; Ramford and Ash 1966). This belief persisted despite demonstration that producing muscle spasms in pain-free volunteers

R. A. Dionne (✉)
Department of Cell Biology, University of Connecticut School of Medicine, UConn Health, Farmington, CT, USA
e-mail: radionne@uchc.edu

S. M. Gordon

Department of Public Health Sciences, University of Connecticut School of Medicine, Farmington, CT, USA

© Springer Nature Switzerland AG 2020
S. G. Dorsey, A. R. Starkweather (eds.), *Genomics of Pain and Co-Morbid Symptoms*, https://doi.org/10.1007/978-3-030-21657-3_8

results in pain, changes in occlusion, and function movements and that these changes were reversible as the spasms dissipated (Obrez and Stohler 1996). Repositioning of the mandible to a putative physiologic rest position identified by an electronic diagnostic device and treatment with a variety of intraoral appliances to establish new maxillary-mandibular relationships was also proposed to relieve chronic oral pain. Displacement of the temporomandibular disc was also proposed as a cause of pain and dysfunction that needed to be corrected with surgical procedures, such as repositioning or replacement of the disc or replacement of the condyle or TMJ with prostheses that resulted in significant iatrogenic injury. The belated recognition that therapeutic practices and their putative mechanisms need to be validated by well-controlled clinical trials that meet standards for validity, minimize patient and observer bias, and be replicated by independent observers (evidence-based dentistry) has finally eroded the clinical management of chronic orofacial pain based on historical beliefs (Aboalnaga et al. 2019; Manfreddi et al. 2017).

Clinical research methodology has advanced from the use of articulating paper to identify occlusal discrepancies and EMG recordings of masticatory muscles to measurement of biomarkers and imaging methods that herald greater understanding of the pathophysiology associated with patients' reports of chronic orofacial pain. Exploring the genetic basis of human variation in pain is also based on a research strategy that understanding the molecular-genetic basis of pain sensitivity and variable responses to therapeutic interventions will provide insight into the individualized treatment of pain, both as a symptom and a disease state. However, the published results over the past two decades of the relationship of genetic factors and pain have generated a mixed picture. This is not surprising given the complexity of pain biology, the evolving knowledge of the human genome, the variability in study designs, sample heterogeneity, phenotype definition, and variable statistical approaches. The seminal observations by Mogil and colleagues demonstrated clear strain differences ranging from 1.2- to 54-fold ranges in nociceptive sensitivity across 12 inbred-mouse strains using several common measures of nociception (thermal, mechanical, and chemical) and different types of pain hypersensitivity (neuropathic, inflammatory) (Mogil et al. 1999a). These results suggested that there are at least 5 genetically distinct types of nociception and that hypersensitivity influences pain responses to different types of stimuli but may have only partial overlapping genetic underpinnings (Mogil et al. 1999b). Animal models have been successfully extrapolated to human models, e.g., the prediction of pain phenotype in humans based on animal models demonstrating the role of MC1R gene in animals (Mogil et al. 2005). But there is no a priori reason that genetic associations present in inbred rodent models will always be reproduced in related human pain states. Moreover, while acute inflammatory pain in response to a controlled, replicable tissue injury (e.g., oral surgery) is a relatively simple clinical pain state, local gene expression profiling indicates that hundreds of different molecules are involved over the first 48 h after surgery and likely contribute differentially to the pain phenotype (Wang et al. 2007). Adding to the genetic factors influencing the expression and function of local mediators are genetic influences on nociceptive signaling pathways, CNS plasticity, mechanisms influencing mood, gender-related mechanisms, and the underlying disease process. The composite complexity of the multiple steps from tissue injury to the development of persistent pain, including temporal changes and individual variability, requires caution to avoid pitfalls when using genomic methods that can generate millions of possible targets leading to fallacious associations.

8.1 Evaluation of Factors Associated with Chronic Orofacial Pain

A large prospective study of risk factors for the development of painful temporo-mandibular disorders (Orofacial Pain: Prospective Evaluation and Risk Assessment OPPERA) has improved understanding of the etiology of chronic orofacial pain by combining clinical and genetic observations. OPPERA recruited and examined 3258 community-based TMD-free adults and assessed genetic and phenotypic measures of biological, psychosocial, clinical, and health status characteristics across 202 baseline risk factors over a median follow-up period of 2.8 years. Over the course of the study, 4% of participants annually developed clinically verified TMD (Slade et al. 2016). The most influential predictors of clinical TMD symptoms were distilled to simple checklists of comorbid health conditions and nonpainful orofa-cial symptoms. Self-reports of jaw parafunction were identified as stronger predic-tors of TMD than examiner's clinical assessments. Among the psychosocial factors examined, the frequency of somatic symptoms was the strongest predictor of TMD onset. Examination of 41 clinical orofacial characteristics thought to predict the onset of TMD in the same subjects revealed that pain upon jaw opening, and pain from palpation of masticatory, neck, and body muscles were significant predictors for the onset of TMD (Ohrbach et al. 2013). Conversely, examiner assessments of temporomandibular joint noise and tooth wear facets did not predict incidence, indi-cating that clinician-observed findings had modest influence on TMD incidence. This dichotomy between etiologic hypotheses based on clinical observations (occlu-sion, clenching, disc displacements) and careful prospective observations that con-trolled for observer bias points to overall health status and psychosocial factors as important predictors of TMD onset. Factors such as these are part of the genetic vulnerability that is initiated or mediated locally by masticatory tissues and pain-regulatory systems within the nervous system that results in a complex pattern of TMD etiology. The important role of genetic factors has prompted genomic research to identify genes that are associated with the etiology or responsiveness of TMD to treatment, which are reviewed next.

The genetic contribution to TMD etiology is supported by twin studies that dem-onstrate 27% heritability (Plesh et al. 2012) and numerous studies using candidate genes related to nociception, inflammation, immunology, and affective processes. A case controlled study using a candidate panel of 350 genes (Slade et al. 2011) found that TMD was associated with single nucleotide polymorphisms (SNPs) in 7 genes: a catecholamine catabolizing enzyme (COMT), a serotonin receptor (HTR2A), the glucocorticoid receptor (NR3C1), the calcium/calmodulin-dependent protein kinase (CAMK4), the muscarinic acetylcholine receptor (CHRN2), the interferon-related developmental regulator (IFRD1), and the G-protein-coupled receptor kinase (GRK5). Another association study evaluated polymorphisms in the estrogen receptor in three groups of women with muscular TMD, articular TMD or without TMD but with systemic arthralgia compared to a group of women without TMD or pain (Quinelato et al. 2018). A genotype for the estrogen receptor-1 was strongly associated with the groups with muscle TMD and articular TMD. An association was also observed for the estrogen-related receptor beta and the group of subjects with articular TMD. The authors conclude that these observations are consistent with the hypothesis that changes in estrogen receptor genes influence the presence of TMD associated with chronic joint pain in females.

A recently published genome-wide association study (GWAS) based on the large OPPERA data set ($N = 3104$) used a randomly selected discovery cohort, replication with meta-analysis in 7 independent cohorts, followed by functional and behavioral validation (Smith et al. 2019). A total of 6 SNPs exceeded the whole-genome corrected threshold for significance in the discovery genome-wide association scan. Five of these SNPs were from distinct loci and were further examined in the 7 replication cohorts, but only 3 of these SNPs, all on chromosome 3, were independently replicated. The functional significance of the genetic variants that were associated with painful TMD in the discovery or replication association studies were further evaluated by bioinformatic analysis. Three of the SNPs associated with TMD in males only showed significant association with the expression of the gene muscle RAS oncogene homolog (MRAS), resulting in lower expression of MRAS mRNA. This association was further evaluated by producing a mutant knockout mouse in which functional Mras protein was reduced or eliminated. The knockout mice were compared to wildtype following injection of CFA (complete Freund's adjuvant) as an assay for chronic pain conditions, especially those with an inflammatory neuroimmune component. In the male knockout mice, but not female mutant mice, the duration of CFA-induced mechanical allodynia was significantly longer, lasting for 14–21 days compared to the usual 3–7 days duration demonstrated in the wildtype mice.

These findings in a GWAS with replication and both functional and behavioral validation illustrate many important factors that differ from conclusions based on clinical observations alone: the findings were unexpected and not based on prior assumptions or the bias often associated with poorly controlled clinical observations. Not only was the association with Mras expression unexpected, it occurred in males only rather than in females who exhibit a much higher prevalence of painful TMD. The authors suggest that these findings may indicate that males and females develop TMD through different pathophysiologic pathways that may be related to the differences in healing that protects males from developing chronic TMD, possibly mediated by the immune system in neurological tissues. While this hypothesis requires replication in independent cohorts, the findings provide pathophysiologic insights that are amenable to the evaluation of novel interventions, further exploration for other relevant genes located near the gene-coding region, and the use of genomic discovery studies for other TMD susceptibility variants with larger sample sizes and improved phenotypes. The magnitude and scientific sophistication of this study minimizes the potential for spurious findings based on clinical observations in under-powered samples using variable outcome measures (Kim et al. 2009, 2013).

A similar GWAS of painful TMD conducted by the OPPERA group focused on subjects from the Hispanic Community Health Study/Study of Latinos (Sanders et al. 2017). An initial discovery GWAS study of TMD in 10,153 participants of the HCHS/SOL (769 cases, 9384 controls), followed by meta-analysis of the most promising SNPs in 4 independent cohorts from the USA, Germany, Finland, and Brazil. A locus near the sarcoglycan alpha (SGCA) gene was identified in the discovery analysis and replicated in the Brazilian cohort. Another locus near the relaxin/insulin-like family peptide receptor 2 (RXP2) was identified in the sex-stratified discovery cohort and replicated among females in the replication meta-analysis. A novel locus identified at the genome-wide level in the intron of the dystrophin gene DMD which is responsible for Duchenne muscular dystrophy. The SGCA gene encodes SGCA which is involved in the cellular structure of muscle fibers; DMD forms part of the dystrophin-glycoprotein complex along with SGCA. The two associations near DMD and SGCA suggest that the dystrophin-glycoprotein pathway contributes to the pathophysiology of muscle fibers in orofacial pain (Sanders et al. 2017).

8.2 Conclusions and Clinical Considerations

The past two decades of genomic research on the biosocial characteristics of TMD and its the molecular-genetic basis have been remarkable. Our understanding of the genomic contributions to chronic orofacial pain, the importance of multiple clinical, psychosocial, and biological factors along with the emergence of evidence-based dentistry heralds a new era of scientifically informed and validated treatments as well as the possibility of individualized pain management. Yet there are still many barriers: even for highly heritable phenotypes such as height (heritability estimates ~90%), the most significant genotype only explains less than 1% of the individual variation (Weedon et al. 2007; Yang et al. 2010). The small contribution of a genetic variation for a highly heritable phenotype stems from the influences of multiple genetic factors as well as environmental factors. Considering the heritability estimates for pain along with other influential factors and the likelihood that pain is a more complex phenotype than height, caution is needed before extrapolating from even the most robust genomic findings. The findings of the OPPERA study, the continuing development of improved genomic methods, and the growing recognition that most of the empirically derived treatments based on clinical observations represent "fools gold" hold promise for safer and more effective treatments for chronic orofacial pain. Meanwhile between the promise of new findings and their translation to validated treatments for chronic orofacial pain we still need to heed the caution "do no harm," resist the urge to treat clinical findings as if they are causative factors, and carefully weigh possible iatrogenic injury when using non-validated treatments. Given the natural history of most temporomandibular disorders to flux over time, palliative interventions combined with a short course of an appropriate analgesic may be sufficient to alleviate symptoms with minimal risk of harm.

References

Aboalnaga AA, Amer NM, Elnahas MO. Malocclusion and temporomandibular disorders: verification and controversy. J Oral Facial Pain Headache. 2019;33:440–50.

Berry DC. Facial pain related to muscle dysfunction. Br J Oral Surg. 1967;4:222–6.

Guichet NF. Applied gnathology, why and how. Dent Clin N Am. 1969;13:687.

Kim H, Clark D, Dionne RA. Genetic contribution to clinical pain and analgesia: avoiding pitfalls in genetic research. J Pain. 2009;10:663–93.

Kim H, Gordon S, Dionne R. Genome-wide approaches (GWA) in oral and craniofacial diseases research. Oral Dis. 2013;19(2):111–20.

Manfreddi D, Lombardo L, Siciliani G. Temporomandibular disorders and dental occlusion. A systematic review of association studies: end of an era? J Oral Rehabil. 2017;44:908–23.

Mogil JS, Wilson SG, Bon K, Lee SE, Chung K, et al. Heritability of nociception I: responses of 11 inbred strains on 12 measures of inflammation. Pain. 1999a;80:67–82.

Mogil JS, Wilson JB, Bon K, Lee SE, Chung K, et al. Heritability of nociception II. 'Types' of nociception revealed by genetic correlation analysis. Pain. 1999b;80:83–93.

Mogil JS, Ritchie J, Strasburg K, Kaplan L, Wallace MR, et al. Melanocortin-1 receptor gene variants affect pain and mu-opioid analgesia in mice and humans. J Med Genet. 2005;42:583–7.

Obrez A, Stohler CS. Jaw muscle pain and its effect on gothic arch tracings. J Prosth Dent. 1996;75:393–8.

Ohrbach R, Bair E, Fillingham RB, Gonzalez Y, Gordon SM, et al. Clinical orofacial characteristics associated with risk of first-onset TMD: the OPPERA prospective cohort study. J Pain. 2013;14(12 Suppl):T33–50.

Plesh O, Noonan C, Buchwald DS, Goldberg J, Alan N. Temporomandibular disorder-type pain and migraine headaches in women: a preliminary twin study. J Orofac Pain. 2012;26(2):91–8.

Quinelato V, Bonato LL, Vierira AR, Granjeiro JM, Tesch R, Casado PL. Association between polymorphisms in the genes of estrogen receptors and the presence of temporomandibular disorders and chronic arthralgia. J Oral Maxillofac Surg. 2018;76:314e1–9.

Ramford SP, Ash MM. Occlusion. Philadelphia, PA: WB Saunders; 1966;161–3.

Sanders AE, Jain D, Sofer T, Kerr KF, Laurie CC, Shaffer JR, et al. GWAS identifies new loci for painful temporomandibular disorder: Hispanic Community Health Study/Study of Latinos. J Dent Res. 2017;96:277–84.

Slade GD, Maixner DW, Greenspan JD, et al. Potential genetic risk factors for chronic TMD: genetic associations from the OPPERA case control study. J Pain. 2011;12(11 Suppl):T92–101.

Slade GD, Ohrbach R, Greenspan JD, Fillingim RB, Bair E, et al. Painful temporomandibular disorders: decade of discovery from OPPERA studies. J Dent Res. 2016;95:1084–92.

Smith SB, Parisien M, Bair E, Belfer I, et al. Genome-wide association reveals contribution of MRAS to painful temporomandibular disorder in males. Pain. 2019;160:579–91.

Wang X-M, Wu T-X, Hamza M, et al. Rofecoxib modulates multiple gene expression pathways in a clinical model of acute inflammatory pain. Pain. 2007;128:136–47.

Weedon MN, Lettre G, Freathy RM, Lindgren CM, Voight BF, et al. A common variant of HMGA2 is associated with adult and childhood height in the general population. Nat Genet. 2007;39:1245–50.

Yang J, Benyamin B, McEvoy BP, Gordon S, Henders AK, Nyholt DR, et al. Common SNPs explain a large proportion of heritability for human height. Nat Genet. 2010;42(7):565–9.

Genomics of Visceral Pain

9

Wendy A. Henderson and Bridgett Rahim-Williams

Contents

9.1 Introduction

The genomics of visceral pain refers to the study of not only the structure but also the evolution, function, and mapping of organisms genes with respect to the phenomenon of perceived pain in the visceral cavity. Visceral pain is a heterogenous condition that typically arises from a known precipitating event. Up to 25% of the US population report digestive symptoms, and visceral pain/abdominal pain is the most common gastrointestinal diagnosis for outpatient visits (Grundmann and Yoon 2010; Mearin et al. 2016; Palsson et al. 2020). Visceral pain is a symptom that affects a large number of individuals and often leads to chronic gastrointestinal

W. A. Henderson (✉)
University of Connecticut School of Nursing, Storrs, CT, USA
e-mail: Wendy.Henderson@uconn.edu

B. Rahim-Williams
University of North Florida, Jacksonville, FL, USA

© Springer Nature Switzerland AG 2020
S. G. Dorsey, A. R. Starkweather (eds.), *Genomics of Pain and Co-Morbid Symptoms*, https://doi.org/10.1007/978-3-030-21657-3_9

disorders and conditions (Palsson et al. 2020). Persons suffering from visceral pain may experience the symptom chronically or chronically intermittently and/or in response to innocuous stimuli (allodynia). The severity of the symptom increases through the application of noxious stimuli (hyperalgesia). Such hypersensitivity and chronic pain may be caused by either or both peripheral and central sensitization of nociceptive neurons (Marchand et al. 2005). Inflammatory mediators activate primed neurons (Gold and Gebhart 2010; Stein et al. 2009). In such cases, the nociceptive sensitivity usually resolves as the inflammation resolves, but in many cases, the hypersensitivity may persist for some time or become chronic. One such disorder is irritable bowel syndrome (IBS). A principal feature of IBS is visceral hypersensitivity and hyperalgesia. Visceral pain, self-reported as abdominal pain, has been universally endorsed by those who suffer from IBS and alone affects approximately 12% of Americans and has an estimated global incidence of 7–20% (Lovell and Ford 2012). In the USA the condition results in an estimated $1.2 billion in medical costs and loss of productivity (Lovell and Ford 2012). Abdominal pain, a principal characteristic of IBS and several other GI conditions, is one of the primary non-trauma reasons for emergency room visits. Unlike related conditions that may present with similar symptoms to IBS, no organic cause for IBS is known. Clinical indices of inflammation, immunity, digestive organ function, and GI permeability are usually normal or cannot be consistently linked to IBS. However, symptoms are chronic and treatment outcomes are inconsistent (Drossman et al. 2002). The lack of gross inflammation, lack of macro- and micro-anatomical pathology, and absence of infection make it a convenient model system in which to investigate the relationship between visceral pain and genetics/genomics in a patient population.

Visceral pain specifically refers to pain located in the trunk area of the human body that includes the lungs, heart, abdominal and pelvic organs (Grundy et al. 2019). Common organic causes of visceral pain include pancreatitis, appendicitis, cholecystitis, diverticulitis, and pelvic pain. Pain that is visceral in nature is generally difficult to localize and diffuse and often accompanied by other symptoms, including but not limited to nausea, vomiting, diarrhea, changes in temperature, blood pressure, and heart rate. Visceral pain can be due to structural lesions, inflammation, biochemical alterations, or functional disorders. The most common functional gastrointestinal disorders are functional dyspepsia and irritable bowel syndrome. For the purposes of this chapter, the focus will be on genomics in chronic visceral pain that is gastrointestinal in origin. Pain lasting greater than 3 months with or without an injurious event without resolution defines chronic visceral pain. There often is no identified biologic marker or organic cause. Approximately 15–20% of children (Petersen et al. 2006) and adults living in the USA suffer from chronic visceral pain (Creed et al. 2001; Russo et al. 1999). There has been speculation that chronic visceral pain may be due to inflammatory changes in the gastrointestinal tract. Other potential mechanisms are altered function at the level of neurotransmission receptors of serotonin and altered colonic mucosal permeability. Chronic visceral pain is an unpleasant physical experience that if left untreated can lead to functional impairment and clinical decline. Health care professionals who work with patients report difficulty when assessing pain, particularly when pain is chronic in nature. The location and type of symptoms underlying the molecular pathology are not fully understood in chronic visceral pain. Pain that is long lasting and has no identifiable cause is particularly difficult to assess and treat. Chronic abdominal pain is divided into several subsets including functional dyspepsia and irritable bowel syndrome (IBS) according to the Rome diagnostic criteria (Hahn et al. 1997). Unspecified chronic visceral pain that has no identified biologic marker or known organic cause and has occurred for greater than 6 months may be diagnosed by Rome IV criteria as IBS if the following criteria are met. Symptoms of abdominal pain must occur once a week on average for the past 3 months with associated (2 out of 3 of the following) improvement of pain with stooling; change in bowel frequency; or change in bowel consistency/form (Mearin et al. 2016). These criteria assume that no metabolic or structural cause is related to the chronic symptoms. Adults and children with IBS experience lessened quality of life in comparison to their healthy peers that leads to numerous absences from school for medical care (Youssef et al. 2006; Whitehead et al. 1996).

Persons with chronic abdominal pain have been shown to have a decreased quality of life (Whitehead et al. 1996). In the USA and other western industrialized countries, more women than men seek health care services for IBS, although this gender imbalance remains poorly understood. Women have different symptom profiles than men with IBS (Lee et al. 2001). IBS symptoms include: (1) abdominal pain relieved by a bowel movement or associated with changes in stool consistency and (2) fewer or more frequent stools, harder or looser stools, straining, urgency, feeling of incomplete evacuation or passage of mucus, and bloating or feeling of abdominal distention (Talley and Phillips 1989). Individuals with chronic abdominal pain also self-report a number of non-GI symptoms.

9.2 Visceral Pain Measures

There are subjective (patient-reported) and objective (clinically observed) measures for visceral pain, more specifically abdominal pain. Subjective Indicators: Most research finds that clinicians underestimate pain intensity in their patients (Mantyselka et al. 2001). Abdominal pain severity is particularly difficult to monitor and assess without standardized measures as it is subjectively reported. Typically, abdominal pain related symptom severity is assessed by a zero to ten pain intensity rating scale. In children and adults self-reported pain measures are the most common (Duff et al. 2001). However, the location of abdominal pain is documented at the discretion of the healthcare provider as subjectively reported by the patient. When patients present with chronic abdominal pain of unknown origin without any significant signs of organic disease (i.e., bloody stools, weight loss, surgical abdomen, etc.) the pain is often localized to the periumbilical region. Shulman et al. (2007) supported subjective reports of localized periumbilical pain in a pediatric cohort and further showed the relationship with interference in activity. Objective Indicators: The autonomic nervous system is the primary pathway between the gut and the brain for portrayal of pain perceptions. Objective physiologic correlates of pain have included heart rate, blood pressure, and skin temperature, as well as 24-h monitoring for heart rate variability. Women with IBS have an increased resting heart rate and blood pressure when compared to a control group of women without IBS (Levine et al. 1997; Heitkemper et al. 2001). A real-time assessment of both subjective and objective symptoms in individuals with chronic abdominal pain of unknown origin has been assessed with the Gastrointestinal Pain Pointer (Henderson et al. 2017).

9.3   Phenotyping Visceral Pain

Abdominal pain has been universally endorsed by those who suffer from irritable bowel syndrome (IBS) with descriptors of the experience include cramping, aching, discomfort, or specifically abdominal pain (Palsson et al. 2020). For persons with IBS or other digestive disorders, chronic abdominal pain disrupts daily life and negatively impacts quality of life (Sherwin et al. 2016, 2017a, b). Although other studies have assessed abdominal pain through subjective patient response questionnaires, pain assessment persists to be challenging, with studies suggesting that clinicians have difficultly both assessing and documenting changes in pain over time.

Electronic pain assessment tools have been developed to address these mandates. Various tools have been developed to measure pain intensity, including blinding features, language descriptors, body model/image, spatial pain indicator, and physiologic pain (Henderson et al. 2017; Eberlein et al. 2000; McCahon et al. 2005; Melzack 1975; North et al. 1998; Fanciullo et al. 2007) (See Table 9.1). Despite these developments, few of the currently available tools have alerting mechanisms for pain assessment reminders, or an option to choose body gender or type such as normal weight or overweight images. Therefore, the Gastrointestinal Pain Pointer (GIPP) was developed to address these unmet needs (Henderson et al. 2017). The GIPP is a novel, electronic tool designed to assess in real time self-reported

Table 9.1 Subjective self-reported pain assessment tools

	Pain intensity rating/ Numerical rating scale	Pain intensity rating blinded	Pain language descriptors	Body model/ Image	Spatial pain indicator	Heart rate/ Blood pressure readings
GIPP, 2017	✔	✔	✔	✔	✔	✔
McGill, 1975	✔	✔	✔	✔	✔	
Eberlein, 2000	✔	✔	✔	✔	✔	
Fanciullo, 2007	✔	✔	✔	✔	✔	
McCahon, 2005	✔	✔	✔	✔	✔	
North, 1998	✔	✔	✔	✔	✔	

abdominal pain. The validation study tested the GIPP in a limited cohort of participants with and without gastrointestinal pain that were a relatively young population of individuals. Further studies are needed to validate the GIPP in other cohorts. GI symptoms and visceral pain may vary depending on age, including children, aging individuals, post-menopausal women, and individuals living with co-morbidities. Noting the aforementioned limitations, pain is not a one-dimensional experience, particularly for patients suffering from chronic GI symptoms including visceral pain. Therefore, tools such as the GIPP are valuable for use in both clinical care and research settings.

There continues to be an unmet clinical and research need for expanded clinical assessment of GI symptoms and for improved characterization of visceral pain phenotypes that include a real-time assessment of both subjective (patient-reported) and objective (measureable and quantifiable) pain. Furthermore, because the GIPP is able to characterize pain intensity, location, and qualitative description, it may be utilized in various diseases, particularly in those that have no clearly defined etiology. As such, the GIPP is a valuable asset and adds a unique contribution to patient-reported outcome of visceral pain assessment.

9.4   Inflammation

Recognition of the role of inflammation and its interaction with the neuro-immune system of the gastrointestinal tract is an emerging area of interest in patients with IBS (De Giorgio and Barbara 2008). Inflammation is a known mechanism related to acute and then chronic visceral pain (Fig. 9.1). The mechanisms of visceral pain have been shown to be related to interactions between the immune and nervous system in the gut thereby leading to visceral hypersensitivity of the intestinal mucosa (De Giorgio and Barbara 2008). Specifically, there is evidence that mast cells interact with colonic nerve endings by secretion of tryptase, histamine, and possibly serotonin (5-hydroxytryptamine, 5-HT) in humans (Barbara et al. 2004). Additionally, the proximity of mast cells to nerve fibers in the colon of IBS patients leads to augmented visceral sensitivity (Barbara et al. 2007). A possible relationship in humans between the mucosal mast cell numbers and rectal sensitivity has also been demonstrated (Park et al. 2006). Increased mast cell numbers in patients with IBS is shown (Piche et al. 2008). Higher mast cell numbers in the colonic mucosa directly correlate with abdominal pain severity in IBS patients (Akbar et al. 2008; Mahjoub et al. 2009) (Fig. 9.2). Inflammation and increased permeability of GI tract mucosa may also induce pain in pediatric patients with functional abdominal pain (Shulman et al. 2008).

There is mounting evidence that chronic visceral pain symptoms of abdominal pain may be due to low-grade inflammatory changes in the GI tract (Barbara 2006; Spiller and Garsed 2009). Sub-clinical inflammation may result in increased intestinal permeability across the mucosal barrier, which in turn permits the entrance of small molecules, toxins, antigens, genetic signals into the intestinal wall (Barbara 2006). Other mechanisms with recent evidence include altered function at the level

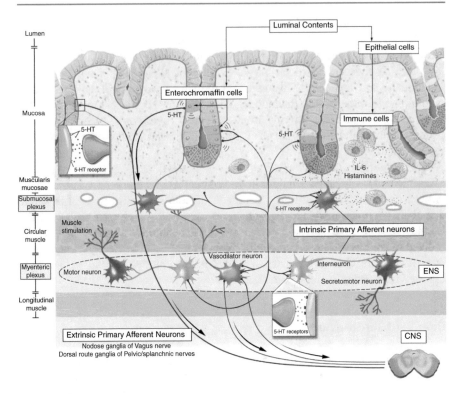

Fig. 9.1 Gastrointestinal-neuronal visceral pain communication. *CNS* Central nervous system, *ENS* Enteric nervous system, *5-HT* 5-hydroxytryptamine

of neurotransmitter receptors of 5-HT, increased visceral hypersensitivity, and impaired colonic mucosal permeability (Crowell 2004). Alleviation of visceral symptoms with medications that target 5-HT receptors suggests that 5-HT is involved in gut motility and visceral pain regulation (Costedio et al. 2007).

Increased gastrointestinal permeability and its role in pediatric abdominal pain have been the focus of work by Shulman et al. (2008). They explored the relationship of sub-clinical inflammation to chronic pain of the GI tract. In a well-designed prospective controlled study, they investigated the difference in GI permeability and fecal calprotectin (protein marker for intestinal inflammation) concentration in children with abdominal pain versus control (Shulman et al. 2008). Proximal GI permeability, colonic permeability, and fecal calprotectin are found to be significantly greater in the abdominal pain group compared to the control group. Fecal calprotectin concentrations correlated with pain interference with activities. However, there was no correlation between GI permeability and pain-related symptoms. This study highlighted that more research is needed to examine the interactions at the molecular genetic level between mast cells and 5-HT.

Intestinal inflammation markers (fecal calprotectin, intestinal permeability, and serum cytokine levels) are associated with chronic abdominal pain. Prior to 2010,

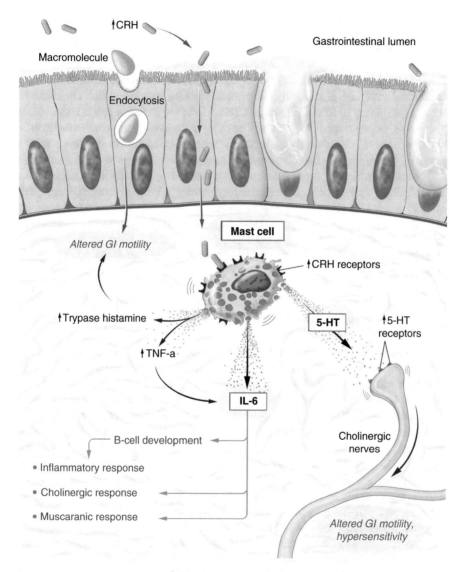

Fig. 9.2 Gastrointestinal lumen pathways to altered gastrointestinal motility and visceral pain (hypersensitivity). *5-HT* 5-hydroxytryptamine, *CRH* Corticotrophin-releasing hormone, *GI* Gastrointestinal, *IL-6* Interleukin-6, *TNF-a* Tumor necrosis factor alpha

there was little evidence that inflammatory changes in the gastrointestinal tract which are considered "microscopic" or at the cellular level without gross mucosal changes are manifested in chronic abdominal pain of unknown origin. Chronic abdominal pain of unknown origin may be due to inflammatory changes in the gastrointestinal tract.

Epithelial inflammation is commonly associated with a range of gastrointestinal disorders including disorders that are explicitly inflammatory in nature such as colitis and conditions in which the inflammation may be sub-clinical such as IBS. The colon is often a primary site of epithelial inflammation and dysregulation. These conditions share various chronic symptoms such as bloating, chronic visceral/ abdominal pain and discomfort, and altered stooling patterns which significantly impact the quality of life of patients. Effective interventions that can alleviate these symptoms, improve quality of life, and possibly permanently improve gastrointestinal health are highly desirable, especially when viewed within the scope of the problem. Research has identified possible mechanisms through which colonic epithelial cell integrity is compromised in persons suffering from chronic visceral pain. These include dysregulation of inflammatory pathways induced through overgrowth of acid producing bacteria, exposure to lipopolysaccharide. Further, molecular genetic pathways (e.g., AKT-pathway, (Fourie et al. 2014)) which are implicated in induced visceral pain of patients have been identified in IBS patients. There is a clear link between chronic abdominal pain and stress (Mayer 2000; Naliboff et al. 2004; Whitehead et al. 1992, 1996).

9.5   Chronic Stress-Associated Visceral Pain (Hyperalgesia)

In the human, chronic stress is associated with enhanced abdominal pain (visceral pain hyperalgesia) and altered bowel function (IBS), a common but poorly understood clinical presentation that consumes significant health care resources. While it is not clear whether stress initiates visceral pain development and abdominal pain symptoms, in 75–80% of IBS patients stress is reported to trigger or exacerbate symptoms (Mayer 2000; Whitehead et al. 1992). Recent evidence indicates that epigenetic regulatory pathways are involved in chronic stress-induced visceral hyperalgesia in the rodent. The genes involved in chronic stress-associated visceral hyperalgesia are poorly understood. Stress is associated with negative chronic health effects, many of which may be associated with stress-induced inflammation. A significantly higher number of mast cells have been shown in the colonic mucosa of pediatric patients with histologically defined non-inflammatory phenotype as compared to controls with inflammatory phenotypes despite similar levels of GI symptoms. Interestingly, a similar finding was not found when the upper gut was compared to colon biopsies (Henderson et al. 2012). Thus, visceral site specificity (the colon) may be important in testing of the relationship of chronic stress to gastrointestinal symptoms.

We prospectively tested this with an intestinal permeability test solution in patients who had chronic abdominal pain and who met inclusion criteria for IBS based on medical history, physical exam, and laboratory evaluation. For the testing procedure, participants (including controls) fasted prior to the clinic visit. They were then asked to describe their GI symptom intensity, sensation, and location with the Gastrointestinal Pain Pointer (GIPP) (Henderson et al. 2017), before and after ingestion of a 100 ml intestinal permeability test solution (sucrose—10 g/dl, lactulose—5 g/dl, mannitol—1 g/dl, and sucralose—0.1 g/dl) that is absorbable at various sites along the GI tract. We administered the test solution to acutely induce GI symptoms of abdominal pain/discomfort. Participants reported the intensity of their abdominal pain/discomfort using a virtual dial on an app-like electronic interface, the GIPP, on a 0–100 point scale. Abdominal pain/discomfort was measured 30 min prior to ingestion, at the time of ingestion, and at 15, 30, 45, 60, 90, 120, 180, 240, and 300 min post ingestion. The maximum abdominal pain scores were used for data analyses. Five hours post ingestion, urine samples were obtained for analysis of the four sugar probes and raffinose (an internal standard). The samples were analyzed by mass spectrometry. We found that patients with IBS (Rome III defined; 55% diarrhea) did not have differences in sucrose urine output, representing gastric permeability, nor changes in lactulose-to-mannitol ratio, representing small intestinal permeability, however did have significantly decreased colonic permeability, represented by sucralose output, compared to controls (Del Valle-Pinero et al. 2013). This provides support to our hypothesis of colonic site specificity for the reported chronic GI symptoms in persons with IBS.

Subsequent gene expression analysis of this patient group showed upregulation of the IL1A gene, which coded for the pro-inflammatory cytokine interleukin-1 alpha (IL1-α), in those with high Perceived Stress Score (PSS) compared to those with low PSS (1.91-fold). Those with high PSS and IBS had IL1A mRNA levels that were 3.47-fold higher compared to participants with low PSS and IBS. These findings provided evidence for a potential inflammatory regulatory mechanism behind stress-related changes in chronic GI symptoms (Peace et al. 2012).

Following the discovery of microRNAs (miRNAs) as regulators of mRNA stability and gene expression, we characterized the miRNA profiles associated with chronic GI symptoms and stress in IBS patients. Our miRNA profiling revealed elevated levels of the hsa-miR-342-3p and hsa-miR-150, compared to controls with similar age, BMI, and ethnicity (Fourie et al. 2014, 2016a), U.S. Patent Application No. 14/892,999. Interestingly, hsa-miR-150 targets IL1A mRNA at multiple sites, suggesting a link between inflammatory gene expression and miRNA.

Collectively, mRNA and miRNA observations suggest cell migration as the common pathway. Supporting this is an earlier finding of significantly elevated levels of the liver chemokine C-C motif ligand 16 (CCL-16) and the vasoactive intestinal peptide (VIP) mRNA in patients with chronic visceral pain compared to controls (Del Valle-Pinero et al. 2011; Del Valle-Pinero et al. 2015). Based on these findings, we began to conduct functional studies focusing on cytoskeletal and motility signaling as both relate to gut epithelial barrier dynamics and potentially microbial translocation and microbe–host interactions. In addition, given the link between gut epithelial barrier dynamics and brain–gut interactions, neuro-related behaviors such as sleep alterations in IBS were studied (Heitkemper et al. 2016; Reddy et al. 2014).

One potential genomic marker related to visceral pain recently identified is vasoactive intestinal peptide (VIP). It may serve as a promising therapeutic target given its role in facilitating overall gut motility and digestion. VIP, which is a member of the glucagon–secretin superfamily, has a broad distribution in the body, such as the heart, lung, brain, and digestive tract. VIP has many physiological effects including exocrine and endocrine secretions, neuro-protection, cell differentiation, and immune response. VIP gene expression from peripheral whole blood of patients ($n = 12$; 45.8% male) with chronic GI symptoms and stress (IBS) is significantly upregulated (2.91-fold) compared to matched (age, gender, race, and BMI) controls ($n = 12$; $P < 0.00001$; CI 95%). VIP plasma protein was not significantly different when compared with controls ($P = 0.193$). However, in the rat model VIP plasma concentration was significantly increased treated with trinitrobenzene sulfonic acid compared to naive rats ($P < 0.05$). TNBS-treated rats had colitis and inflammation as evident from weight changes and elevated myeloperoxidase levels in the intracellular granules of neutrophils. These observations provide additional evidence of a role for VIP in stress and IBS. The results provide a basis for future studies to understand the physiology and pharmacology of VIP for the treatment of chronic visceral pain (Del Valle-Pinero et al. 2015).

9.6 Chronic Stress-Associated Visceral Hyperalgesia: Animal Model

Several reports have established that daily chronic stress predicts the intensity and severity of subsequent symptoms in IBS patients. This observation led to the testing of a variety of rodent models involving chronic intermittent exposure to stress. One of the first chronic stress models to be adapted to the study of visceral hypersensitivity was repeated exposure to psychological stress of water avoidance (Larauche et al. 2011). Initial studies using this model indicated that male rats exposed to 10 consecutive days of water avoidance stress for 1-h daily developed visceral hypersensitivity to colorectal distension and increased stool output. The visceral pain response was monitored using electromyographic recording that entails surgical implantation of electrodes and subsequent single housing of animals. However, when naïve male and female Wistar rats were exposed to a similar water avoidance stress schedule, they developed visceral analgesia to CRD as monitored by intraluminal colonic solid-state manometry. Therefore, the impact of repeated mild stress such as water avoidance stress in modulating visceral pain response is influenced by the basal state condition of the animals such as surgery and housing conditions before (Barbara et al. 2007; Piche et al. 2009) response.

Stress and peripheral administration of corticotrophin-releasing factor (CRF) induce mast cell degranulation in the colon in experimental animals and humans resulting in the development of visceral hypersensitivity via the release of several potential mediators including histamine, tryptase, prostaglandin E2, and nerve growth factor that can activate or sensitize sensory afferents (Larauche et al. 2008, 2012; Wallon et al. 2008). It is well established that stress and administration of corticosterone in rodents can disrupt the intestinal epithelial barrier which may increase the penetration of soluble factors (antigens) into the lamina propria, resulting in nociceptor sensitization (Ait-Belgnaoui et al. 2005; Matricon et al. 2012). Increased intestinal permeability appears as a prerequisite for the development of visceral hypersensitivity in both humans and rodents (Barbara et al. 2007; Piche et al. 2009). There is mounting evidence that stress can affect the intestinal microbiota of rodents which display changes in composition, diversity, and number of gut microorganisms (Bailey et al. 2011; O'Mahony et al. 2009). These alterations in the microbiota can have significant impact on the host to affect behavior, visceral sensitivity, and inflammatory susceptibility (Bercik 2011; Cryan and O'Mahony 2011). A role for aberrant microRNA (miRNA) expression in gastrointestinal disorders has been postulated. miRNAs are cleaved from 70- to 100-nucleotide hairpin pre-miRNA precursors which then bind through partial sequence complementarity to the 3'-UTR of target mRNAs. This leads to a block of translation as well as some degree of mRNA degradation. A pathophysiologic role for microRNAs in patients with IBS has been reported that postulates links between miRNA-29a and inhibition of a target gene, glutamine synthetase (Zhou et al. 2010). Glutamine has trophic effects that help maintain gut mucosal barrier function (Zhou and Verne 2011). Thus, profiling of miRNA may be useful as a molecular marker for IBS patients and other gastrointestinal diseases (such as ulcerative colitis) that demonstrate an increase in intestinal permeability.

9.7 Emerging Role of Epigenetic Regulation in Stress and Visceral Pain

Epigenetics has emerged rapidly from groundbreaking studies demonstrating genome-wide distribution of DNA methylation and histone acetylation sites in human cell lines and primary cells and functional correlation in multiple key physiological processes, including cell cycle regulation (Barski et al. 2007; Shi et al. 2004; Wang et al. 2008). Recent studies have assessed the role of epigenetic pathways to regulation of behavior to include memory, mood, and pain (Denk and McMahon 2012; Gao et al. 2010; Szyf and Meaney 2008; Weaver et al. 2004). The upstream promoter sites in the glucocorticoid gene recently demonstrated increased methylation associated with decreased glucocorticoid expression in the amygdala in the water avoidance stress rat model (Tran et al. 2013). The miR-29 family (29a, 29b, 29c) has significant relationship to two key enzymes involved in DNA methylation, the 3'-UTRs of DNA methyltransferase 3a and -3b (de novo methyltransferases) (Fabbri et al. 2007). Little is known about the role of epigenetic mechanisms in the modulation of peripheral pain pathways and intestinal barrier function. Studies of the effects of histone deacetylase (HDAC) inhibitors suggest that epigenetic gene regulation modulates pain perception in mouse models of inflammatory or neuropathic pain (Zhou and Verne 2011; Denk and McMahon 2012; Chiechio et al. 2009; Geranton 2012; Uchida et al. 2010). Zheng et al. (2013) report that visceral hyperalgesia in a chronic stress rat model was associated with decreased levels of intestinal epithelial tight junction protein levels and increased permeability for macromolecules in the colon (Zheng et al. 2013). This study supports the potentially pivotal role for epigenetic regulatory pathways in chronic stress-induced hyperalgesia in peripheral pain pathways and intestinal barrier function including activation of DNA methyltransferase 3 alpha (DNMT3A) in the colonic mucosal. DNA methylation is a key function in the determination of whether signals from specific segments of the DNA are transmitted and complete the intended function or are silenced. Such signals may include relay of neurotransmitters or reactions involving proteins and or fats. Both DNMT3A and DNMT3B are mostly responsible for de novo methylation. DNMT3A is shown to establish patterns in early development before birth.

9.8   Microbiome

Recent research has highlighted the importance of the human microbiome in mediating a variety of brain–gut interactions (Mayer et al. 2014) and specifically its role in various GI conditions and symptoms, including visceral pain, through mediating inflammatory processes. Commensal GI bacteria fulfill a large array of beneficial functions to the host, assisting in digestion, regulating biological function, playing a role in immunity, and even affecting mental processes, at the same time these microbes also play critical roles in the pathogenesis of many GI diseases and disorders (Arrieta et al. 2014). Furthermore, a growing body of research has yielded several promising lines of inquiry, which shed light onto the etiology of IBS and through which consistent trends maybe identified. These include altered permeability of the GI epithelium, specifically in the colon (Del Valle-Pinero et al. 2013; Camilleri and Gorman 2007; McOmber et al. 2020; Turcotte et al. 2013) and sub-clinical inflammation identified via genomic lines of inquiry, both systemic and colonic (Fourie et al. 2014, 2016a; Henderson et al. 2012; Peace et al. 2012; Del Valle-Pinero et al. 2011, 2015; Matricon et al. 2012; Taylor et al. 2010), both possible mechanisms underlying visceral pain. The microbiome undoubtedly interacts with and regulates many of the physiologic and metabolic pathways in the gut as well as in other neurophysiological axes which may be implicated in the etiology of visceral pain (Lew et al. 2019). Studying the microbial community profiles and bacterial richness and abundance in IBS while exploiting IBS associated hyperalgesia to specifically look at bacterial associations of visceral pain severity in relation to a noxious stressor may identify novel diagnostic genomic and microbial markers, and may reveal genomic mechanisms underlying visceral pain, which may form the basis of simple and targeted therapeutic interventions.

Traditional methodologies for studying of microbial community variation in GI disorders have focused on the luminal microbiome (feces) or colonic mucosal microbiome. Fecal microbial profiling offers a non-invasive approach to studying the colonic (the GI segment thought to be most affected in IBS) microbiome. The fecal microbiome consists mostly of two components, a luminal non-mucosa-adherent bacterial fraction as well shed mucosa-adherent bacteria (Durban et al. 2011). As the bacterial composition of stool is inclusive of both luminal and adherent bacterial communities it is not known in what proportions of either is reflected in stool. Evidence indicates that the mucosal microbiome is poorly represented in fecal samples and that the fecal microbiome differs substantially in composition from the mucosa-adherent microbiome of the colon (Durban et al. 2011). Profiling of the colonic mucosal microbiome provides an arguably more functionally relevant look at the colonic microbial community as it relates to the health and function of the colonic epithelium and mucosa. Mucosa-adherent bacteria are likely to interact more intimately with the epithelium. Dysregulation in these microbes may affect epithelial function, they are the most likely to translocate across the compromised epithelium and therefore initiate or contribute to the pathogenesis of sustained visceral pain and hypersensitivity.

The oral buccal mucosal microbiome of 38 participants were characterized using PhyloChip microarrays. The severity of abdominal pain was assessed by administering an oral test solution. Participants self-reported their induced abdominal pain. Pain severity was highest in IBS participants ($P = 0.0002$), particularly IBS-overweight participants ($P = 0.02$), and was robustly correlated to the abundance of 60 OTUs, 4 genera, 5 families, and 4 orders of bacteria ($r^2 > 0.4$, $P < 0.001$). IBS-overweight participants showed decreased richness in the phylum Bacteroidetes ($P = 0.007$) and the genus *Bacillus* ($P = 0.008$). Analysis of beta-diversity found significant separation of the IBS-overweight group ($P < 0.05$). Our oral microbial results are concordant with described fecal and colonic microbiome-IBS and -weight associations. Visceral pain severity was strongly correlated to the abundance of many taxa, suggesting the potential of the oral microbiome in diagnosis and patient phenotyping. The oral microbiome has potential as a source of microbial information in IBS. IBS is a symptom-defined condition and symptom severity may differ widely among individuals with IBS. The variation in the severity of IBS symptoms provides a more nuanced interpretation of biological data than symptom categories. The results of prior investigations provide the basis for future research aimed at expanding research on the microbiome and visceral pain symptoms and intestinal health (Fourie et al. 2016b).

9.9 Sleep

Diagnostic assessment and treatment of visceral pain is a challenge due to its often unknown origin, lack of biomarkers, and comorbid symptoms. Up to 20% of high school and college-age persons in the USA suffer from chronic abdominal pain. In addition, of the over three million people identified in the National Health Interview Survey as having chronic abdominal pain, 34% rated their health as fair to poor health, had an average of 30 sick days a year, and over 20% reported limitations in their daily activity (Hahn et al. 1997). Many patients with chronic abdominal pain report a significant precipitating illness with associated abdominal pain. Proposed causes of the pain include inflammatory changes in the gastrointestinal tract (Taylor et al. 2010), altered function of serotonin receptors, motility, and colonic mucosal permeability (Bellini et al. 2003; Camilleri et al. 2002; Chen et al. 2001). The molecular pathology underlying the comorbid symptoms is not known. There is a link between chronic abdominal pain and sleep, in that when sleep is dysregulated patients have an increase in abdominal pain symptoms (Burr et al. 2009; Gold et al.

2009; Wu et al. 2010). Likewise, genetic expression may also be altered. Data show that intrinsic differences in genetic expression of inflammatory chemokines differentiate patients with chronic abdominal pain and sleep disturbances.

There is a growing science of sleep disturbances, specifically the study of rhythms in living organisms, known as chronobiology, which has significantly evolved since the early studies (Ritschel and Forusz 1994; Wetterberg 1994). Biological rhythms, such as circadian (18–24 h), allow nearly all organisms to maintain homeostasis. Sleep disturbances affect the sleep–wake cycle which is controlled jointly by a circadian process regulating the timing of sleep and by a homeostatic process regulating sleep drive (and/or sleep "intensity") (Borbely 1982; Daan et al. 1984). About one third of the adult population experience sleep problems weekly (Bixler et al. 2002; Singleton et al. 2003). Chronotype refers to the diurnal preference of an individual preferring to be awake in the morning (i.e., morningness) or in the evening (i.e., eveningness). The sleep/activity rhythm is an endogenous genetically automatic oscillator that completes a cycle every 24 h. The oscillating periodicity repeats itself by increasing and decreasing sleep/activity movements in time to the daily clock (Touitou and Haus 1992). Measuring the sleep/activity rhythm is an accepted way of approximating the sleep/wake cycle. It also can be used to estimate the degree of disruption of the sleep drive. Locomotor activity was found to be a "marker" rhythm for the clock (Turek et al. 1987). Nearly all living organisms express circadian rhythms. Circadian rhythms are generated by transcriptional/translational feedback loops consisting of clock genes (Dunlap 1999; Reppert and Weaver 2002; Shearman et al. 2000).

There is an interaction of gastrointestinal visceral pain and sleep disturbances. Patients with chronic abdominal pain had an increased sensitivity to environmental, physical, and visceral stimuli. Females, specifically, have a dysregulation of noradrenaline, adrenaline, and cortisol levels throughout their sleep interval (Burr et al. 2009). Patients with chronic abdominal pain also have differences in apnea-hypopnea index when compared with controls (Gold et al. 2009). Moreover, patients with chronic abdominal pain have an increased risk of sleep disturbances and poor sleep quality. The cluster of the aforementioned symptoms exacerbate sleep disorders in patients with chronic abdominal pain (Wu et al. 2010).

Sleep is necessary to well-being and health and, too much or too little sleep may increase the risk of developing chronic comorbid conditions (Buxton and Marcelli 2010; Cirelli and Tononi 2008). Disturbances in sleep are also associated with increased IBS symptoms (Nicholl et al. 2008; Maneerattanaporn and Chey 2009; Heitkemper et al. 1998; Chen et al. 2011). In IBS patients, abdominal pain scores are significantly correlated with sleep (Chen et al. 2011). There is evidence that there is a genetic predisposition to alterations in both sleep architecture and pain (Buxton and Marcelli 2010; Cirelli and Tononi 2008; Roehrs 2009; Bachmann et al. 2012). The brain-derived neurotrophic factor (BDNF) gene encodes BDNF, which contains a single nucleotide polymorphism (SNP) Val66Met (rs6265), wherein adenine and guanine alleles vary, yielding an alteration of valine to methionine at codon 66. Bachmann et al. (2012) show in healthy humans that BDNF Val66Met polymorphism modulates sleep intensity. The study found pronounced variations in slow wave sleep, measured by electroencephalogram, activity specific to BDNF. Another study demonstrated elevated BDNF expression in IBS patients compared to healthy controls was significantly correlated with higher abdominal pain scores (Yu et al. 2012). Although there are limited studies involving IBS and BDNF, a 10-biomarker index that included BDNF was used to differentiate IBS from non-IBS subjects and found to be sensitive and specific (Lembo et al. 2009).

9.10 Discussion

Recent advances on the role of systemic and circulating genomic signals, the enteric nervous systems and its relationship to visceral pain have contributed significantly to the understanding of IBS and comorbid symptoms. This increased knowledge has helped shift the paradigm that these disorders are exclusively behavioral in nature and that pathophysiologic disturbances at the cellular and genomic level exist. To date, the consideration of inflammation and visceral pain has been traditionally reserved for inflammatory conditions such as Crohn's disease and ulcerative colitis. The findings presented therein indicate that similar relationships may exist in previously termed functional gastrointestinal disorders. Future research is needed to uncover the associated causes and genomic underpinning of visceral pain. Novel biobehavioral treatments will arise by unveiling the role of genomics in the pathophysiology and genomics of chronic visceral pain.

Acknowledgements Alan Hoofring, National Institutes of Health, Medical Arts for illustrations. Conceptual contributions from prior Henderson Lab members: Tara J.Taylor, Nicolaas H. Fourie, LeeAnne B. Sherwin, Kristen R. Weaver, Arsemia Del Valle Pinero. Acknowledgments and mentoring support from Margaret M. Heitkemper and John W. Wiley.

The authors acknowledge funding from the United States Department of Health and Human Services, National Institutes of Health, National Institute of Nursing Research, Division of Intramural Research to Wendy A. Henderson, 1ZIANR000018; National Institute on Minority Health & Health Disparities, NIH fellowship DREAM award to Bridgett Rahim-Williams, 1K22MD006143; Bethesda, Maryland, USA.

The opinions expressed herein and the interpretation and reporting of these data are the responsibility of the author(s) and should not be seen as an official recommendation, interpretation, or policy of the National Institutes of Health or the United States Government.

References

Ait-Belgnaoui A, Bradesi S, Fioramonti J, Theodorou V, Bueno L. Acute stress-induced hypersensitivity to colonic distension depends upon increase in paracellular permeability: role of myosin light chain kinase. Pain. 2005;113(1–2):141–7.

Akbar A, Yiangou Y, Facer P, Walters JR, Anand P, Ghosh S. Increased capsaicin receptor TRPV1-expressing sensory fibres in irritable bowel syndrome and their correlation with abdominal pain. Gut. 2008;57(7):923–9.

Arrieta MC, Stiemsma LT, Amenyogbe N, Brown EM, Finlay B. The intestinal microbiome in early life: health and disease. Front Immunol. 2014;5:427.

Bachmann V, Klein C, Bodenmann S, Schafer N, Berger W, Brugger P, et al. The BDNF Val66Met polymorphism modulates sleep intensity: EEG frequency- and state-specificity. Sleep. 2012;35(3):335–44.

Bailey MT, Dowd SE, Galley JD, Hufnagle AR, Allen RG, Lyte M. Exposure to a social stressor alters the structure of the intestinal microbiota: implications for stressor-induced immunomodulation. Brain Behav Immun. 2011;25(3):397–407.

Barbara G. Mucosal barrier defects in irritable bowel syndrome. Who left the door open? Am J Gastroenterol. 2006;101(6):1295–8.

Barbara G, Stanghellini V, De Giorgio R, Cremon C, Cottrell GS, Santini D, et al. Activated mast cells in proximity to colonic nerves correlate with abdominal pain in irritable bowel syndrome. Gastroenterology. 2004;126(3):693–702.

Barbara G, Wang B, Stanghellini V, de Giorgio R, Cremon C, Di Nardo G, et al. Mast cell-dependent excitation of visceral-nociceptive sensory neurons in irritable bowel syndrome. Gastroenterology. 2007;132(1):26–37.

Barski A, Cuddapah S, Cui K, Roh TY, Schones DE, Wang Z, et al. High-resolution profiling of histone methylations in the human genome. Cell. 2007;129(4):823–37.

Bellini M, Rappelli L, Blandizzi C, et al. Platelet 5-HT transporter in patients with diarrhea-predominant IBS both before and after treatment with alosetron. Am J Gastroenterol. 2003;98:2705–11.

Bercik P. The microbiota-gut-brain axis: learning from intestinal bacteria? Gut. 2011;60(3):288–9.

Bixler EO, Vgontzas AN, Lin HM, Vela-Bueno A, Kales A. Insomnia in Central Pennsylvania. J Psychosom Res. 2002;53(1):589–92.

Borbely AA. Sleep regulation. Introduction. Hum Neurobiol. 1982;1(3):161–2.

Burr RL, Jarrett ME, Cain KC, Jun SE, Heitkemper MM. Catecholamine and cortisol levels during sleep in women with irritable bowel syndrome. Neurogastroenterol Motil. 2009;21(11):1148–e97.

Buxton OM, Marcelli E. Short and long sleep are positively associated with obesity, diabetes, hypertension, and cardiovascular disease among adults in the United States. Soc Sci Med. 2010;71(5):1027–36.

Camilleri M, Gorman H. Intestinal permeability and irritable bowel syndrome. Neurogastroenterol Motil. 2007;19(7):545–52.

Camilleri M, Atanasova E, Carlson P, et al. Serotonin-transporter polymorphism pharmacogenetics in diarrhea-predominant irritable bowel syndrome. Gastroenterology. 2002;123:425–32.

Chen J, Li Z, Pan H, et al. Maintenance of 5-HT in the intestinal mucosa & ganglia of mice that lack the high-affinity 5-HT transporter: abnormal intestinal motility & the expression of cation transporters. J Neurosci. 2001;21:6348–61.

Chen CL, Liu TT, Yi CH, Orr WC. Evidence for altered anorectal function in irritable bowel syndrome patients with sleep disturbance. Digestion. 2011;84(3):247–51.

Chiechio S, Zammataro M, Morales ME, Busceti CL, Drago F, Gereau RWT, et al. Epigenetic modulation of mGlu2 receptors by histone deacetylase inhibitors in the treatment of inflammatory pain. Mol Pharmacol. 2009;75(5):1014–20.

Cirelli C, Tononi G. Is sleep essential? PLoS Biol. 2008;6(8-e216):1605–11.

Costedio MM, Hyman N, Mawe GM. Serotonin and its role in colonic function and in gastrointestinal disorders. Dis Colon Rectum. 2007;50(3):376–88.

Creed F, Ratcliffe J, Fernandez L, Tomenson B, Palmer S, Rigby C, et al. Health-related quality of life and health care costs in severe, refractory irritable bowel syndrome. Ann Intern Med. 2001;134(9_Part_2):860–8.

Crowell MD. Role of serotonin in the pathophysiology of the irritable bowel syndrome. Br J Pharmacol. 2004;141(8):1285–93.

Cryan JF, O'Mahony SM. The microbiome-gut-brain axis: from bowel to behavior. Neurogastroenterol Motil. 2011;23(3):187–92.

Daan S, Beersma DG, Borbely AA. Timing of human sleep: recovery process gated by a circadian pacemaker. Am J Physiol Regul Integr Comp Physiol. 1984;246(2):R161–83.

De Giorgio R, Barbara G. Is irritable bowel syndrome an inflammatory disorder? Curr Gastroenterol Rep. 2008;10(4):385–90.

Del Valle-Pinero AY, Martino AC, Taylor TJ, Majors BL, Patel NS, Heitkemper MM, et al. Pro-inflammatory chemokine C-C motif ligand 16 (CCL-16) dysregulation in irritable bowel syndrome (IBS): a pilot study. Neurogastroenterol Motil. 2011;23(12):1092–7.

Del Valle-Pinero AY, Van Deventer HE, Fourie NH, Martino AC, Patel NS, Remaley AT, et al. Gastrointestinal permeability in patients with irritable bowel syndrome assessed using a four probe permeability solution. Clin Chim Acta. 2013;418:97–101.

Del Valle-Pinero AY, Sherwin LB, Anderson EM, Caudle RM, Henderson WA. Altered vasoactive intestinal peptides expression in irritable bowel syndrome patients and rats with trinitrobenzene sulfonic acid-induced colitis. World J Gastroenterol. 2015;21(1):155–63.

Denk F, McMahon SB. Chronic pain: emerging evidence for the involvement of epigenetics. Neuron. 2012;73(3):435–44.

Drossman DA, Camilleri M, Mayer EA, Whitehead WE. AGA technical review on irritable bowel syndrome. Gastroenterology. 2002;123(6):2108–31.

Duff L, Louw G, Loftus-Hills A, Morrell C. Clinical practice guidelines: the recognition and assessment of acute pain in children. London: Royal College of Nursing; 2001. 1–33 p.

Dunlap JC. Molecular bases for circadian clocks. Cell. 1999;96(2):271–90.

Durban A, Abellan JJ, Jimenez-Hernandez N, Ponce M, Ponce J, Sala T, et al. Assessing gut microbial diversity from feces and rectal mucosa. Microb Ecol. 2011;61(1):123–33.

Eberlein R, Kruse JT, Wiehe S. Education plan on the subject of "pain." 1: understanding of patient care affects the purpose of nursing education. Pflege Z. 2000;53(6):409–14.

Fabbri M, Garzon R, Cimmino A, Liu Z, Zanesi N, Callegari E, et al. MicroRNA-29 family reverts aberrant methylation in lung cancer by targeting DNA methyltransferases 3A and 3B. Proc Natl Acad Sci U S A. 2007;104(40):15805–10.

Fanciullo GJ, Cravero JP, Mudge BO, McHugo GJ, Baird JC. Development of a new computer method to assess children's pain. Pain Med. 2007;8(Suppl 3):S121–8.

Fourie NH, Peace RM, Abey SK, Sherwin LB, Rahim-Williams B, Smyser PA, et al. Elevated circulating miR-150 and miR-342-3p in patients with irritable bowel syndrome. Exp Mol Pathol. 2014;96(3):422–5.

Fourie NH, Peace RM, Abey SK, Sherwin LB, Wiley JW, Henderson WA. Perturbations of circulating miRNAs in irritable bowel syndrome detected using a multiplexed high-throughput gene expression platform. J Vis Exp. 2016a;(117):e54693.

Fourie NH, Wang D, Abey SK, Sherwin LB, Joseph PV, Rahim-Williams B, et al. The microbiome of the oral mucosa in irritable bowel syndrome. Gut Microbes. 2016b;7(4):286–301.

Gao J, Wang WY, Mao YW, Graff J, Guan JS, Pan L, et al. A novel pathway regulates memory and plasticity via SIRT1 and miR-134. Nature. 2010;466(7310):1105–9.

Geranton SM. Targeting epigenetic mechanisms for pain relief. Curr Opin Pharmacol. 2012;12(1):35–41.

Gold MS, Gebhart GF. Nociceptor sensitization in pain pathogenesis. Nat Med. 2010;16(11):1248–57.

Gold AR, Broderick JE, Amin MM, Gold MS. Inspiratory airflow dynamics during sleep in irritable bowel syndrome: a pilot study. Sleep Breath. 2009;13(4):397–407.

Grundmann O, Yoon SL. Irritable bowel syndrome: epidemiology, diagnosis and treatment: an update for health-care practitioners. J Gastroenterol Hepatol. 2010;25(4):691–9.

Grundy L, Erickson A, Brierley SM. Annual review of physiology: visceral pain. Ann Rev. 2019;81:261–84.

Hahn B, Saunders W, Maier W. Differences between individuals with self-reported irritable bowel syndrome (IBS) and IBS-like symptoms. Dig Dis Sci. 1997;42(12):2585–90.

Heitkemper M, Charman AB, Shaver J, Lentz MJ, Jarrett ME. Self-report and polysomnographic measures of sleep in women with irritable bowel syndrome. Nurs Res. 1998;47(5):270–7.

Heitkemper M, Jarrett M, Cain K, et al. Autonomic nervous system function in women with irritable bowel syndrome. Dig Dis Sci. 2001;46:1276–84.

Heitkemper MM, Han CJ, Jarrett ME, Gu H, Djukovic D, Shulman RJ, et al. Serum tryptophan metabolite levels during sleep in patients with and without irritable bowel syndrome (IBS). Biol Res Nurs. 2016;18(2):193–8.

Henderson WA, Shankar R, Taylor TJ, Del Valle-Pinero AY, Kleiner DE, Kim KH, et al. Inverse relationship of interleukin-6 and mast cells in children with inflammatory and non-inflammatory abdominal pain phenotypes. World J Gastroint Pathophysiol. 2012;3(6):102–8.

Henderson WA, Rahim-Williams B, Kim KH, Sherwin LB, Abey SK, Martino AC, et al. The gastrointestinal pain pointer: a valid and innovative method to assess gastrointestinal symptoms. Gastroenterol Nurs. 2017;40(5):357–63.

Larauche M, Bradesi S, Million M, McLean P, Tache Y, Mayer EA, et al. Corticotropin-releasing factor type 1 receptors mediate the visceral hyperalgesia induced by repeated psychological stress in rats. Am J Physiol Gastrointest Liver Physiol. 2008;294(4):G1033–40.

Larauche M, Mulak A, Tache Y. Stress-related alterations of visceral sensation: animal models for irritable bowel syndrome study. J Neurogastroenterol Motility. 2011;17(3):213–34.

Larauche M, Mulak A, Tache Y. Stress and visceral pain: from animal models to clinical therapies. Exp Neurol. 2012;233(1):49–67.

Lee O, Mayer E, Schmulson M, et al. Gender-related differences in IBS symptoms. Am J Gastroenterol. 2001;96:2184–93.

Lembo AJ, Neri B, Tolley J, Barken D, Carroll S, Pan H. Use of serum biomarkers in a diagnostic test for irritable bowel syndrome. Aliment Pharmacol Ther. 2009;29(8):834–42.

Levine B, Jarrett M, Cain K, Heitkemper M. Psychophysiological response to a laboratory challenge in women with and without diagnosed IBS. Res Nurs Health. 1997;20:431–41.

Lew KN, Starkweather A, Cong X, Judge M. A mechanistic model of gut-brain axis perturbation and high-fat diet pathways to gut microbiome homeostatic disruption, systemic inflammation, and type 2 diabetes. Biol Res Nurs. 2019;21(4):384–99.

Lovell RM, Ford AC. Prevalence of gastro-esophageal reflux-type symptoms in individuals with irritable bowel syndrome in the community: a meta-analysis. Am J Gastroenterol. 2012;107(12):1793–801. quiz 802.

Mahjoub FE, Farahmand F, Pourpak Z, Asefi H, Amini Z. Mast cell gastritis: children complaining of chronic abdominal pain with histologically normal gastric mucosal biopsies except for increase in mast cells, proposing a new entity. Diagn Pathol. 2009;4:34.

Maneerattanaporn M, Chey WD. Sleep disorders and gastrointestinal symptoms: chicken, egg or vicious cycle? Neurogastroenterol Motil. 2009;21(2):97–9.

Mantyselka P, Kumpusato E, Ahonen R, Takala J. Patients' verses general practitioners assessment of pain intensity in primary care patients with non-cancer pain. Br Gen Pract. 2001;51:995–7.

Marchand F, Perretti M, McMahon SB. Role of the immune system in chronic pain. Nat Rev Neurosci. 2005;6(7):521–32.

Matricon J, Meleine M, Gelot A, Piche T, Dapoigny M, Muller E, et al. Review article: associations between immune activation, intestinal permeability and the irritable bowel syndrome. Aliment Pharmacol Ther. 2012;36(11–12):1009–31.

Mayer E. The neurobiology of stress and gastrointestinal disease. Gut. 2000;47:861–9.

Mayer EA, Savidge T, Shulman RJ. Brain-gut microbiome interactions and functional bowel disorders. Gastroenterology. 2014;146(6):1500–12.

McCahon S, Strong J, Sharry R, Cramond T. Self-report and pain behavior among patients with chronic pain. Clin J Pain. 2005;21(3):223–31.

McOmber M, Rafati D, Cain K, Devaraj S, Weidler EM, Heitkemper M, et al. Increased gut permeability in first-degree relatives of children with irritable bowel syndrome or functional abdominal pain. Clin Gastroenterol Hepatol. 2020;18(2):375–84 e1.

Mearin F, Lacy BE, Chang L, Chey WD, Lembo AJ, Simren M, et al. Bowel disorders. Gastroenterology. 2016;150(6):1393–407.

Melzack R. The McGill pain questionnaire: major properties and scoring methods. Pain. 1975;1(3):277–99.

Naliboff B, Mayer M, Fass R, et al. The effect of the life stress on symptoms of heartburn. Psychosom Med. 2004;66:426–34.

Nicholl BI, Halder SL, Macfarlane GJ, Thompson DG, O'Brien S, Musleh M, et al. Psychosocial risk markers for new onset irritable bowel syndrome--results of a large prospective population-based study. Pain. 2008;137(1):147–55.

North RB, Sieracki JM, Fowler KR, Alvarez B, Cutchis PN. Patient-interactive, microprocessor-controlled neurological stimulation system. Neuromodulation. 1998;1(4):185–93.

O'Mahony SM, Marchesi JR, Scully P, Codling C, Ceolho AM, Quigley EM, et al. Early life stress alters behavior, immunity, and microbiota in rats: implications for irritable bowel syndrome and psychiatric illnesses. Biol Psychiatry. 2009;65(3):263–7.

Palsson OS, Whitehead W, Tornblom H, Sperber AD, Simren M. Prevalence of Rome IV functional bowel disorders among adults in the United States, Canada, and the United Kingdom. Gastroenterology. 2020;158(5):1262–73.

Park JH, Rhee PL, Kim HS, Lee JH, Kim YH, Kim JJ, et al. Mucosal mast cell counts correlate with visceral hypersensitivity in patients with diarrhea predominant irritable bowel syndrome. J Gastroenterol Hepatol. 2006;21(1 Pt 1):71–8.

Peace RM, Majors BL, Patel NS, Wang D, Valle-Pinero AY, Martino AC, et al. Stress and gene expression of individuals with chronic abdominal pain. Biol Res Nurs. 2012;14(4):405–11.

Petersen S, Brulin C, Bergström E. Recurrent pain symptoms in young schoolchildren are often multiple. Pain. 2006;121(1–2):145–50.

Piche T, Saint-Paul MC, Dainese R, Marine-Barjoan E, Iannelli A, Montoya ML, et al. Mast cells and cellularity of the colonic mucosa correlated with fatigue and depression in irritable bowel syndrome. Gut. 2008;57(4):468–73.

Piche T, Barbara G, Aubert P, Bruley des Varannes S, Dainese R, Nano JL, et al. Impaired intestinal barrier integrity in the colon of patients with irritable bowel syndrome: involvement of soluble mediators. Gut. 2009;58(2):196–201.

Reddy SY, Rasmussen NA, Fourie NH, Berger RS, Martino AC, Gill J, et al. Sleep quality, BDNF genotype and gene expression in individuals with chronic abdominal pain. BMC Med Genet. 2014;7:61.

Reppert SM, Weaver DR. Coordination of circadian timing in mammals. Nature. 2002;418(6901):935–41.

Ritschel WA, Forusz H. Chronopharmacology: a review of drugs studied. Methods Find Exp Clin Pharmacol. 1994;16(1):57–75.

Roehrs TA. Does effective management of sleep disorders improve pain symptoms? Drugs. 2009;69(2 Suppl):5–11.

Russo MW, Gaynes BN, Drossman DA. A national survey of practice patterns of gastroenterologists with comparison to the past two decades. J Clin Gastroenterol. 1999;29(4):339–43.

Shearman LP, Sriram S, Weaver DR, Maywood ES, Chaves IS, Zheng B, et al. Interacting molecular loops in the mammalian circadian clock. Science. 2000;288(5468):1013–9.

Sherwin LB, Leary E, Henderson WA. Effect of illness representations and catastrophizing on quality of life in adults with irritable bowel syndrome. J Psychosoc Nurs Ment Health Serv. 2016;54(9):44–53.

Sherwin LB, Leary E, Henderson WA. The association of catastrophizing with quality-of-life outcomes in patients with irritable bowel syndrome. Qual Life Res. 2017a;26(8):2161–70.

Sherwin LB, Ozoji OM, Boulineaux CM, Joseph PV, Fourie NH, Abey SK, et al. Gender and weight influence quality of life in irritable bowel syndrome. J Clin Med. 2017b;6(11):103.

Shi Y, Lan F, Matson C, Mulligan P, Whetstine JR, Cole PA, et al. Histone demethylation mediated by the nuclear amine oxidase homolog LSD1. Cell. 2004;119(7):941–53.

Shulman RJ, Eakin MN, Jarrett M, Czyzewski DI, Zeltzer LK. Characteristics of pain and stooling in children with recurrent abdominal pain. J Pediatr Gastroenterol Nutr. 2007;44(2):203–8.

Shulman RJ, Eakin MN, Czyzewski DI, Jarrett M, Ou CN. Increased gastrointestinal permeability and gut inflammation in children with functional abdominal pain and irritable bowel syndrome. J Pediatr. 2008;153(5):646–50.

Singleton N, Bumpstead R, O'Brien M, Lee A, Meltzer H. Psychiatric morbidity among adults living in private households, 2000. Int Rev Psychiatry. 2003;15(1–2):65–73.

Spiller R, Garsed K. Postinfectious irritable bowel syndrome. Gastroenterology. 2009;136(6):1979–88.

Stein C, Clark JD, Oh U, Vasko MR, Wilcox GL, Overland AC, et al. Peripheral mechanisms of pain and analgesia. Brain Res Rev. 2009;60(1):90–113.

Szyf M, Meaney MJ. Epigenetics, behaviour, and health. Allergy Asthma Clin Immunol. 2008;4(1):37–49.

Talley N, Phillips S, Melton LJ, et al. A patient questionnaire to identify bowel disease. Ann Intern Med. 1989;111:671–4.

Taylor TJ, Youssef NN, Shankar R, Kleiner DE, Henderson WA. The association of mast cells and serotonin in children with chronic abdominal pain of unknown etiology. BMC Res Notes. 2010;3:265.

Touitou YH, Haus E. Biological rhythms in clinical and laboratory medicine. Berlin: Springer; 1992.

Tran L, Chaloner A, Sawalha AH, Van-Meerveld BG. Importance of epigenetic mechanisms in visceral pain induced by chronic water avoidance stress. Psychoneuroendocrinology. 2013;38(6):898–906.

Turcotte JF, Kao D, Mah SJ, Claggett B, Saltzman JR, Fedorak RN, et al. Breaks in the wall: increased gaps in the intestinal epithelium of irritable bowel syndrome patients identified by confocal laser endomicroscopy (with videos). Gastrointest Endosc. 2013;77(4):624–30.

Turek FW, Losee-Olson S, Swann JM, Horwath K, Van Cauter E, Milette JJ. Circadian and seasonal control of neuroendocrine-gonadal activity. J Steroid Biochem. 1987;27(1–3):573–9.

Uchida H, Ma L, Ueda H. Epigenetic gene silencing underlies C-fiber dysfunctions in neuropathic pain. J Neurosci. 2010;30(13):4806–14.

Wallon C, Yang PC, Keita AV, Ericson AC, McKay DM, Sherman PM, et al. Corticotropin-releasing hormone (CRH) regulates macromolecular permeability via mast cells in normal human colonic biopsies in vitro. Gut. 2008;57(1):50–8.

Wang Z, Zang C, Rosenfeld JA, Schones DE, Barski A, Cuddapah S, et al. Combinatorial patterns of histone acetylations and methylations in the human genome. Nat Genet. 2008;40(7):897–903.

Weaver IC, Cervoni N, Champagne FA, D'Alessio AC, Sharma S, Seckl JR, et al. Epigenetic programming by maternal behavior. Nat Neurosci. 2004;7(8):847–54.

Wetterberg L. Light and biological rhythms. J Intern Med. 1994;235(1):5–19.

Whitehead W, Crowell MD, Robinson JC. Effects of stressful life events on bowel symptoms: subjects with IBS compared with subjects without bowel dysfunction. Gut. 1992;33:825–30.

Whitehead W, Burnett C, Cook E, et al. Impact of IBS on quality of life. Dig Dis Sci. 1996;41:2248–53.

Wu JP, Song ZY, Xu Y, Zhang YM, Shen RH. Probe into sleep quality in the patients with irritable bowel syndrome. Zhonghua Nei Ke Za Zhi. 2010;49(7):587–90.

Youssef NN, Murphy TG, Langseder AL, Rosh JR. Quality of life for children with functional abdominal pain: a comparison study of patients' and parents' perceptions. Pediatrics. 2006;117(1):54–9.

Yu YB, Zuo XL, Zhao QJ, Chen FX, Yang J, Dong YY, et al. Brain-derived neurotrophic factor contributes to abdominal pain in irritable bowel syndrome. Gut. 2012;61(5):685–94.

Zheng G, Wu SP, Hu Y, Smith DE, Wiley JW, Hong S. Corticosterone mediates stress-related increased intestinal permeability in a region-specific manner. Neurogastroenterol Motil. 2013;25(2):e127–39.

Zhou Q, Verne GN. miRNA-based therapies for the irritable bowel syndrome. Expert Opin Biol Ther. 2011;11(8):991–5.

Zhou Q, Souba WW, Croce CM, Verne GN. MicroRNA-29a regulates intestinal membrane permeability in patients with irritable bowel syndrome. Gut. 2010;59(6):775–84.

Genomics of Fracture Pain

10

Mari A. Griffioen, Susan G. Dorsey, and Cynthia L. Renn

Contents

10.1 Introduction

Fracture-related pain at the time of injury serves a protective function and is meant to lessen and disappear as the fracture heals. While nearly everyone with fractures report pain, the intensity, characteristics, extent, and duration of pain vary from person to person and reflect the complex biopsychosocial interaction between genetic, developmental, environmental, and psychological factors along with the severity and extent of the injury (Martelli et al. 2004; Pizzo et al. 2011). The effects

M. A. Griffioen (✉)
University of Delaware School of Nursing, Newark, DE, USA
e-mail: mgriffi@udel.edu

S. G. Dorsey · C. L. Renn
Department of Pain and Translational Symptom Science, University of Maryland School of Nursing, Baltimore, MD, USA
e-mail: SDorsey@umaryland.edu; Renn@umaryland.edu

© Springer Nature Switzerland AG 2020
S. G. Dorsey, A. R. Starkweather (eds.), *Genomics of Pain and Co-Morbid Symptoms*, https://doi.org/10.1007/978-3-030-21657-3_10

of fracture-related pain are substantial. Poorly controlled acute pain adds to an increased stress response, longer hospital stay, higher cost, delayed return to work, and disability (Ahmadi et al. 2016; Wells et al. 2008; Goldsmith and Mccloughen 2016; Rivara et al. 2008; Batoz et al. 2016). Persons with fracture-related chronic pain miss work and seek medical care more frequently than those without chronic pain in addition to reporting higher levels of pain intensity, anxiety, and depression (Stålnacke 2011; Jenewein et al. 2009). Although some factors (e.g. older age, being female, high pain intensity) have been identified as fracture-related chronic pain risk factors (Rivara et al. 2008; Clay et al. 2012; Holmes et al. 2010; Rosenbloom et al. 2013), few studies have ventured beyond collecting self-reported measures at the time of injury or subsequently. Since pain is a complex phenomenon including sensory, emotional, and genomic components, it is crucial that future investigations rigorously incorporate physiological, psychological, and genomic factors that contribute to fracture-related pain.

10.2 Pathophysiology of Fracture-Related Pain

Fracture-related pain is the outcome of deliberate brain activity in response to noxious stimuli from the fracture site and has cognitive, affective, and sensory components (Loeser and Treede 2008). A fracture and surrounding tissue injury initiate a cascade of complex mechanisms that have not been fully uncovered. Briefly, acute pain proceeds the activation of peripheral endings of nociceptors (A-beta, A-delta, and C-fibers) located in muscles, tendons, joints, fasciae, and bones (Katz and Rothenberg 2005; Nencini and Ivanusic 2016). Some of the substances expressed by these sensory nerve fibers are tropomyosin receptor kinase A (TrkA), calcitonin gene-related peptide (CGRP), substance P (SP), and pro-inflammatory cytokines, which all have been implicated in animal fracture pain models (Nencini and Ivanusic 2016; Castañeda-Corral et al. 2011). Therefore, further evaluation of these substances in humans is warranted; however, the complexity of accessing fracture sites at the time of injury might not be feasible. A more practical option would be to draw peripheral blood for circulating biomarkers.

Not only does the nociceptive process contribute to pain, the bone and tissue injury also initiate an inflammatory response (Ren and Dubner 2010; Blunk et al. 1999) where immune cells recruited to the site of injury form a network with peripheral neurons such that activation of an immune response leads to further excitation of pain pathways (Ren and Dubner 2010; Padi and Kulkarni 2008). Sensitization develops so that subsequent stimuli are intensified, which leads to guarding at the site of injury to prevent further damage (Katz and Rothenberg 2005). This heightened peripheral sensitivity is expected to return to baseline as the injury heals (Latremoliere and Woolf 2009). However, with traumatic injuries such as fractures, peripheral-nerve axons can suffer mild-to-severe damage and nerve regeneration can be slow and often incomplete (Grinsell and Keating 2014). In instances where sensation is restored, function returns in a specified order starting with coolness followed by cold pain, heat pain, mechanical pain, touch, and warmth (Van Boven and

Johnson 1994). In other cases, the damage persists and may lead to spontaneous pain and multiple positive and negative somatosensory signs such as thermal and mechanical hyperalgesia, allodynia, hypoesthesia, and hypoalgesia (Maier et al. 2010).

The few studies that have examined sensory aspects of acute fracture pain have revealed that, in the acute phase following i.e. facial fractures, patients report hypoesthesia, loss of discriminant power, and paresthesia in the affected side (Schultze-Mosgau et al. 1999; Benoliel et al. 2005; Okochi et al. 2015) and lowered pressure pain threshold immediately following upper and lower extremity fracture (Hall et al. 2016). These findings correspond with facial fracture patients self-report of abnormal sensation and paresthesia in the affected area (Okochi et al. 2015; Marchena et al. 1998). This would seem counterintuitive to what would be expected as hyperalgesia presents as a protective mechanism immediately following an injury. However this hyposensitivity has been shown to persist in lower extremity fracture patients with chronic pain as they present with higher warmth detection thresholds and increased activation thresholds for Aβ and Aδ fibers (Griffioen et al. 2017a). This hyposensitivity has also been reported in other chronic pain injury conditions, i.e. low back pain, partial nerve injuries, knee pain, and whiplash injuries (Leffler and Hansson 2008; Starkweather et al. 2016; Jensen et al. 2007; Raak and Wallin 2006) But, persistent local hyperalgesia has also been reported in patients with patellar fractures (Larsen et al. 2018). Thus, it is currently not clear which factors contribute to some patients with fractures developing chronic pain with hypoalgesia or hyperalgesia. Carefully designed genotype-phenotype studies including sensory testing and genomics have the potential to illuminate some of the underlying biological mechanisms that might have taken place at one or multiple sites during the nociceptive process.

10.3 Fracture-Related Pain

Pain intensity, or pain severity, is what most healthcare providers and patients refer to when they describe fracture-related pain. This subjective experience of the magnitude or strength a person assigns a painful episode, is only one aspect of pain and ignores other qualities such as sensory and affective components which can guide treatment or the degree of interference and disability that can have significant effects on activities of daily living. While pain intensity is an appropriate measure when assessing for the presence of pain, the effectiveness of treatment for pain, or how pain changes over time in individuals, its usefulness is limited when used for group comparisons. This is because the highly subjective pain intensity score is influenced by how it compares with past painful experiences as well as the pain score immediately preceding the current one (Watkinson et al. 2013). Some of this variability may be due to differences in the severity of the injuries (Nielsen et al. 2009); however, many persons with seemingly identical injuries report very different pain intensities. In experimental acute pain studies with strictly controlled stimuli in normal subjects, variable pain intensity reports between subjects are found (Chen et al.

1989; Fillingim 2004; Kim et al. 2004). Some of these differences can be attributed to environmental influences (Nielsen et al. 2008). However, there is abundant evidence from the literature suggesting that individual differences in pain response may be inherited (Mogil 2012). For example, several twin studies demonstrate that heritable factors significantly contribute to an individual's pain response (Nielsen et al. 2008; Hartvigsen et al. 2009; Norbury et al. 2007). As very few fracture-related pain studies have included genomics, it is important that future studies collect specimens for current and future analyses so as to illuminate biological factors that may drive fracture-related pain.

Currently, a majority of the fracture-related pain literature reports the mean pain intensity score. Regardless of fracture site, age, or gender the mean pain score has been reported to be approximately five (Warren et al. 2016; Archer et al. 2012; Ware et al. 2012; Griffioen et al. 2017b; Pasyar et al. 2018; Storesund et al. 2016; Griffioen et al. 2016; Zanchi et al. 2018; Radinovic et al. 2014). What complicates using the mean pain score for comparisons is that the same mean pain score of five can be referred to as either a high or low pain score in analyses. In lower extremity fractures and traumatic injuries it has been referred to as a high pain score and is used to predict outcomes (Holmes et al. 2010; Griffioen et al. 2017b). In other studies on traumatic injuries it has been referred to as a low pain score (Trevino et al. 2014). Some of this variability can be attributed to the study population under investigation, but condensing multiple pain scores into one ignores the fluctuation of pain over time. In one study on lower extremity fracture pain during hospitalization, the pain either stayed the same or got worse for 54% of patients (Griffioen et al. 2017b). Further analyses in this study showed that persons with a similar pain score (five) had vastly different pain experiences during hospitalization; the pain either improved, stayed the same, or got worse (Griffioen et al. 2016). The ability to create trajectories of how pain changes over time allows for accounting of within- and between-person variability associated with poorer pain outcomes as well as examining predictors for each type of trajectory. Further, if these trajectories were available in real time for healthcare providers, the effectiveness of treatment would be easier to discern and the opportunity to change treatment would present sooner during care.

10.4 Factors that Affect Fracture-Related Pain

10.4.1 Fracture Type and Injury Severity

In general, the more severe and complicated the fracture, the more likely it is that the person will report higher pain (Suzuki et al. 2009; Arinzon et al. 2007). However, individual differences do exist as at times there is no difference in pain based on fracture type (Chou et al. 2018). Fractures are most common in youth and in old age when the skeleton is porous, with weak points at the physes and metaphyses (Hedström et al. 2010). Children are most likely to fracture the upper extremity

(Naranje et al. 2016; Hedström et al. 2010); as children age, lower extremity fractures become more frequent, echoing the most common fracture site reported in adults and older adults (Baig 2017; Somersalo et al. 2014). Hip fractures, the most common fracture site in older adults (Somersalo et al. 2014; Pressley et al. 2011; Baidwan and Naranje 2017), can frequently be attributed to underlying pathological conditions such as osteoporosis and rheumatoid arthritis (Warriner et al. 2011).

Persons most likely to report chronic pain are those who suffered lower extremity fractures (Rivara et al. 2008; Griffioen et al. 2017b; Fuzier et al. 2015; Gerbershagen et al. 2010). A majority of persons with chronic fracture-related pain report the pain to be mild to moderate (Fuzier et al. 2015) with the exception of lower extremity fracture patients who report the pain to be both severe and persistent (Rosenbloom et al. 2013; Warren et al. 2016). It could be that persons with fracture-related chronic pain may have developed a neuropathic pain component (Fuzier et al. 2015; Rbia et al. 2017) as they report the sensations of tenderness, throbbing, sharpness, tingling, and numbness (Griffioen et al. 2017a). Neuropathic pain has also been reported in children with chronic pain following fractures (Batoz et al. 2016). Complex Regional Pain Syndrome, a devastatingly painful condition, occurs at rates of 5.5–26.2 cases per 100,000 following fractures where women and those with complicated fractures are more at risk (Birklein and Dimova 2017). Much of the fracture-related pain literature reports on a specific type of fracture and patient population, which is the most feasible and practical approach; however, there is not a standard set of pain measurement tools, sensory testing, or genomic approach that is applied to examine pain across fracture types and across the life span, which would greatly facilitate the discovery of common factors.

10.4.2 Treatment

Treatment has a significant effect on fracture-related pain. While pharmacological and non-pharmacological therapies can effectively reduce pain intensity (Berben et al. 2008), many patients report that the therapies provided do not adequately control the pain (Archer et al. 2012). Approximately 70% of trauma patients report pain during hospitalization (Jabusch et al. 2015) and nearly 59% of patients undergoing surgery to repair a traumatic lower extremity fracture report moderate to severe pain at the time of discharge from the hospital (Archer et al. 2012). Further, up to 30% of patients with long bone fractures do not receive any pain medication while in the emergency department (Brown et al. 2003; Ware et al. 2012). Intravenous fentanyl, hydromorphone, and morphine are the preferred analgesics for treating acute pain immediately upon arrival in the hospital (Ahmadi et al. 2016; MacKenzie et al. 2016) and provide much improvement in pain (Griffioen et al. 2019), but are not as effective as regional nerve blocks (Griffioen and O'Brien 2018). Oral opioids and non-opioids are much less effective in controlling fracture pain once the patient has been transferred from the emergency room to the in-patient unit (Griffioen et al. 2019). And, children, older adults, and those with cognitive impairment are more likely to be under-treated for their fracture-related pain (Fuentes-Losada et al. 2016; Morrison and Siu 2000; Herrick et al. 2004). While pharmacogenomics has the potential for personalized treatment, the ability to use genomics to predict postoperative pain is still developing (Palada et al. 2018).

10.4.3 Sociodemographic Factors

Disparities in pain prevalence, access to care, and under treatment of pain based on sociodemographic factors (age, race, ethnicity, gender, educational level, health insurance, primary care provider, occupation, residence [urban/rural], smoking, alcohol, controlled substances) have been identified (Pizzo et al. 2011). While the physiological process of pain changes with age as the sensory system matures and then ages (Baccei and Fitzgerald 2013), persons of all ages report fracture-related pain verbally or non-verbally, and warrant having their pain assessed and treated effectively. The treatment of fracture-related pain in children and persons with cognitive impairment is poor (Fuentes-Losada et al. 2016; Morrison and Siu 2000). Up to 60% of children report pain at the time of discharge from the emergency room following fractures (Fuentes-Losada et al. 2016) and older children are less likely to receive medication for their pain (Liao and Reyes 2018). Effective treatment of pain in infants and children is particularly important as persistent and repetitive pain may alter pain plasticity and result in long-term negative consequences such as chronic pain and altered pain threshold and sensitization, not to mention the anxiety that may accompany future medical procedures (Batoz et al. 2016; Loizzo et al. 2009; Wong et al. 2010). Younger children are more sensitive to pain compared to older children, and once puberty becomes apparent, sex differences appear where girls are more sensitive compared to boys (Blankenburg et al. 2010; Blankenburg et al. 2011). In fracture-related pain, girls and younger children are more likely to report higher pain intensity compared to boys and older children (Clapp et al. 2017). Not much is known about the underlying biological mechanisms associated with fracture-related pain in children. Future studies including physiological, psychological, and genomic factors are warranted. Of note is that quantitative sensory testing (QST) has been validated in children above the age of five (Blankenburg et al. 2010) and could provide valuable information regarding the peripheral sensory aspect associated with fracture-related pain. However, the expertise and equipment to conduct QST requires an investment from both scientists, institutions, and funding agencies.

Just as younger persons are less likely to have their fracture-related pain assessed, the same is true for older persons (Boccio et al. 2014). What further complicates assessment and treatment of fracture-related pain in some older persons is the presence of cognitive impairment, and as cognitively impaired people are less likely to have their pain treated, they also have higher pain compared to non-cognitively impaired people (Daniels et al. 2014). For future studies it would be important to use behaviors such as motor movements, communications, and facial expressions that have been found to be associated with pain for assessment (Mills 1989; Lichtner et al. 2014). However these behavioral changes can be difficult to analyze and further research is needed to ensure reliability and validity of existing tools assessing these behaviors (Lichtner et al. 2014). Very few studies have examined fracture-related pain by gender; however, there is anecdotal evidence that women report higher pain intensity scores following ankle fracture surgery (Storesund et al. 2016). It is crucial that future studies include results on the sex differences related to fracture-related pain not only to generate knowledge, but the NIH now requires that sex be included as a biological variable in research.

10.4.4 Co-morbid Symptoms

Depression, anxiety, and sleep disturbances are frequently reported by persons with pain (Pizzo et al. 2011), and all three are correlated with chronic pain following lower extremity injury (Castillo et al. 2006). The literature reports that at the time of fracture, persons with depressive symptoms are more likely to report higher pain scores (Archer et al. 2012; Radinovic et al. 2014; Alexiou et al. 2018) and that persons with fractures are more likely to report increase in depressive symptoms over time (Gironda et al. 2009; Benzinger et al. 2012). Or, as in one study, depressive symptoms improved from the time of fracture (Golkari et al. 2015; Mossey et al. 1989). The literature examining anxiety and sleep disturbance with fracture pain is minimal, although pain catastrophizing is associated with pain intensity, pain-related anxiety, and poor sleep with pain intensity following fractures (Archer et al. 2012; Keogh et al. 2010; Accardi-Ravid et al. 2018).

10.5 Genomics and Fracture Pain

The administration of Trk-inhibitor and monoclonal antibodies against nerve growth factor (anti-NGF) result in a reduction of fracture-induced pain in mice (Ghilardi et al. 2011; Jimenez-Andrade et al. 2007) and in one human study, elevated tumor necrosis factor alpha (TNF-α), soluble TNF II (sTNF-RII), and interleukin 18 (IL-18) plasma levels were associated with increased resting pain following hip fracture surgery (Ko et al. 2018). The few studies that have examined genomics, pain, and depression found BDNF Met/Met (Rawson et al. 2015) and 5HTR1A (-1019) G allele were associated with increased depressive symptoms following hip fracture (Lenze et al. 2008). The shortage of studies that focus fracture-related pain, or co-morbid conditions and fracture pain studies is concerning as interventions will have to be multifactorial and address more than just pain. Additional studies examining lower extremity fracture pain, depression, and anxiety are needed to better understand which genomic factors may be associated in pain and co-morbid symptoms. It is currently not clear what the biological mechanisms are that have depression, anxiety, sleep disturbance, and chronic pain co-occur.

10.6 Summary and Future Directions

The study of fracture pain is challenging and interesting, and an advantage is the onset of acute pain is known; therefore, the opportunity to examine the transition to chronic pain has a clear starting point. It is clear that fracture-related pain results from genetic, developmental, environmental, and psychological factors working alone and/or in combination; however, mechanisms underlying acute pain, chronic pain, and the transition from acute to chronic pain in this population remain unclear. The results from comprehensive studies that include all these factors can then be applied to identify new therapeutic targets to prevent and/or better manage pain following a fracture as well as be used in predictive risk models as biomarkers of susceptibility to chronic fracture-related pain.

References

Accardi-Ravid MC, Dyer JR, Sharar SR, Wiechman S, Jensen MP, Hoffman HG, et al. The nature of trauma pain and its association with catastrophizing and sleep. Int J Behav Med. 2018;25(6):698–705.

Ahmadi A, Bazargan-hejazi S, Zadie ZH, Euasobhon P, Keturman P, Karbasfrushan A, et al. Pain management in trauma: a review study. Inj Violence. 2016;8(2):89–98.

Alexiou KI, Roushias A, Evaritimidis SE, Malizos KN. Quality of life and psychological consequences in elderly patients after a hip fracture: a review. Clin Interv Aging. 2018;13:143–50.

Archer KR, Castillo RC, Wegener ST, Abraham CM, Obremskey WT. Pain and satisfaction in hospitalized trauma patients: the importance of self-efficacy and psychological distress. J Trauma Acute Care Surg. 2012;72(4):1068–77.

Arinzon Z, Gepstein R, Shabat S, Berner Y. Pain perception during the rehabilitation phase following traumatic hip fracture in the elderly is an important prognostic factor and treatment tool. Disabil Rehabil. 2007;29(8):651–8.

Baccei ML, Fitzgerald M. Baccei. In: McMahon SB, Koltzenburg M, Tracey I, Turk DC, editors. Wall & Melzack's textbook of pain. 6th ed. Philadelphia: Elsevier; 2013. p. 143–55.

Baidwan NK, Naranje SM. Epidemiology and recent trends of geriatric fractures presenting to the emergency department for United States population from year 2004–2014. Public Health. 2017;142:64–9.

Baig M. A review of epidemiological distribution of different types of fractures in paediatric age. Cureus. 2017;9(8):1–9.

Batoz H, Semjen F, Bordes-Demolis M, Bnard A, Nouette-Gaulain K. Chronic postsurgical pain in children: prevalence and risk factors. A prospective observational study. Br J Anaesth. 2016;117(4):489–96.

Benoliel R, Birenboim R, Regev E, Eliav E. Neurosensory changes in the infraorbital nerve following zygomatic fractures. Oral Surg Oral Med Oral Pathol Oral Radiol Endod. 2005;99(6):657–65.

Benzinger P, Nicolai S, Hoffrichter R, Becker C, Pfeiffer K, Zijlstra GAR, et al. Depressive symptoms and fear of falling in previously community-dwelling older persons recovering from proximal femoral fracture (Aging Clinical and Experimental Research (2011) 23, 5-6 (450-456)). Aging Clin Exp Res. 2012;24(1):108.

Berben SAA, Meijs THJM, van Dongen RTM, van Vugt AB, Vloet LCM, Mintjes-de Groot JJ, et al. Pain prevalence and pain relief in trauma patients in the Accident & Emergency department. Injury. 2008;39(5):578–85.

Birklein F, Dimova V. Complex regional pain syndrome type I. Pain Clin Updat. 2017;e624(1–8):556–61.

Blankenburg M, Boekens H, Hechler T, Maier C, Krumova E, Scherens A, et al. Reference values for quantitative sensory testing in children and adolescents: developmental and gender differences of somatosensory perception. Pain. 2010;149(1):76–88. https://doi.org/10.1016/j.pain.2010.01.011.

Blankenburg M, Meyer D, Hirschfeld G, Kraemer N, Hechler T, Aksu F, et al. Developmental and sex differences in somatosensory perception - a systematic comparison of 7-versus 14-year-olds using quantitative sensory testing. Pain. 2011;152(11):2625–31. https://doi.org/10.1016/j.pain.2011.08.007.

Blunk J, Osiander G, Nischik M, Schmelz M. Pain and inflammatory intradermal injections leukocytes hyperalgesia induced by of human platelets and leukocytes. Eur J Pain. 1999;3(3):247–59.

Boccio E, Wie B, Pasternak S, Salvador-Kelly A, Ward MF, D'Amore J. The relationship between patient age and pain management of acute long-bone fracture in the ED. Am J Emerg Med. 2014;32(12):1516–9.

Brown JC, Klein EJ, Lewis CW, Johnston BD, Cummings P. Emergency department analgesia for fracture pain. Ann Emerg Med. 2003;42(2):197–205.

Castañeda-Corral G, Jimenez-Andrade JM, Bloom AP, Reid N, Mantyh WG, Kaczmarska MJ, et al. The majority of myelinated and unmyelinated sensory nerve fibers that innervate bone express the tropomyosin receptor kinase A. Neuroscience. 2011;520:196–207.

Castillo RC, MacKenzie EJ, Wegener ST, Bosse MJ. Prevalence of chronic pain seven years following limb threatening lower extremity trauma. Pain. 2006;124(3):321–9.

Chen AC, Dworkin SF, Haug J, Gehrig J. Human pain responsivity in a tonic pain model: psychological determinants. Pain. 1989;37(2):143–60.

Chou LB, Niu EL, Williams AA, Duester R, Anderson SE, Harris AHS, et al. Postoperative pain after surgical treatment of ankle fractures. J Am Acad Orthop Surg Glob Res Rev. 2018;2(9):e021.

Clapp ADM, Thull-Freedman J, Mitra T, Lethebe BC, Williamson T, Stang AS. Patient-reported pain outcomes for children attending an emergency department with limb injury. Pediatr Emerg Care. 2017;1 https://doi.org/10.1097/PEC.0000000000001317.

Clay FJ, Watson WL, Newstead SV, McClure RJ. A systematic review of early prognostic factors for persistent pain following acute orthopedic trauma. Pain Res Manag. 2012;17(1):35–45.

Daniels AH, Daiello LA, Lareau CR, Robidoux KA, Luo W, Ott B, et al. Preoperative cognitive impairment and psychological distress in hospitalized elderly hip fracture patients. Am J Orthop. 2014;43(7):E146–52.

Fillingim RB. Social and environmental influences on pain: implications for pain genetics. In: Mogil JS, editor. The genetics of pain. Seattle: IASP Press; 2004. p. 283–303.

Fuentes-Losada LM, Vergara-Amador E, Laverde-Cortina R. Pain management assessment in children with limb fractures in an emergency service. Rev Colomb Anestesiol. 2016;44(4):305–10.

Fuzier R, Rousset J, Bataille B, Salces-y-Nedeo A, Magues J-P. One half of patients reports persistent pain three months after orthopaedic surgery. Anaesth Crit Care Pain Med. 2015;34:159–64.

Gerbershagen HJ, Dagtekin O, Isenberg J, Martens N, Özgür E, Krep H, et al. Chronic pain and disability after pelvic and acetabular fractures-assessment with the Mainz Pain Staging System. J Trauma. 2010;69(1):128–36.

Ghilardi JR, Freeman KT, Jimenez-Andrade JM, Mantyh WG, Bloom AP, Bouhana KS, et al. Sustained blockade of neurotrophin receptors TrkA, TrkB and TrkC reduces non-malignant skeletal pain but not the maintenance of sensory and sympathetic nerve fibers. Bone. 2011;48(2):389–98. https://doi.org/10.1016/j.bone.2010.09.019.

Gironda MW, Der-Martirosian C, Belin TR, Black EE, Atchison KA. Predictors of depressive symptoms following mandibular fracture repair. J Oral Maxillofac Surg. 2009;67(2):328–34. https://doi.org/10.1016/j.joms.2008.06.007.

Goldsmith H, Mccloughen A. Review article analgesic adherence in recently discharged trauma patients: an integrative literature review. Pain Manag Nurs. 2016;17(1):63–79.

Golkari S, Teunis T, Ring D, Vranceanu AM. Changes in depression, health anxiety, and pain catastrophizing between enrollment and 1 month after a radius fracture. Psychosomatics. 2015;56(6):652–7. https://doi.org/10.1016/j.psym.2015.03.008.

Griffioen MA, O'Brien G. Analgesics administered for pain during hospitalization following lower extremity fracture: a review of the literature. J Trauma Nurs. 2018;25(6):360–5.

Griffioen MA, Johantgen M, Von Rueden K, Greenspan JD, Dorsey SG, Renn CL. Characteristics of patients with lower extremity trauma with improved and not improved pain during hospitalization: a pilot study. Pain Manag Nurs. 2016;17(1):3–13.

Griffioen MA, Greenspan JD, Johantgen M, Von Rueden K, O'Toole RV, Dorsey SG, et al. Quantitative sensory testing and current perception threshold testing in patients with chronic pain following lower extremity fracture. Biol Res Nurs. 2017a;20(1):16–24.

Griffioen MA, Greenspan JD, Johantgen M, Von Rueden K, O'Toole RV, Dorsey SG, et al. Acute pain characteristics in patients with and without chronic pain following lower extremity injury. Pain Manag Nurs. 2017b;18(1):33–41.

Griffioen MA, Ziegler ML, O'Toole RV, Dorsey SG, Renn CL. Change in pain score after administration of analgesics for lower extremity fracture pain during hospitalization. Pain Manag Nurs. 2019;20:158–63.

Grinsell D, Keating CP. Peripheral nerve reconstruction after injury: a review of clinical and experimental therapies. Biomed Res Int. 2014;2014:1–13.

Hall J, Llewellyn A, Palmer S, Rowett-Harris J, Atkins RM, McCabe CS. Sensorimotor dysfunction after limb fracture – an exploratory study. Eur J Pain (United Kingdom). 2016;20(9):1402–12.

Hartvigsen J, Nielsen J, Kyvik KO, Fejer R, Vach W, Iachine I, et al. Heritability of spinal pain and consequences of spinal pain: a comprehensive genetic epidemiologic analysis using a population-based sample of 15,328 twins ages 20-71 years. Arthritis Rheum. 2009;61(10):1343–51.

Hedström EM, Svensson O, Bergström U, Michno P. Epidemiology of fractures in children and adolescents: increased incidence over the past decade: a population-based study from northern Sweden. Acta Orthop. 2010;81(1):148–53.

Herrick C, Steger-May K, Sinacore DR, Brown M, Schechtman KB, Binder EF. Persistent pain in frail older adults after hip fracture repair. J Am Geriatr Soc. 2004;52(12):2062–8.

Holmes A, Williamson O, Hogg M, Arnold C, Prosser A, Clements J, et al. Predictors of pain 12 months after serious injury. Pain Med. 2010;238:1599–611.

Jabusch KM, Lewthwaite BJ, Mandzuk LL, Schnell-Hoehn KN, Wheeler BJ. The pain experience of inpatients in a teaching hospital: revisiting a strategic priority. Pain Manag Nurs. 2015;16(1):69–76.

Jenewein J, Moergeli H, Wittmann L, Büchi S, Kraemer B, Schnyder U. Development of chronic pain following severe accidental injury. Results of a 3-year follow-up study. J Psychosom Res. 2009;66(2):119–26.

Jensen R, Hystad T, Kvale A, Baerheim A. Quantitative sensory testing of patients with long lasting patellofemoral pain syndrome. Eur J Pain. 2007;11(6):665–76.

Jimenez-Andrade JM, Martin CD, Koewler NJ, Freeman KT, Sullivan LJ, Halvorson KG, et al. Nerve growth factor sequestering therapy attenuates non-malignant skeletal pain following fracture. Pain. 2007;133(1–3):183–96.

Katz WA, Rothenberg R. The nature of pain: pathophysiology. J Clin Rheumatol. 2005;11(2 Suppl):11–5.

Keogh E, Book K, Thomas J, Giddins G, Eccleston C. Predicting pain and disability in patients with hand fractures: comparing pain anxiety, anxiety sensitivity and pain catastrophizing. Eur J Pain. 2010;14(4):446–51. https://doi.org/10.1016/j.ejpain.2009.08.001.

Kim H, Neubert JK, San Miguel A, Xu K, Krishnaraju RK, Iadarola MJ, et al. Genetic influence on variability in human acute experimental pain sensitivity associated with gender, ethnicity and psychological temperament. Pain. 2004;109(3):488–96.

Ko FC, Rubenstein WJ, Lee EJ, Siu AL, Morrison RS. TNF-α and sTNF-RII are associated with pain following hip fracture surgery in older adults. Pain Med (United States). 2018;19(1):169–77.

Larsen P, Vedel JO, Vistrup S, Elsoe R. Long-lasting hyperalgesia is common in patients following patella fractures. Pain Med (United States). 2018;19(3):429–37.

Latremoliere A, Woolf CJ. Central sensitization: a generator of pain hypersensitivity by central neural plasticity. J Pain. 2009;10(9):895–926.

Leffler AS, Hansson P. Painful traumatic peripheral partial nerve injury-sensory dysfunction profiles comparing outcomes of bedside examination and quantitative sensory testing. Eur J Pain. 2008;12(4):397–402.

Lenze EJ, Shardell M, Ferrell RE, Orwig D, Yu-Yahiro J, Hawkes W, et al. Association of serotonin-1A and 2A receptor promoter polymorphisms with depressive symptoms and functional recovery in elderly persons after hip fracture. J Affect Disord. 2008;111(1):61–6.

Liao L, Reyes L. Evaluating for racial differences in pain management of long-bone fractures in a pediatric rural population. Pediatr Emerg Care. 2018;1 https://doi.org/10.1097/PEC.0000000000001696.

Lichtner V, Dowding D, Esterhuizen P, Closs SJ, Long AF, Corbett A, et al. Pain assessment for people with dementia: a systematic review of systematic reviews of pain assessment tools. BMC Geriatr. 2014;14(138):1–19.

Loeser JD, Treede R-D. The Kyoto protocol of IASP basic pain terminology. Pain. 2008;137(3):473–7.

Loizzo A, Loizzo S, Capasso A. Neurobiology of pain in children: an overview. Open Biochem J. 2009;3:18–25.

MacKenzie M, Zed PJ, Ensom MHH. Opioid pharmacokinetics-pharmacodynamics. Ann Pharmacother. 2016;50(3):209–18. https://doi.org/10.1177/1060028015625659.

Maier C, Baron R, Tölle TR, Binder A, Birbaumer N, Birklein F, et al. Quantitative sensory testing in the German research network on neuropathic pain (DFNS): somatosensory abnormalities in 1236 patients with different neuropathic pain syndromes. Pain. 2010;150(3):439–50.

Marchena JM, Padwa BL, Kaban LB. Sensory abnormalities associated with mandibular fractures: incidence and natural history. J Oral Maxillofac Surg. 1998;56(7):822–6.

Martelli MF, Zasler ND, Bender MC, Nicholson K. Psychological, neuropsychological, and medical considerations in assessment and management of pain. J Head Trauma Rehabil. 2004;19(1):10–28.

Mills NM. Pain behaviors in infants and toddlers. J Pain Symptom Manag. 1989;4(4):184–90.

Mogil JS. Pain genetics: past, present and future. Trends Genet. 2012;28(6):258–66.

Morrison RS, Siu AL. A comparison of pain and its treatment in advanced dementia and cognitively intact patients with hip fracture. J Pain Symptom Manag. 2000;19(4):240–8.

Mossey JM, Mutran E, Knott K, Craik R. Determinants of recovery 12 months after hip fracture: the importance of psychosocial factors. Am J Public Health. 1989;79(3):279–86.

Naranje SM, Erali RA, Warner WC, Sawyer JR, Kelly DM. Epidemiology of pediatric fractures presenting to emergency departments in the United States. J Pediatr Orthop. 2016;36(4):e45–8.

Nencini S, Ivanusic JJ. The physiology of bone pain. How much do we really know? Front Physiol. 2016;7(Apr):1–15.

Nielsen CS, Stubhaug A, Price DD, Vassend O, Czajkowski N, Harris JR. Individual differences in pain sensitivity: genetic and environmental contributions. Pain. 2008;136(1–2):21–9.

Nielsen CS, Staud R, Price DD. Individual differences in pain sensitivity: measurement, causation, and consequences. J Pain. 2009;10(3):231–7.

Norbury TA, MacGregor AJ, Urwin J, Spector TD, McMahon SB. Heritability of responses to painful stimuli in women: a classical twin study. Brain. 2007;130(Pt 11):3041–9.

Okochi M, Ueda K, Mochizuki Y, Okochi H. How can paresthesia after zygomaticomaxillary complex fracture be determined after long-term follow-up? A new and quantitative evaluation method using current perception threshold testing. J Oral Maxillofac Surg. 2015;73(8):1554–61.

Padi SSV, Kulkarni SK. Minocycline prevents the development of neuropathic pain, but not acute pain: possible anti-inflammatory and antioxidant mechanisms. Eur J Pharmacol. 2008;601(1–3):79–87.

Palada V, Kaunisto MA, Kalso E. Genetics and genomics in postoperative pain and analgesia. Curr Opin Anaesthesiol. 2018;31(5):569–74.

Pasyar N, Rambod M, Kahkhaee FR. The effect of foot massage on pain intensity and anxiety in patients having undergone a tibial shaft fracture surgery: a randomized clinical trial. J Orthop Trauma. 2018;32(12):e482–6.

Pizzo PA, Clark NM, Pokras OC. Relieving pain in America: a blueprint for transforming prevention, care, education, and research. Washington, D.C.: National Academies Press; 2011. 382 p. www.nap.edu. Accessed 3 Nov 2015

Pressley JC, Kendig TD, Frencher SK, Barlow B, Quitel L, Waqar F. Epidemiology of bone fracture across the age span in blacks and whites. J Trauma. 2011;71(5 Suppl 2):S541–8.

Raak R, Wallin M. Thermal thresholds and catastrophizing in individuals with chronic pain after whiplash injury. Biol Res Nurs. 2006;8(2):138–46.

Radinovic K, Milan Z, Markovic-Denic L, Dubljanin-Raspopovic E, Jovanovic B, Bumbasirevic V. Predictors of severe pain in the immediate postoperative period in elderly patients following hip fracture surgery. Injury. 2014;45(8):1246–50. https://doi.org/10.1016/j.injury.2014.05.024.

Rawson KS, Dixon D, Nowotny P, Ricci WM, Binder EF, Rodebaugh TL, et al. Association of functional polymorphisms from brain-derived neurotrophic factor and serotonin-related genes with depressive symptoms after a medical stressor in older adults. PLoS One. 2015;10(3):1–17. https://doi.org/10.1371/journal.pone.0120685.

Rbia N, van der Vlies CH, Cleffken BI, Selles RW, Hovius SER, Nijhuis THJ. High prevalence of chronic pain with neuropathic characteristics after open reduction and internal fixation of ankle fractures. Foot Ankle Int. 2017;38(9):987–96.

Ren K, Dubner R. Interactions between the immune and nervous systems in pain. Nat Med. 2010;16(11):1267–76.

Rivara FP, Mackenzie EJ, Jurkovich GJ, Nathens AB, Wang J, Scharfstein DO. Prevalence of pain in patients 1 year after major trauma. Arch Surg. 2008;143(3):282–7; discussion 288.

Rosenbloom BN, Khan S, McCartney C, Katz J. Systematic review of persistent pain and psychological outcomes following traumatic musculoskeletal injury. J Pain Res. 2013;6:39–51.

Schultze-Mosgau S, Erbe M, Rudolph D, Ott R, Neukam FW. Prospective study on post-traumatic and postoperative sensory disturbances of the inferior alveolar nerve and infraorbital nerve in mandibular and midfacial fractures. J Craniomaxillofac Surg. 1999;27(2):86–93.

Somersalo A, Paloneva J, Kautiainen H, Lönnroos E, Heinänen M, Kiviranta I. Incidence of fractures requiring inpatient care. Acta Orthop. 2014;85(5):525–30.

Stålnacke B-M. Life satisfaction in patients with chronic pain - relation to pain intensity, disability, and psychological factors. Neuropsychiatr Dis Treat. 2011;7:683–9.

Starkweather AR, Lyon DE, Kinser P, Heineman A, Sturgill JL, Deng X, et al. Comparison of low Back pain recovery and persistence: a descriptive study of characteristics at pain onset. Biol Res Nurs. 2016;18(4):401–10.

Storesund A, Krukhaug Y, Olsen MV, Rygh LJ, Nilsen RM, Norekvål TM. Females report higher postoperative pain scores than males after ankle surgery. Scand J Pain. 2016;12:85–93.

Suzuki N, Ogikubo O, Hansson T. The prognosis for pain, disability, activities of daily living and quality of life after an acute osteoporotic vertebral body fracture: its relation to fracture level, type of fracture and grade of fracture deformation. Eur Spine J. 2009;18(1):77–88.

Trevino C, Harl F, Deroon-Cassini T, Brasel K, Litwack K. Predictors of chronic pain in traumatically injured hospitalized adult patients. J Trauma Nurs. 2014;21(2):50–6.

Van Boven RW, Johnson KO. A psychophysical study of the mechanisms of sensory recovery following nerve injury in humans. Brain. 1994;117:149–67.

Ware LJ, Epps CD, Clark J, Chatterjee A. Do ethnic differences still exist in pain assessment and treatment in the emergency department? Pain Manag Nurs. 2012;13(4):194–201.

Warren AM, Jones AL, Bennett M, Solis JK, Reynolds M, Rainey EE, et al. Prospective evaluation of posttraumatic stress disorder in injured patients with and without orthopaedic injury. J Orthop Trauma. 2016;30(9):e305–11.

Warriner AH, Patkar NM, Curtis JR, Delzell E, Gary L, Kilgore M, et al. Which fractures are most attributable to osteoporosis? J Clin Epidemiol. 2011;64(1):46–53. https://doi.org/10.1016/j.jclinepi.2010.07.007.

Watkinson P, Wood AM, Lloyd DM, Brown GD. Pain ratings reflect cognitive context: a range frequency model of pain perception. Pain. 2013;154(5):743–9.

Wells N, Pasero C, Mccaffery M. Improving the quality of care through pain assessment and management. In: Hughes RH, editor. Patient safety and quality: an evidence-based handbook for nurses. Rockville: Agency for Healthcare Research and Quality; 2008. https://www.ncbi.nlm.nih.gov/books/NBK2658/.

Wong C, Lau E, Palozzi L, Campbell F. Pain management in children: part 1 — pain assessment tools and a brief review of nonpharmacological and pharmacological treatment options. Can Pharm J. 2010;145(5):222–5.

Zanchi C, Giangreco M, Ronfani L, Germani C, Giorgi R, Calligaris L, et al. Pain intensity and risk of bone fracture in children with minor extremity injuries. Pediatr Emerg Care. 2018;00(00):1. https://doi.org/10.1097/PEC.0000000000001418.

Genomics of Fibromyalgia

11

Nada Lukkahatai and Leorey N. Saligan

Contents

11.1   Introduction and Significance

Fibromyalgia is a chronic and often debilitating condition that is characterized by widespread pain, persistent fatigue, cognitive impairment, sleep disturbance, and other somatic symptoms (Wolfe 2010). Increased prevalence of fibromyalgia has led to more primary care providers tasked with the assessment and management of this complex condition (Arnold et al. 2012). Unfortunately, about 36% of primary care providers report very limited knowledge about fibromyalgia (Perrot et al. 2012), resulting in many patients with fibromyalgia un- or misdiagnosed and untreated (Arnold et al. 2011). The management of this syndrome can be an

N. Lukkahatai
School of Nursing, Johns Hopkins University, Baltimore, MD, USA
e-mail: nada.lukkahatai@jhu.edu

L. N. Saligan (✉)
Division of Intramural Research, National Institute of Nursing Research, National Institutes of Health, Bethesda, MD, USA
e-mail: saliganl@mail.nih.gov

© Springer Nature Switzerland AG 2020
S. G. Dorsey, A. R. Starkweather (eds.), *Genomics of Pain and Co-Morbid Symptoms*, https://doi.org/10.1007/978-3-030-21657-3_11

economic burden. In the USA, persons with fibromyalgia spend an annual average of nearly $10,000 on health care cost, which is three times higher than patients without FM (Berger et al. 2007).

11.1.1 Prevalence Rates

An epidemiologic study estimated the prevalence of fibromyalgia to be approximately 2–6% in the world population (Hauser et al. 2015; Cabo-Meseguer et al. 2017). The prevalence is higher in Europe (2.31%) compared to North America (1.90%), Asia (1.64%), and South America (1.12%) (Mas et al. 2008). By gender, women have more than 20 times higher prevalence than men (Mas et al. 2008). The highest prevalence peak is found among women within the 40–49 age group (Cabo-Meseguer et al. 2017). Non-Caucasian females, between 49 and 63 years, are more at risk than men in developing fibromyalgia (Aparicio et al. 2012; Lee et al. 2018).

11.2 Diagnostic Criteria

Fibromyalgia diagnosis relies on patients' self-report of symptoms. The criteria that define the condition have been repeatedly changed over the years. The subjective nature of the diagnostic criteria for fibromyalgia often makes an accurate and timely diagnosis of the condition a difficult task to achieve. In the absence of any measurable physiological change, clinicians must base their diagnosis on their patient's self-reported symptoms, which often leads to misdiagnosis.

The first diagnostic criteria for fibromyalgia was proposed in 1981 which included tender points and the presence of generalized aches and pains in 3 or more anatomic sites (Yunus et al. 1981). Then, the American College of Rheumatology (ACR) revised the diagnostic criteria in 1990 based on the results of a multi-center study which reduced the number of tender points at 11 (out of a possible 18) and clarified the widespread pain criterion to 4-quadrant plus axial pain (Wolfe et al. 1990). Then in 2010, the tender point and widespread pain requirements were dropped from the criteria, replacing them with a count of 19 painful regions and the patients' symptom severity reports (Wolfe et al. 2010). In 2011, the 2010 criteria were modified and recommended that fibromyalgia diagnosis be accomplished entirely by the patient's symptom reports (Wolfe et al. 2015). Then in 2016, the diagnostic criteria were revised to include (Wolfe et al. 2016; Heymann et al. 2017):

1. Generalized pain, defined as pain in at least 4 of 5 regions, is present.
2. Symptoms have been present at a similar level for at least 3 months.
3. Widespread pain index (WPI) ≥ 7 and symptom severity scale (SSS) score ≥ 5 OR WPI of 4–6 and SSS score ≥ 9.
4. A diagnosis of fibromyalgia is valid irrespective of other diagnoses. A diagnosis of fibromyalgia does not exclude the presence of other clinically important illnesses.

11.3   Predisposing Factors of Fibromyalgia

Stress, obesity, and a sedentary lifestyle have been suggested as predisposing factors of fibromyalgia. Stress related to physical and emotional traumas and stressful life events could alter stress mechanisms, leading to increased vulnerability to develop the disorder (Gonzalez et al. 2013). In fibromyalgia patients, studies showed dysregulation of stress response manifested by decrease in hypothalamus–pituitary–adrenal (HPA) reactivity, baroreflect and low sympathetic and parasympathetic regulations (Galvez-Sanchez et al. 2019; Ablin et al. 2008; Van Houdenhove and Egle 2004). This abnormal stress response mechanisms may lead to inappropriate responses to physical and psychological stressors (Galvez-Sanchez et al. 2019).

Obesity and a sedentary lifestyle were associated with the risk to develop fibromyalgia (Mork et al. 2010; Segura-Jimenez et al. 2016a; Kang et al. 2016; Gota et al. 2015; de Araujo et al. 2015). The prevalence of overweight individuals with fibromyalgia ranges from 50 to 70% (Okifuji et al. 2009). The underlying mechanism is unclear. Studies suggested the possible involvement of adipocytes, specifically leptin level, in modulating pain perception and fibromyalgia symptoms though the HPA axis (Paiva et al. 2017; Ablin et al. 2012). Furthermore, evidence suggests that fibromyalgia demonstrated a clear familial aggregation (Pellegrino et al. 1989; Ablin and Buskila 2015; Buskila et al. 2007, 1996). However, the mode of inheritance in fibromyalgia is unknown, but it is most probably polygenic.

11.3.1 Symptom Clusters

Fibromyalgia in adults is seen as a syndrome characterized by moderate-to-severe symptoms which include widespread musculoskeletal pain and point tenderness, fatigue, poor sleep, cognitive impairment, and hyperesthesia with or without co-morbidities such as depression, anxiety, and migraine that compromise quality of life (Waylonis and Heck 1992). The co-occurrence of widespread pain with other symptoms contributes to the challenges associated with its management. Some observed a syndrome overlap between different but similar comorbid conditions. For example, fibromyalgia patients who meet the 1990 or 2010/2011 diagnostic criteria are indistinguishable in symptom reports from others with chronic fatigue and irritable bowel syndromes (Wolfe et al. 2016; Heymann et al. 2017). Thus in the past, fibromyalgia has been labeled or clumped with other syndromes such as somatoform disorder, functional somatic syndrome, and bodily distress syndrome (Fink and Schroder 2010; Sharpe and Carson 2001; Henningsen and Creed 2010; Creed et al. 2010).

11.4   Impact of Fibromyalgia

Fibromyalgia negatively affects the individual's quality of life (Wolfe et al. 2019; Carta et al. 2018; Beyazal et al. 2018; Zhang et al. 2018). Studies found the association of fibromyalgia with higher level of stress, (Carta et al. 2018;

Conversano et al. 2019) catastrophizing, (Lukkahatai et al. 2016) aversive emotions such as sadness, anger, fear, and guilt (Gota et al. 2015; Carta et al. 2018; Alciati et al. 2018). Cognitive disturbance is often reported by patients with fibromyalgia, complaining of having difficulty in planning, attention, memory, executive function, and processing speed (Samartin-Veiga et al. 2019; Blanco et al. 2019; Elkana et al. 2019; Segura-Jimenez et al. 2016b). Moreover, patients with fibromyalgia often experience low self-esteem and negative self-image, which can be related to the poor cognitive performance, pain experience, and negative health care experiences (Manivannan 2017; Boyington et al. 2015; Akkaya et al. 2012).

11.5   Management of Fibromyalgia

The management of fibromyalgia is complex. Currently, there are three medications approved by the United States Food and Drug Administration for fibromyalgia: pregabalin, duloxetine, and milnacipran (White et al. 2018). Unfortunately, many patients with fibromyalgia do not receive the recommended dosage of these medications, driven by several factors such as individuals' tolerability and other chronic comorbidities (White et al. 2018). Some regulatory bodies recommend other pharmacologic agents to treat symptoms associated with fibromyalgia. The association of the Scientific Medical Societies' guidelines recommend the daily use of amitriptyline (10–50 mg), pregabalin (150–450 mg/day), and gabapentin in fibromyalgia patients, where the concomitant use of duloxetine (60 mg/day), fluoxetine (20–40 mg/day), and paroxetine (20–40 mg/day) have also been reported (Kia and Choy 2017). The European League Against Rheumatism guidelines also recommend the use of milnacipran (Kia and Choy 2017). These therapies come with several side-effects.

11.6   Underlying Mechanisms

Pathophysiology of fibromyalgia is very composite and multifactorial. Proposed etiologies assert neurochemical disparity in the brain and peripheral tissues along with some autonomic, neuroendocrine, and immune dysregulations (Sarchielli et al. 2007; Wallace et al. 2001). There have been numerous studies that have attempted to locate a biological marker that can predict the risk of development or receipt of a definitive diagnosis of fibromyalgia (Buskila et al. 2007; Jones et al. 2016; Lukkahatai et al. 2015; Docampo et al. 2014). Presently, there is still no accepted marker that can reliably replace or complement patient symptom reports.

The systemic nature of fibromyalgia suggests that a single gene is most likely not the sole determining factor for this condition; it is more likely that there is a synergistically working group of genes contributing to the symptoms of fibromyalgia. It is generally accepted that the etiology of the condition involves a combination of genetic and environmental factors. Currently, there is no consensus on any definitive prognostic or diagnostic biomarker for fibromyalgia.

If the etiology of fibromyalgia involves the interplay of several genes, then it could be hypothesized that this is translated to differences in the types and levels of proteins in the body fluids of individuals with fibromyalgia. Hence, various teams have investigated several proteins and proposed the role of mitochondrial dysfunction and oxidative stress in fibromyalgia as seen by an increase in lipid peroxidation and a decrease in the concentrations of vitamins A and E in the peripheral blood of patients with fibromyalgia (Castro-Marrero et al. 2013; Akkus et al. 2009). Hypovitaminosis, specifically reduction of vitamin A is proposed to be a manifestation in the imbalance between reactive oxygen species production and antioxidant defense system observed in patients with fibromyalgia (Ruggiero et al. 2014). Another group discovered increased number of mast cells in the papillary dermis of patients with fibromyalgia and proposed that enhanced pain sensitivity reported by fibromyalgia patients may be explained by the increased local release of chemicals from mastocytes in the skin (Blanco et al. 2010). Most of these studies are initial investigations involving small sample sizes. Further investigations are warranted to validate or explore the mechanisms that can explain the etiology of fibromyalgia.

11.6.1 Genomic Correlates

Minor alleles of catechol-*O*-methyltransferase (*COMT*) single nucleotide polymorphisms (SNPs) rs4680, rs4818, rs4633, and rs6269 were overrepresented in the fibromyalgia population in one study (Lee et al. 2018). However, a research team from Spain observed a lack of associations between the rs4680 and rs4818 polymorphisms of the *COMT* gene with fibromyalgia, either susceptibility or pain, but found an increased frequency of the TT genotype and the T allele of the rs6860 polymorphism of the charged multivesicular body protein 1A (*CHMP1A*) gene to the genetic susceptibility to fibromyalgia, but not with the pain levels experienced by patients with fibromyalgia (Estevez-Lopez et al. 2018a). Another study by the same research group from Spain described the associations of the rs841 (guanosine triphosphate cyclohydrolase 1 gene) and rs2097903 (*COMT*) SNPs with higher risk of fibromyalgia susceptibility and confirmed that the rs1799971 SNP (opioid receptor μ1 gene) might confer genetic risk for fibromyalgia (Estevez-Lopez et al. 2018b). One study observed that fibromyalgia has a distinct hypomethylation DNA pattern, which is enriched in genes implicated in stress response and DNA repair/free radical clearance (Ciampi de Andrade et al. 2017). Our group observed differential expressions of genes related to B-cell development, primary immunodeficiency signaling, and mitotic roles of polo-like kinase in patients with fibromyalgia (Lukkahatai et al. 2015).

Serotonergic metabolism and neurotransmission is another pathway that had receive attention as possible underlying mechanism for fibromyalgia. The genetic mutation affecting serotonin receptors and the serotonin transporter (5-HTT) can be potential markers to determine susceptibility to fibromyalgia and understand its pathophysiology. Genes included 5-hydroxytryptamine receptor 2A (*5-HTR2A*) and solute carrier family 6 member 4 (*SLC6A4*) were identified as fibromyalgia candidate genes (Ablin and Buskila 2015; Al-Nimer et al. 2018; Kuzelova et al. 2010; Bondy et al. 1999; Eison and Mullins 1996). The possible mechanism involves the 5-hydroxytryptamine receptor 2A (*5-HT2A*) found in the central nervous system, which is encoded by the *5-HTR2A*. The 5-HT2A and the 5-HTT regulate the serotonin system (Eison and Mullins 1996). The 5-HTT or the sodium-dependent serotonin transporter, encoded by the *SLC6A4*, regulates the transport of serotonin from synaptic spaces into the presynaptic neurons during serotonergic neurotransmission (Kuzelova et al. 2010). However, to date, the role of serotonergic metabolism in the etiology of fibromyalgia remains inconclusive.

All these reports were from cross-sectional studies, making it difficult to conclude how the variability of symptoms experienced by patients with fibromyalgia is influenced by these genetic vulnerabilities, transcriptional differences, and epigenetic modifications. In addition, the diagnostic criteria used by these studies varied from the use of the International Statistical Classification of Diseases and Related Health Problems (ICD) series codes (ICD-9 and 10) (Lee et al. 2018), and the 1990 or 2010 American College of Rheumatology (ACR) criteria for fibromyalgia (Lukkahatai et al. 2015; Estevez-Lopez et al. 2018a, b). These factors may contribute to the inconsistencies in the reports; hence, a reliable marker for fibromyalgia continues to be a research goal.

11.7 Summary and Future Directions

Fibromyalgia is a complex, multicausal, and multidimensional symptom. Individuals with fibromyalgia report widespread debilitating pain clustered with other symptoms and psychological distress; substantial medical costs and high rates of disability. Previous reports failed to establish a reliable marker for the condition. Future investigations must address the inconsistencies in the use of diagnostic criteria for fibromyalgia in order to build consensus and reproducibility of data, as well as conduct appropriately powered, exploratory "omic" inquiries to identify functional pathways that can explain the biologic underpinnings of this condition.

References

Ablin JN, Buskila D. Update on the genetics of the fibromyalgia syndrome. Best Pract Res Clin Rheumatol. 2015;29(1):20–8.

Ablin JN, et al. Coping styles in fibromyalgia: effect of co-morbid posttraumatic stress disorder. Rheumatol Int. 2008;28(7):649–56.

Ablin JN, et al. Evaluation of leptin levels among fibromyalgia patients before and after three months of treatment, in comparison with healthy controls. Pain Res Manag. 2012;17(2):89–92.

Akkaya N, et al. Relationship between the body image and level of pain, functional status, severity of depression, and quality of life in patients with fibromyalgia syndrome. Clin Rheumatol. 2012;31(6):983–8.

Akkus S, et al. Levels of lipid peroxidation, nitric oxide, and antioxidant vitamins in plasma of patients with fibromyalgia. Cell Biochem Funct. 2009;27(4):181–5.

Alciati A, et al. Features of mood associated with high body weight in females with fibromyalgia. Compr Psychiatry. 2018;80:57–64.

Al-Nimer MSM, Mohammad TAM, Alsakeni RA. Serum levels of serotonin as a biomarker of newly diagnosed fibromyalgia in women: its relation to the platelet indices. J Res Med Sci. 2018;23:71.

Aparicio VA, et al. Are there gender differences in quality of life and symptomatology between fibromyalgia patients? Am J Mens Health. 2012;6(4):314–9.

Arnold LM, Clauw DJ, McCarberg BH, FibroCollaborative. Improving the recognition and diagnosis of fibromyalgia. Mayo Clin Proc. 2011;86(5):457–64.

Arnold LM, et al. A framework for fibromyalgia management for primary care providers. Mayo Clin Proc. 2012;87(5):488–96.

Berger A, et al. Characteristics and healthcare costs of patients with fibromyalgia syndrome. Int J Clin Pract. 2007;61(9):1498–508.

Beyazal MS, et al. The impact of fibromyalgia on disability, anxiety, depression, sleep disturbance, and quality of life in patients with migraine. Noro Psikiyatr Ars. 2018;55(2):140–5.

Blanco I, et al. Abnormal overexpression of mastocytes in skin biopsies of fibromyalgia patients. Clin Rheumatol. 2010;29(12):1403–12.

Blanco S, et al. Olfactory and cognitive functioning in patients with fibromyalgia. Psychol Health Med. 2019;24(5):530–41.

Bondy B, et al. The T102C polymorphism of the 5-HT2A-receptor gene in fibromyalgia. Neurobiol Dis. 1999;6(5):433–9.

Boyington JE, Schoster B, Callahan LF. Comparisons of body image perceptions of a sample of black and white women with rheumatoid arthritis and fibromyalgia in the US. Open Rheumatol J. 2015;9:1–7.

Buskila D, et al. Familial aggregation in the fibromyalgia syndrome. Semin Arthritis Rheum. 1996;26(3):605–11.

Buskila D, Sarzi-Puttini P, Ablin JN. The genetics of fibromyalgia syndrome. Pharmacogenomics. 2007;8(1):67–74.

Cabo-Meseguer A, Cerda-Olmedo G, Trillo-Mata JL. Fibromyalgia: prevalence, epidemiologic profiles and economic costs. Med Clin (Barc). 2017;149(10):441–8.

Carta MG, et al. The impact of fibromyalgia syndrome and the role of comorbidity with mood and post-traumatic stress disorder in worsening the quality of life. Int J Soc Psychiatry. 2018;64(7):647–55.

Castro-Marrero J, et al. Could mitochondrial dysfunction be a differentiating marker between chronic fatigue syndrome and fibromyalgia? Antioxid Redox Signal. 2013;19(15):1855–60.

Ciampi de Andrade D, et al. Epigenetics insights into chronic pain: DNA hypomethylation in fibromyalgia-a controlled pilot-study. Pain. 2017;158(8):1473–80.

Conversano C, et al. Potentially traumatic events, post-traumatic stress disorder and post-traumatic stress spectrum in patients with fibromyalgia. Clin Exp Rheumatol. 2019;116(1):39–43.

Creed F, et al. Is there a better term than "medically unexplained symptoms"? J Psychosom Res. 2010;68(1):5–8.

de Araujo TA, Mota MC, Crispim CA. Obesity and sleepiness in women with fibromyalgia. Rheumatol Int. 2015;35(2):281–7.

Docampo E, et al. Genome-wide analysis of single nucleotide polymorphisms and copy number variants in fibromyalgia suggest a role for the central nervous system. Pain. 2014;155(6):1102–9.

Eison AS, Mullins UL. Regulation of central 5-HT2A receptors: a review of in vivo studies. Behav Brain Res. 1996;73(1–2):177–81.

Elkana O, et al. Does the cognitive index of the symptom severity scale evaluate cognition? Data from subjective and objective cognitive measures in fibromyalgia. Clin Exp Rheumatol. 2019;116(1):51–7.

Estevez-Lopez F, et al. The TT genotype of the rs6860 polymorphism of the charged multivesicular body protein 1A gene is associated with susceptibility to fibromyalgia in southern Spanish women. Rheumatol Int. 2018a;38(3):531–3.

Estevez-Lopez F, et al. Identification of candidate genes associated with fibromyalgia susceptibility in southern Spanish women: the al-Andalus project. J Transl Med. 2018b;16(1):43.

Fink P, Schroder A. One single diagnosis, bodily distress syndrome, succeeded to capture 10 diagnostic categories of functional somatic syndromes and somatoform disorders. J Psychosom Res. 2010;68(5):415–26.

Galvez-Sanchez CM, Duschek S, Reyes Del Paso GA. Psychological impact of fibromyalgia: current perspectives. Psychol Res Behav Manag. 2019;12:117–27.

Gonzalez B, et al. Fibromyalgia: antecedent life events, disability, and causal attribution. Psychol Health Med. 2013;18(4):461–70.

Gota CE, Kaouk S, Wilke WS. Fibromyalgia and obesity: the association between body mass index and disability, depression, history of abuse, medications, and comorbidities. J Clin Rheumatol. 2015;21(6):289–95.

Hauser W, et al. Fibromyalgia. Nat Rev Dis Primers. 2015;1:15022.

Henningsen P, Creed F. The genetic, physiological and psychological mechanisms underlying disabling medically unexplained symptoms and somatisation. J Psychosom Res. 2010;68(5):395–7.

Heymann RE, et al. New guidelines for the diagnosis of fibromyalgia. Rev Bras Reumatol Engl Ed. 2017;57(Suppl 2):467–76.

Jones KD, et al. Genome-wide expression profiling in the peripheral blood of patients with fibromyalgia. Clin Exp Rheumatol. 2016;34(2 Suppl 96):S89–98.

Kang JH, et al. Severity of fibromyalgia symptoms is associated with socioeconomic status and not obesity in Korean patients. Clin Exp Rheumatol. 2016;34(2 Suppl 96):S83–8.

Kia S, Choy E. Update on treatment guideline in fibromyalgia syndrome with focus on pharmacology. Biomedicine. 2017;5(2) https://doi.org/10.3390/biomedicines5020020.

Kuzelova H, Ptacek R, Macek M. The serotonin transporter gene (5-HTT) variant and psychiatric disorders: review of current literature. Neuro Endocrinol Lett. 2010;31(1):4–10.

Lee C, et al. Association of catechol-O-methyltransferase single nucleotide polymorphisms, ethnicity, and sex in a large cohort of fibromyalgia patients. BMC Rheumatol. 2018;2:38.

Lukkahatai N, et al. Comparing genomic profiles of women with and without fibromyalgia. Biol Res Nurs. 2015;17(4):373–83.

Lukkahatai N, et al. Understanding the association of fatigue with other symptoms of fibromyalgia: development of a cluster model. Arthritis Care Res (Hoboken). 2016;68(1):99–107.

Manivannan V. What we see when we digitize pain: the risk of valorizing image-based representations of fibromyalgia over body and bodily experience. Digit Health. 2017;3:2055207617708860.

Mas AJ, et al. Prevalence and impact of fibromyalgia on function and quality of life in individuals from the general population: results from a nationwide study in Spain. Clin Exp Rheumatol. 2008;26(4):519–26.

Mork PJ, Vasseljen O, Nilsen TI. Association between physical exercise, body mass index, and risk of fibromyalgia: longitudinal data from the Norwegian Nord-Trondelag Health Study. Arthritis Care Res (Hoboken). 2010;62(5):611–7.

Okifuji A, Bradshaw DH, Olson C. Evaluating obesity in fibromyalgia: neuroendocrine biomarkers, symptoms, and functions. Clin Rheumatol. 2009;28(4):475–8.

Paiva ES, et al. Serum levels of leptin and adiponectin and clinical parameters in women with fibromyalgia and overweight/obesity. Arch Endocrinol Metab. 2017;61(3):249–56.

Pellegrino MJ, Waylonis GW, Sommer A. Familial occurrence of primary fibromyalgia. Arch Phys Med Rehabil. 1989;70(1):61–3.

Perrot S, et al. Survey of physician experiences and perceptions about the diagnosis and treatment of fibromyalgia. BMC Health Serv Res. 2012;12:356.

Ruggiero V, et al. A preliminary study on serum proteomics in fibromyalgia syndrome. Clin Chem Lab Med. 2014;52(9):e207–10.

Samartin-Veiga N, Gonzalez-Villar AJ, Carrillo-de-la-Pena MT. Neural correlates of cognitive dysfunction in fibromyalgia patients: reduced brain electrical activity during the execution of a cognitive control task. Neuroimage Clin. 2019;23:101817.

Sarchielli P, et al. Increased levels of neurotrophins are not specific for chronic migraine: evidence from primary fibromyalgia syndrome. J Pain. 2007;8(9):737–45.

Segura-Jimenez V, et al. The association of total and central body fat with pain, fatigue and the impact of fibromyalgia in women; role of physical fitness. Eur J Pain. 2016a;20(5):811–21.

Segura-Jimenez V, et al. Gender differences in symptoms, health-related quality of life, sleep quality, mental health, cognitive performance, pain-cognition, and positive health in Spanish fibromyalgia individuals: the Al-Andalus project. Pain Res Manag. 2016b;2016:5135176.

Sharpe M, Carson A. "Unexplained" somatic symptoms, functional syndromes, and somatization: do we need a paradigm shift? Ann Intern Med. 2001;134(9 Pt 2):926–30.

Van Houdenhove B, Egle UT. Fibromyalgia: a stress disorder? Piecing the biopsychosocial puzzle together. Psychother Psychosom. 2004;73(5):267–75.

Wallace DJ, et al. Cytokines play an aetiopathogenetic role in fibromyalgia: a hypothesis and pilot study. Rheumatology (Oxford). 2001;40(7):743–9.

Waylonis GW, Heck W. Fibromyalgia syndrome. New associations. Am J Phys Med Rehabil. 1992;71(6):343–8.

White C, et al. Analysis of real-world dosing patterns for the 3 FDA-approved medications in the treatment of fibromyalgia. Am Health Drug Benefits. 2018;11(6):293–301.

Wolfe F. New American College of Rheumatology criteria for fibromyalgia: a twenty-year journey. Arthritis Care Res (Hoboken). 2010;62(5):583–4.

Wolfe F, et al. The American College of Rheumatology 1990 Criteria for the classification of fibromyalgia. Report of the multicenter criteria committee. Arthritis Rheum. 1990;33(2):160–72.

Wolfe F, et al. The American College of Rheumatology preliminary diagnostic criteria for fibromyalgia and measurement of symptom severity. Arthritis Care Res (Hoboken). 2010;62(5):600–10.

Wolfe F, et al. The use of polysymptomatic distress categories in the evaluation of fibromyalgia (FM) and FM severity. J Rheumatol. 2015;42(8):1494–501.

Wolfe F, et al. 2016 revisions to the 2010/2011 fibromyalgia diagnostic criteria. Semin Arthritis Rheum. 2016;46(3):319–29.

Wolfe F, et al. Primary and secondary fibromyalgia are the same: the universality of polysymptomatic distress. J Rheumatol. 2019;46(2):204–12.

Yunus M, et al. Primary fibromyalgia (fibrositis): clinical study of 50 patients with matched normal controls. Semin Arthritis Rheum. 1981;11(1):151–71.

Zhang Y, et al. Clinical, psychological features and quality of life of fibromyalgia patients: a cross-sectional study of Chinese sample. Clin Rheumatol. 2018;37(2):527–37.

Genomics of Neuropathic Pain

12

Katerina Zorina-Lichtenwalter

Contents

12.1 Introduction

Neuropathic pain is a debilitating maladaptive condition. It manifests as recurring, physiologically unjustified painful sensations, resulting either from a genetic mutation or another primary disease. While their causes converge on a chronically damaged nociceptive network, neuropathic pain conditions are heterogeneous in their etiology and symptoms. As such, they are difficult to diagnose and treat. Recent literature reflects an increased interest in identification of specific mechanisms at the root of neuropathic pain (Baron et al. 2017). Genetic studies have been a contribution to this effort, given their potential to shed light on the underlying pathophysiology at the molecular level.

Originally defined as "pain initiated or caused by a primary lesion or dysfunction in the nervous system" by the International Association for the Study of Pain (IASP), neuropathic pain was redefined in a 2011 publication (Jensen et al. 2011). The new definition—"pain arising as a direct consequence of a lesion or disease affecting the somatosensory system"—replaces the term "dysfunction" with

K. Zorina-Lichtenwalter (✉)

McGill University, Alan Edwards Centre for Research on Pain, Montreal, QC, Canada

e-mail: katerina.lichtenwalter@mail.mcgill.ca

© Springer Nature Switzerland AG 2020

S. G. Dorsey, A. R. Starkweather (eds.), *Genomics of Pain and Co-Morbid Symptoms*, https://doi.org/10.1007/978-3-030-21657-3_12

"disease" and imposes a narrower anatomical restriction on the source of pain. This definition, now adopted by the IASP, was put forth by the authors to aid in classification, epidemiology, diagnosis, and treatment of these conditions. However, the goal of this writing is a better understanding of the molecular pathophysiology of conditions involving the dysfunction of the somatosensory system. Therefore, laid out here are all findings relevant to conditions characterised by such dysfunction, whether they present as the primary disease or as abnormal sensitivity to paracrine signalling or physical interaction with nerve fibres. In the latter case, neuropathic pain may have been caused by events outside the somatosensory system, but there is sufficient reason to suspect that it is precipitated into chronicity by somatosensory mediators.

This chapter is divided into two sections. The first describes hereditary conditions, for which the responsible genetic variants are very rare and highly—if not completely—penetrant, meaning that individuals who carry them have a very high chance of expressing the associated neuropathic pain disease. The second describes acquired conditions, for which the responsible genetic variants are common. These variants merely contribute to the risk of neuropathic pain onset given a specific event or another primary disease. Notwithstanding the wide discrepancy in their frequency and effects, both types of risk factors are important to our understanding of the molecular pathophysiology of neuropathic pain.

While the contribution of genetic variants to hereditary conditions is well established, there is also evidence that susceptibility to acquired chronic neuropathic pain conditions varies beyond the variability explainable by environmental factors (Zorina-Lichtenwalter et al. 2018). Additionally, animal models of neuropathic pain have shown a substantial genetic contribution (Mogil et al. 1999). Therefore, the search for molecular risk factors and potential treatment targets through genetic studies is ongoing. A comprehensive review with extensive citations for genetic variant associations with neuropathic pain discussed below may be found in (Zorina-Lichtenwalter et al. 2018).

12.2   Hereditary Monogenic Disorders and Channelopathies

A defining symptom for hereditary monogenic neuropathic pain disorders is chronic pain, usually with early onset. Monogenic implies that they are caused by variants in one gene. These disorders are rare, and causal genetic variants are usually identified using linkage analysis in multi-generation pedigrees. In many of these conditions the responsible gene encodes a sodium channel, giving rise to the nickname channelopathies. Sodium channels are vital mediators of neurotransmission, therefore their abnormal sensitivity significantly disrupts the nociceptive network. Variants in sodium channel-encoding genes lead to either increased or decreased (or even annulled) nociception, which in turn leads to painful or painless conditions, respectively. While painless conditions share insensitivity to pain as their defining symptom, they are still included in discussion here, because their clinical phenotype is a manifestation of chronically aberrant nociceptive signalling, and their causal genetic factors are important to our understanding of the somatosensory system as it pertains to neuropathic pain. All genes discussed here and their associated hereditary neuropathic pain conditions are listed in Table 12.1, grouped by functional pathway.

Table 12.1 Genes reported in association studies of hereditary neuropathic pain conditions

Functional pathway	Gene	Protein	Condition(s)
Cell component digestion and recycling	*GLA*	α-galactosidase A	Small fibre neuropathy
	NACLU	α-N-acytelglucosamidase	Late-onset hereditary peripheral neuropathy
	RETREG1	Reticulophagy regulator 1	HSAN, type II
Development and survival of pain-transmitting neurons	*NTRK1*	Neurotrophic receptor tyrosine kinase 1	HSAN, type IV
	NGF	Nerve growth factor	HSAN, type V
Heme transporting	*FLVCB1*	Feline leukemia virus subgroup C cellular receptor 1	HSAN, unclassified
Myelin formation	*MPZ*	Myelin protein zero	Neuropathic pain and demyelination
Neuronal maturation, differentiation, migration, and networking	*DNMT1*	DNA methyltransferase 1	HSAN, type I
Neurotransmission	*SCN9A*	Sodium channel NaV1.7	Erythromelalgia; Idiopathic painful small fibre neuropathies; Paroxysmal extreme pain disorder; HSAN, type II
	SCN10A	Sodium channel NaV1.8	Idiopathic painful small fibre neuropathies
	SCN11A	Sodium channel NaV1.9	Idiopathic painful small fibre neuropathies; HSAN, type VII
Neurotransmission regulation	*WNK1*	WNK lysine deficient protein kinase 1	HSAN, type II
Neuronal organelle transport	*KIF5A*	Kinesin family member 5A	Hereditary spastic paraplegia with axonal neuropathy and pain
	KIF1A	Kinesin family member IA	HSAN, type II
Neuronal growth and brain localisation	*ELP1*	Elongator complex protein	HSAN, type III
Protein processing and transport	*SPTLC1*	Serine palmitoyltransferase, long chain base subunit 1	HSAN, type I
Transcriptional regulator of sensory neuronal specification	*PRDM12*	PR domain zinc finger protein 12	HSAN, type VIII

Abbreviations: *HSAN* hereditary sensory and autonomic neuropathy

12.2.1 Painful Hereditary Disorders

Among rare monogenic disorders with neuropathic pain as a defining symptom, erythromelalgia has the longest history and is consequently the best characterised. It is attributed to hypersensitivity of C fibres (unmyelinated nerves that transmit slower-response, primarily thermal or chemical noxious stimuli), manifesting as redness and painful swelling of hands and feet. Primary erythromelalgia is hereditary, caused by rare hyperfunctional variants in sodium channel NaV1.7 (*SCN9A*). The variants in *SCN9A* change the electrophysiological properties—activation, deactivation, and current amplitude—of dorsal root ganglion (DRG) neurons, thus affecting nociceptive signalling. The timing of disease onset appears to be related to the magnitude of effect on these properties (Han et al. 2009). In other words, a variant with a smaller effect may lead to later onset. An alternative theory proposes that *SCN9A* variants with effects on different electrophysiological properties lead to different neuropathic conditions. In cellular assays, one group has demonstrated that alleles responsible for erythromelalgia disrupted fast inactivation in nociceptors, while alleles that lower firing thresholds, slow deactivation, and potentiate currents result in paroxysmal extreme pain disorder (Estacion et al. 2008). The latter is also a rare neuropathic disorder caused by hyperfunctional *SCN9A* variants, and it manifests as rectal, periocular, and perimandibular pain (Estacion et al. 2008).

Another set of conditions that fall into the painful hereditary monogenic disorders category is known as idiopathic painful small fibre neuropathies. Variants responsible for these conditions have been found in sodium channels NaV 1.7, NaV 1.8 and NaV 1.9 (*SCN9A*, *SCN10A* and *SCN11A*, respectively). With clinical manifestations such as sudden bouts of pain originating in the extremities and propagating inward, these conditions affect small-diameter A-delta (which transmit acute, fast-response pain signals of different modalities) and C fibers. As shown in electrophysiology assays in DGR neurons, the associated variants in these sodium channels contribute to hyperexcitability by raising the resting membrane potential and reducing firing thresholds, which would explain their associated clinical manifestation of idiopathic pain (Han et al. 2015).

While sodium channels dominate studies of painful peripheral neuropathies, variants in four other genes have been implicated. α-galactosidase A, *GLA*, was discovered in a small fibre neuropathy patient (de Greef et al. 2016). Myelin protein zero, *MPZ*, has been found in a family with debilitating neuropathic pain and demyelination. A variant in a subunit of kinesin, *KIF5A*, a protein involved in intracellular motility, has been reported as causal in hereditary spastic paraplegia with axonal neuropathy and pain (Rinaldi et al. 2015). Lastly, a variant in α-*N*-acetyl-glucosaminidase, *NAGLU*, has been reported in individuals with a rare case of late-onset hereditary peripheral neuropathy (Tétreault et al. 2015).

12.2.2 Painless Hereditary Disorders

While the above described disorders are defined by increased pain, a class of conditions characterised by insensitivity to pain have also been linked to sodium channel variants, specifically hypofunctional variants in *SCN9A* and hyperfunctional variants in *SCN11A*. The electrophysiological properties of the two sodium channels explain the direction of effect on their function. Increased activity in NaV1.9 prolongs neuronal depolarisation, which abrogates NaV1.7 and NaV1.8 activity in nociceptors, effectively shutting down pain signalling.

A group of disorders known as hereditary sensory and autonomic neuropathies, HSAN, are also characterised by insensitivity to pain. Variations in their autonomic symptoms and genetic causes segregate these disorders into seven major subtypes. Variants of 11 genes have been implicated, marking the importance of their corresponding pathways in pain insensitivity. These are *SPTLC1*, *DMNT1*, *WNK1*, *KIF1A*, *RETREG1*, *SCN9A*, *ELP1*, *NTRK1*, *NGF*, *SCN11A*, and *PRDM12*.

SPTLC1 encodes a subunit of serine palmitoyltransferase; disturbance to its function contributes to neuronal toxicity and death. *DNMT1* encodes a DNA methyltransferase, whose impairment disrupts neuronal maintenance. These 2 genes' variants have been reported as causal in HSAN, type I (Houlden et al. 2005; Baets et al. 2015). *WNK1* encodes WNK lysine deficient protein kinase 1, and its variants lead to a reduction in the number of sensory neurons, although the exact mechanism for this is unknown. *KIF1A* encodes kinesin family member 1A, an axonal transporter of synaptic vesicles. Its variants lead to impaired neuronal function. *RETREG1* encodes reticulophagy regulator 1, and its variants disrupt its autophagy function, leading to neuronal toxicity and death. Different subtypes of HSAN, type II, are brought about by variants in *WNK1*, *KIF1A*, and *RETREG1* (Shekarabi et al. 2008; Rivière et al. 2011; Murphy et al. 2012) as well as variants in *SCN9A* (Yuan et al. 2013). *ELP1* encodes elongator complex protein 1, a scaffolding protein, variants wherein have been found in patients with HSAN, type III (Anderson et al. 2001). HSAN, type IV, which is also known as congenital insensitivity to pain with anhidrosis (CIPA), features insensitivity to pain as its primary symptom. Variants in *NTRK1*, which encodes neurotrophic receptor tyrosine kinase 1, affect its role in neuronal cell maintenance leading to CIPA (Shatzky et al. 2000). Nerve growth factor beta, encoded by *NGF*, is the binding partner of NTRK1, and its variants lead to HSAN, type V (Capsoni et al. 2011). HSAN subtypes VII and VIII are more recently-described conditions; their genetic causes have been determined to be variants in *SCN11A* (Phatarakijnirund et al. 2016) and *PRDM12* (which encodes PR domain zinc finger protein 12) (Chen et al. 2015), respectively. A recent study showed variants in a heme transporter-encoding *FLVCR1* in a patient with an unclassified HSAN (Chiabrando et al. 2016). All of these genes, in their pathological variant forms, ablate pain sensitivity by reducing the number of viable sensory neurons.

12.3 Acquired Neuropathic Pain Disorders

A lack of clear and consistent phenotyping continues to plague studies of common neuropathic pain conditions. The definition and grading system of neuropathic pain has undergone several iterations during the past 20 years; yet accurate diagnoses—and consequently effective therapies—have remained a challenge. While conditions such as diabetic neuropathy, radicular pain, trigeminal neuralgia, and viral infection-related sensory neuropathies lend themselves better to phenotyping, others, such as cancer pain and postoperative pain often display mixed phenotypes of neuropathic and nociceptive pain, thus rendering identification of those among them with neuropathic pain difficult. Additionally, genetic reports do not adhere to standardised terminology and diagnostic procedures, with some studies reporting associations with chronic pain in cancer or post-surgery patients without characterising it. Nevertheless, all these conditions carry a substantial neuropathic component and are included here in an effort to provide a more complete overview of genetic risk factors for neuropathic pain.

Common conditions have complex etiologies and often see contribution from a large number of genetic loci, each with a small-to-negligible effect; common neuropathic pain conditions are no exception. In order to identify these many low-risk loci, large cohorts of unrelated individuals (founders) are needed. The recent publication of three genome-wide association studies (GWAS) in neuropathic pain has been informative, although the top findings in these studies fall just short of the threshold of genome-wide significance. Additionally, targeted gene panel studies have confirmed or dismissed loci as modulators of susceptibility to developing neuropathic pain given a primary traumatic event, such as physical injury or progressive disease onset.

Genes with variants identified in acquired neuropathic pain conditions are listed in Table 12.2, grouped by functional pathway.

12.3.1 Diabetic Neuropathy

Like some rare hereditary conditions, diabetic neuropathy may manifest as painful or as insensitivity to pain. Its prevalence is high, with some estimates showing 50% in diabetes patients (Tesfaye and Selvarajah 2012). As causes, prolonged glycemic mismanagement and disrupted nerve microvasculature have been proposed (Tesfaye and Selvarajah 2012). Given that neuropathic pain penetrance is incomplete in diabetic patients, genetic susceptibility in those who are at risk is suspected. The first neuropathic pain GWAS to be published was in fact on diabetic neuropathic pain (Meng et al. 2015a). Shortly thereafter a second report from the same group followed (Meng et al. 2015b). These two studies were conducted in the same cohort, almost 7000 diabetic patients. In the first study, neuropathic pain cases were defined as individuals taking at least one specified diabetic peripheral neuropathy drug and testing positive on a monofilament test, which indicates sensory neuropathy. A nearly genome-wide significant association was reported for *GRFA2*, which encodes GDNF, glial cell-derived neurotrophic factor, family receptor alpha 2. In the second study, the monofilament test was not considered, but a minimum of 2 diabetic neuropathy targetting prescriptions was requisite. This study reported the same level of significance for a wide region on chr1p35.1, gated by zinc-finger and a conserved N-terminal motif, SCAN domain encoding *ZSCAN20* and by toll-like receptor 12 pseudogene *TLR12P*, in females. In males a region on chr8p23.1, carrying a high-mobility group box 1 pseudogene 46, *HMGB1P46*, was associated.

Table 12.2 Genes reported in association studies of acquired neuropathic pain conditions

Functional pathway	Gene	Protein	Condition(s)
Apoptosis	CASP9	Caspase-9	Radicular pain
	TNFHSF1B	Tumour necrosis factor receptor superfamily member 1B	Cancer pain
Biosynthesis	MAT2B	Methionine adenosyltransferase 2B	PSP
Cellular processes regulation	MAPK1	Mitogen-activated protein kinase 1	Cancer pain
	PRKCA	Protein kinase C-α	PSP
Extracellular matrix, remodeling	MMP1	Matrix metalloproteinase 1	Radicular pain
Immune regulation	HLA-A	Major histocompatibility complex, class I A	HIV-SN
	HLA-B	Major histocompatibility complex; class II B	HIV-SN
	HLA-C	Major histocompatibility complex, class III C	HIV-SN
	HLA-DRB1	Major histocompatibility complex, class II DR Beta 1	HIV SN
	IL1A	Interleukin 1 alpha	Radicular pain
	IL4R	Interleukin 4 receptor	DNP
	IL6	Interleukin 6	Radicular pain
	IL10RB	Interleukin 10 receptor subunit beta	Cancer pain
	IL13	Interleukin 13	Cancer pain
	IL1R1	Interleukin 1 receptor type 1	Cancer pain
	IL1R2	Interleukin 1 receptor type 2	PSP
	IL1RN	Interleukin 1 receptor antagonist	Radicular pain
	NFKBIA	Nuclear factor kappa-B inhibitor alpha	Cancer pain
	TNF	Tumour necrosis factor	Cancer pain; HIV-SN
Immune regulation, cell growth, and differentiation	IL10	Interleukin 10	PSP
Immune response, apoptosis	LTA	Lymphotoxin alpha	Cancer pain
	P2RX7	Purinergic receptor 7	DNP; PSP
Inflammation and mitogonesis	PTGS2	Prostaglandin-endoperoxidc synthase 2	Cancer pain
Inflammatory response mediator, cell proliferation, differentiation, and apoptosis	IL1B	Interleukin 1-beta	Cancer pain

(continued)

Table 12.2 (continued)

Functional pathway	Gene	Protein	Condition(s)
Neuronal survival and differentiation	CFRA2	Glial cell-derived neurotrophic factor, family receptor alpha 2	DNP
Neurotransmission	CACNG2	Calcium voltage-gated channel auxiliary subunit gamma 2	PSP
	KCNJ3	Potassium voltage-gated channel subfamily J member 3	Cancer pain
	KCNJ6	Potassium voltage-gated channel subfamily J member 6	Cancer pain
	KCNJ9	Potassium voltage-gated channel subfamily J member 9	Cancer pain
	KCNS1	Potassium voltage-gated channel modifier subfamily S member 1	PSP; HIV-SN
	SCN8A	Sodium channel NaV1.6	Trigeminal neuralgia
	SCN9A	Sodium channel NaV1.7	Acquired neuropathic pain (combined phenotype); DNP
Neurotransmission regulation	COMT	Catechol-O-methyl transferase	Cancer pain; Radicular pain
	DRD2	Dopamine receptor D2	Acquired neuropathic conditions (combined)
	NOS3	Nitric oxide synthase 3	Cancer pain
	SLC6A4	Solute carrier family 6 member 4	Trigeminal neuralgia
Neurotransmitter synthesis	GCH1	GTP cyclohydrolase 1	Cancer pain; PSP
Neurtrophil chemokine	CXCL8	C-X-C motif chemochine ligand 8	Cancer pain
Pain and reward processing	OPRM1	Opioid receptor mu 1	DNP; PSP; Radicular pain
Pharmacokinetics	ABCB1	ATP-binding cassette subfamily B member 1	Cancer pain
Sensory neurotransmission	TRPA1	Transient receptor potential cation channel subfamily A member 1	Acquired neuropathic conditions (combined)
	TRPV1	Transient receptor potential cation channel subfamily V member 1	Acquired neuropathic conditions (combined)
Structural component of cartilage	COL9A3	Collagen type IX alpha 3 chain	Radicular pain

Abbreviations: *DNP* diabetic neuropathic pain, *PSP* post-operative pain, *HIV-SN* HIV-related sensory neuropathy

In addition to GWAS, targeted association studies have examined the effects of genetic variants selected a priori, based on their implication in related diseases. One such study has reported the well-known hypofunctional variant, A188G, in μ-opioid receptor, *OPRM1*, to be associated in diabetic patients with foot ulcer pain (Cheng et al. 2010). Another study reported 2 variants in purinergic receptor 7, *P2RX7*, both hyperfunctional, to be associated with increased pain in diabetic neuropathic women (Ursu et al. 2014). Interleukin-4 receptor, encoded by *IL4R* has been shown to have its variable number of tandem repeats correlated with diabetic neuropathy (Basol et al. 2013). Lastly, a study in a cohort of 1000 individuals with diabetes has reported several gain-of-function variants in sodium channel NaV1.7, *SCN9A*, to be associated with neuropathic pain (Li et al. 2015).

12.3.2 Radicular Pain

Spinal disc herniation or prolapse often leads to neuropathic pain via a combination of nerve compression and inflammation. Pain intensity and duration vary, and several genetic risk modifiers have been found. Several inflammatory mediators have been reported in association with herniated disc-related pain intensity, specifically interleukin 1-α (*IL1A*) (Moen et al. 2014), interleukin 1 receptor antagonist (*IL1RN*) (Moen et al. 2014), and interleukin 6 (*IL6*) (Karppinen et al. 2008). Likewise, associations have been published for genetic variants in *OPRM1* (Olsen et al. 2012), catechol-O-methyltransferase (*COMT*) (Jacobsen et al. 2012), *COL9A3* (encoding a chain of type IX collagen) (Paassilta et al. 2001), *MMP1* (encoding matrix metalloproteinase 1) (Jacobsen et al. 2013), and *CASP9* (encoding caspase-9) (Jacobsen et al. 2013).

12.3.3 Trigeminal Neuralgia

Paroxysmal bursts of pain following the trigeminal nerve innervation pathway characterise the condition known as trigeminal neuralgia. Recently proposed diagnostic criteria have defined its onset as: (1) idiopathic, (2) sequelae to an underlying condition, or (3) pressure on the trigeminal nerve root exerted by surrounding blood vessels (Cruccu et al. 2016). Genetic studies of this condition have been few. One report has suggested a variant in serotonin transporter, *SLC6A4*, and another has put forward sodium channel NaV1.6, encoded by *SCN8A* (Tanaka et al. 2016). Both are involved in neurotransmission and implicate the nociceptive pathway. The report of NaV1.6 is unique, because it is the first mention of this channel's role in pain. Its expression in high-frequency firing neurons, however, has previously made it a target in epilepsy studies (Tanaka et al. 2016).

12.3.4 Viral Infection-Related Sensory Neuropathies

Painful neuropathy is a common accompaniment to the HIV infection, and it has been under study by several groups. An excellent review of HIV-associated painful neuropathy genetics in Africa was published in 2017 (Mbenda et al. 2017). Two genes with variants in HIV-carrying South Africans have been reported as associated with pain intensity, namely *KCNS1* (encoding a voltage-gated potassium channel subunit) and *TNF* (encoding tumour necrosis factor). The involvement of *GCH1* (GTP cyclohydrolase) has also been investigated in HIV-associated neuropathic pain in Africans, but no association was found.

Another neuropathic painful condition that develops secondarily to a viral infection is postherpetic neuralgia, characterised by recurring spontaneous or light stimulus-evoked pain. Several studies of postherpetic neuralgia in East Asian patients have been published that examined the role of variants in the human leukocyte antigen (HLA) region (Chung et al. 2016). Both class I HLA complex molecules, encoded by *HLA-A*, *HLA-B*, and *HLA-C*, and a class II HLA complex molecule, encoded by *HLADRB1* have been reported as implicated. The proposed pathway whereby variants in these genes contribute to postherpetic neuropathic pain is an inadequate immune system response to the initial viral infection leading to nerve damage (Sumiyama et al. 2008).

12.3.5 Cancer Pain

It has been estimated that a substantial portion of all cancer pain, up to 40%, has a neuropathic component (Bennett et al. 2012). In addition to chemotherapy and cancer-related surgery, the cancer itself may lead to neuropathic pain either via the tumour's invasion of nociceptors or by inflammatory cytokines leaking from cancerous cells. Initially, cytokines act on nociceptors resulting in nociceptive (non-neuropathic) pain. With protracted inflammation, however, sustained nociceptor activation may lead to persistent changes in neuronal connectivity, modulating the response thresholds and intensities and conveying innocuous stimuli as painful (Reyes-Gibby et al. 2009). Cancer pain is therefore often classified as a mixed phenotype of nociceptive and neuropathic pain.

Genetic studies have linked susceptibility to pain in cancer patients to the immune system. Associations with severe cancer pain have been reported for variants in prostaglandin-endoperoxide synthase 2, *PTGS2*, *TNF*, and NFκB inhibitor-α *NFKBIA*, (Reyes-Gibby et al. 2009), and tumour necrosis factor-β, *LTA*, (Rausch et al. 2012). Another study has reported an aggregate of phenotypes, including high pain intensity, to be associated with the joint effect of genetic variants in nitric oxide synthase-3, *NOS3*; interleukin-1β, *IL1B*; TNF receptor super-family member 1B, *TNFRSF1B*; *PTGS2*; and interleukin-10 receptorβ, *IL10RB* (Reyes-Gibby et al. 2013). Furthermore, variants in *GCH1*, involved in nitric oxide production, have been reported to confer a reduced risk of pain in patients with advanced cancer (Lötsch et al. 2010).

In addition to immune system mediators, there have been findings of association with neurotransmission mediators and even specifically pain inhibition. Variants in several genes encoding voltage-gated potassium ion channels—*KCNJ3*, *KCNJ6*, and *KCNK9*—have been reported in breast cancer patients prior to surgery (Langford et al. 2014). Genes *COMT* and *ABCB1* (which encodes membrane bounded P-glycoprotein, in charge of clearing exogenous opioids) have also been reported in cancer patients to be associated with pain (Wang et al. 2015).

Finally, mitogen-activated protein kinase 1 (*MAPK1*), which is a broad-spectrum regulator, has been shown to have a role in cancer pain in an association study (Reyes-Gibby et al. 2016).

12.3.6 Postoperative Pain

Postoperative acute pain is a common phenomenon. In some cases it may outlast the healing period and turn persistent, at which point it is considered a chronic pain condition. The lower boundary for chronic postoperative pain is 3–6 months after surgery. The neuropathic component of postoperative pain is highest (68%) in patients who have undergone thoracotomy and mastectomy (Haroutiunian et al. 2013). Putative causal mechanisms include central sensitisation resultant from painful surgical procedures (Coderre et al. 1993) and nerve damage clean-up by immune cells (Moalem and Tracey 2006). Pain signalling registered during surgery has been shown to contribute to chronic postoperative pain, attenuated by preoperative analgesics (Coderre et al. 1993). An alternative, or perhaps complementary, theory posits that immune cells recruited to the site of damage engender a prolonged inflamed state, originating during the acute postoperative period. Excessive inflammatory cytokine activity leads to lasting nociceptor activity, which may lead to a permanently altered pain transmission system (Moalem and Tracey 2006).

In support of the intraoperative central sensitisation theory are several reports of the involvement of molecules directly participating in neurotransmission as risk modifiers in persistent postoperative pain. Specifically, this includes (*OPRM1*) in a post-abdominal surgery cohort, (*KCNS1*) in two limb amputation cohorts and one post-mastectomy cohort, and stargazin (*CACNG2*)—involved in the trafficking of AMPA receptors—in a post-mastectomy cohort (Nissenbaum et al. 2010). Variants in GTP hydrolase, encoded by *GCH1*, have also been reported to modulate postoperative pain (Belfer et al. 2015).

In support of inflammation-mediated neuropathic pain are two studies of postmastectomy patients, which have reported associations between persistent pain in the breast and variants in interleukin receptors 2 (*IL1R2*) and 10 (*IL10*) (Stephens et al. 2014) and purinergic receptor 7, *P2RX7* (Sorge et al. 2012).

A recent publication reports the first genome-wide association study in a postoperative pain cohort, specifically in individuals after knee and hip replacement surgery (Warner et al. 2017). The top reported association is for a variant just short of genome-wide significance in *PRKCA*, encoding protein kinase C-α. The next best association is for a variant in *MAT2B*, which encodes methionine adenosyltransferase 2B. Both associations were replicated in at least one of their two replication cohorts consisting of people with joint-related neuropathic pain.

12.3.7 Other Conditions

Some studies have grouped common neuropathic pain conditions into one phenotype, operating under the assumption that a common pathophysiology underlies these conditions, regardless of their specific origin. One such study has reported a dopamine receptor *DRD2* variant as a risk factor in neuropathic pain resulting from any of the following primary conditions: atypical facial pain burning mouth syndrome, nerve injury, and trigeminal neuropathy (Jääskeläinen et al. 2014). Transient receptor potential channels, *TRPA1* and *TRPV1*, have also been shown to increase sensitivity in neuropathic pain patients with a variety of neuropathic conditions (Binder et al. 2012). Lastly, a variant in *SCN9A* has been reported to be associated with pain in five different cohorts with neuropathic pain (Reimann et al. 2010).

12.4 Conclusion

This survey of genes involved in neuropathic pain conditions offers insights into molecular mediators of pain signalling and the extent of their impact on pain pathophysiology. The frequency of the variants and their associated disease penetrance (high for hereditary rare conditions and low for common conditions) are informative about the degree to which they are operative or redundant in nociceptive processing.

Variants in sodium ion channels are involved in rare hereditary conditions (painful and painless) as well as acquired neuropathic pain conditions. It is likely that a more drastic cellular phenotype for these variants corresponds to the more certain clinical outcome, while risk for acquired neuropathic pain is merely increased by milder cellular phenotype-changing variants. On the other hand, the rare hereditary painless conditions appear to be driven by genetic variants that disrupt maintenance and compromise vitality of nociceptive neurons, suggesting that proper nociception relies on availability of optimally functioning nociceptive fibres more than on individual channels participating in signal propagation.

Common neuropathic pain conditions present a wider variety of implicated molecules with small individual effects, albeit dominated by neuroimmune interactions. As informed by genetic association studies, the role of the immune system appears to be such that both an inadequate inflammatory response and an overactive inflammatory response lead to some acquired neuropathic pain: viral infection-related pain in the former case, and cancer and postoperative pain in the latter. These findings offer that timely treatment with attention to the immune system may have a role in preventing or reducing neuropathic pain.

Like other complex conditions, genetic studies of common neuropathic pain conditions suffer from a lack of uniform phenotyping, underpowered studies, spurious findings, and heterogeneity in discovery and replication cohorts (Van Hecke et al. 2015). An added level of complexity for neuropathic pain comes from the fact that pain is often a secondary symptom and pain conditions are comorbid with each other, making it difficult to disentangle the one under study. It is furthermore important to acknowledge the paucity of studies done in non-Caucasian populations, which is regrettable not only because of differential disease prevalence in different regions of the world but also because, as Ngassa Mbenda and colleagues point out in their review (Mbenda et al. 2017), lower linkage disequilibrium in Africans compared to non-Africans makes studies in the former population more useful to the latter than vice versa.

Nevertheless, genetic studies have played and continue to play an important role in gaining a better understanding of neuropathic pain. While still arguably in their nascent stage, these studies strive toward two important goals: facilitating accurate diagnoses by including genetic risk factors in pain patient assessment and identifying or ascertaining treatment targets at the molecular level for high effectiveness and specificity.

References

Anderson SL, Coli R, Daly IW, Kichula EA, Rork MJ, Volpi SA, Ekstein J, Rubin BY. Familial dysautonomia is caused by mutations of the IKAP gene. Am J Hum Genet. 2001;68(3):753–8.

Baets J, Duan X, Wu Y, Smith G, Seeley WW, Mademan I, McGrath NM, Beadell NC, Khoury J, Botuyan M-V, et al. Defects of mutant DNMT1 are linked to a spectrum of neurological disorders. Brain. 2015;138(4):845–61.

Baron R, Maier C, Attal N, Binder A, Bouhassira D, Cruccu G, Finnerup NB, Haanpää M, Hansson P, Huellemann P, et al. Peripheral neuropathic pain. Pain. 2017;158(2):261.

Basol N, Inanir A, Yigit S, Karakus N, Kaya SU. High association of IL-4 gene intron 3 VNTR polymorphism with diabetic peripheral neuropathy. J Mol Neurosci. 2013;51(2):437–41.

Belfer I, Dai F, Kehlet H, Finelli P, Qin L, Bittner R, Aasvang EK. Association of functional variations in COMT and GCH1 genes with posthernotomy pain and related impairment. Pain. 2015;156(2):273–9.

Bennett MI, Rayment C, Hjermstad M, Aass N, Caraceni A, Kaasa S. Prevalence and aetiology of neuropathic pain in cancer patients: a systematic review. Pain. 2012;153(2):359–65.

Binder A, May D, Baron R, Maier C, Tölle TR, Treede R-D, Berthele A, Faltraco F, Flor H, Gierthmühlen J, et al. Transient receptor potential channel polymorphisms are associated with the somatosensory function in neuropathic pain patients, vol. vol. 6. Erlangen: Friedrich-Alexander-Universität Erlangen-Nürnberg (FAU); 2012.

Capsoni S, Covaceuszach S, Marinelli S, Ceci M, Bernardo A, Minghetti L, Ugolini G, Pavone F, Cattaneo A. Taking pain out of NGF: a "painless" NGF mutant, linked to hereditary sensory autonomic neuropathy type V, with full neurotrophic activity. PLoS One. 2011;6(2):e17321.

Chen Y-C, Auer-Grumbach M, Matsukawa S, Zitzelsberger M, Themistocleous AC, Strom TM, Samara C, Moore AW, Cho LT-Y, Young GT, et al. Transcriptional regulator PRDM12 is essential for human pain perception. Nat Genet. 2015;47(7):803–8.

Cheng K-I, Lin S-R, Chang L-L, Wang J-Y, Lai C-S. Association of the functional A118G polymorphism of OPRM1 in diabetic patients with foot ulcer pain. J Diabetes Complicat. 2010;24(2):102–8.

Chiabrando D, Castori M, di Rocco M, Ungelenk M, Gießelmann S, Di Capua M, Madeo A, Grammatico P, Bartsch S, Hübner CA, et al. Mutations in the heme exporter FLVCR1 cause sensory neurodegeneration with loss of pain perception. PLoS Genet. 2016;12(12):e1006461.

Chung HY, Song EY, Yoon JA, Suh DH, Lee SC, Kim YC, Park MH. Association of human leuko-cyte antigen with postherpetic neuralgia in Koreans. APMIS. 2016;124(10):865–71.

Coderre TJ, Katz J, Vaccarino AL, Melzack R. Contribution of central neuroplasticity to pathologi-cal pain: review of clinical and experimental evidence. Pain. 1993;52(3):259–85.

Cruccu G, Finnerup NB, Jensen TS, Scholz J, Sindou M, Svensson P, Treede R-D, Zakrzewska JM, Nurmikko T. Trigeminal neuralgia new classification and diagnostic grading for practice and research. Neurology. 2016;87(2):220–8.

de Greef BT, Hoeijmakers JG, Wolters EE, Smeets HJ, van den Wijngaard A, Merkies IS, Faber CG, Gerrits MM. No fabry disease in patients presenting with isolated small fiber neuropathy. PLoS One. 2016;11(2):e0148316.

Estacion M, Dib-Hajj S, Benke P, te Morsche RH, Eastman E, Macala L, Drenth J, Waxman S. NaV1. 7 gain-of-function mutations as a continuum: A1632E displays physiological changes associated with erythromelalgia and paroxysmal extreme pain disorder mutations and produces symptoms of both disorders. J Neurosci. 2008;43:11079–88.

Han C, Dib-Hajj SD, Lin Z, Li Y, Eastman EM, Tyrrell L, Cao X, Yang Y, Waxman SG. Early- and late-onset inherited erythromelalgia: genotype– phenotype correlation. Brain. 2009;132:1–12.

Han C, Yang Y, de Greef BT, Hoeijmakers JG, Gerrits MM, Verhamme C, Qu J, Lauria G, Merkies IS, Faber CG, et al. The domain II S4-S5 linker in Nav1. 9: a missense mutation enhances activation, impairs fast inactivation, and produces human painful neuropathy. Neuromol Med. 2015;17(2):158–69.

Haroutiunian S, Nikolajsen L, Finnerup NB, Jensen TS. The neuro-pathic component in persistent postsurgical pain: a systematic literature review. Pain. 2013;154(1):95–102.

Houlden H, King R, Blake J, Groves M, Love S, Woodward C, Hammans S, Nicoll J, Lennox G, O'donovan DG, et al. Clinical, pathological and genetic characterization of hereditary sensory and autonomic neuropathy type 1 (HSAN I). Brain. 2005;129(2):411–25.

Jääskeläinen SK, Lindholm P, Valmunen T, Pesonen U, Taiminen T, Virtanen A, Lamusuo S, Forssell H, Hagelberg N, Hietala J, et al. Variation in the dopamine D2 receptor gene plays a key role in human pain and its modulation by transcranial magnetic stimulation. Pain. 2014;155(10):2180–7.

Jacobsen L, Schistad E, Storesund A, Pedersen L, Rygh L, Røe C, Gjerstad J. The COMT rs4680 met allele contributes to long-lasting low back pain, sciatica and disability after lumbar disc herniation. Eur J Pain. 2012;16(7):1064–9.

Jacobsen LM, Schistad EI, Storesund A, Pedersen LM, Espeland A, Rygh LJ, Røe C, Gjerstad J. The MMP1 rs1799750 2G allele is associated with increased low back pain, sciatica, and disability after lumbar disk herniation. Clin J Pain. 2013;29(11):967–71.

Jensen TS, Baron R, Haanpää M, Kalso E, Loeser JD, Rice AS, Treede R-D. A new definition of neuropathic pain. Pain. 2011;152(10):2204–5.

Karppinen J, Daavittila I, Noponen N, Haapea M, Taimela S, Vanharanta H, Ala-Kokko L, Männikkö M. Is the interleukin-6 haplotype a prognostic factor for sciatica? Eur J Pain. 2008;12(8):1018–25.

Langford DJ, West C, Elboim C, Cooper BA, Abrams G, Paul SM, Schmidt BL, Levine JD, Merriman JD, Dhruva A, et al. Variations in potassium channel genes are associated with breast pain in women prior to breast cancer surgery. J Neurogenet. 2014;28(1-2):122–35.

Li QS, Cheng P, Favis R, Wickenden A, Romano G, Wang H. SCN9A variants may be implicated in neuropathic pain associated with diabetic peripheral neuropathy and pain severity. Clin J Pain. 2015;31(11):976.

Lötsch J, Klepstad P, Doehring A, Dale O. A GTP cyclohydrolase 1 genetic variant delays cancer pain. Pain. 2010;148(1):103–6.

Mbenda HGN, Wadley A, Lombard Z, Cherry C, Price P, Kamer-man P. Genetics of HIV-associated sensory neuropathy and related pain in Africans. J Neurovirol. 2017;23:511–9.

Meng W, Deshmukh H, Zuydam N, Liu Y, Donnelly L, Zhou K, Morris A, Colhoun H, Palmer C, Smith B. A genome-wide association study suggests an association of Chr8p21. 3 (GFRA2) with diabetic neuropathic pain. Eur J Pain. 2015a;19(3):392–9.

Meng W, Deshmukh HA, Donnelly LA, Wellcome Trust Case Control Consortium 2 (WTCCC2), Surrogate markers for Micro- and Macro-vascular hard endpoints for Innovative diabetes Tools (SUMMIT) study group, Torrance N, Colhoun HM, Palmer CN, Smith BH, et al. A genome-wide association study provides evidence of sex-specific involvement of Chr1p35. 1 (ZSCAN20-TLR12P) and Chr8p23. 1 (HMGB1P46) with diabetic neuropathic pain. EBioMedicine. 2015b;2(10):1386–93.

Moalem G, Tracey DJ. Immune and inflammatory mechanisms in neuropathic pain. Brain Res Rev. 2006;51(2):240–64.

Moen A, Schistad EI, Rygh LJ, Røe C, Gjerstad J. Role of IL1A rs1800587, IL1B rs1143627 and IL1RN rs2234677 genotype regarding development of chronic lumbar radicular pain; a prospective one-year study. PLoS One. 2014;9(9):e107301.

Mogil JS, Wilson SG, Bon K, Lee SE, Chung K, Raber P, Pieper JO, Hain HS, Belknap JK, Hubert L, et al. Heritability of nociception I:responses of 11 inbred mouse strains on 12 measures of nociception. Pain. 1999;80(1):67–82.

Murphy SM, Davidson GL, Brandner S, Houlden H, Reilly MM. Mutation in FAM134B causing severe hereditary sensory neuropathy. J Neurol Neurosurg Psychiatry. 2012;83(1):119–20.

Nissenbaum J, Devor M, Seltzer Z, Gebauer M, Michaelis M, Tal M, Dorfman R, Abitbul-Yarkoni M, Lu Y, Elahipanah T, et al. Susceptibility to chronic pain following nerve injury is genetically affected by CACNG2. Genome Res. 2010;20(9):1180–90.

Olsen MB, Jacobsen LM, Schistad EI, Pedersen LM, Rygh LJ, Røe C, Gjerstad J. Pain intensity the first year after lumbar disc herniation is associated with the A118G polymorphism in the opioid receptor Mu 1 gene: evidence of a sex and genotype interaction. J Neurosci. 2012;32(29):9831–4.

Paassilta P, Lohiniva J, Göring HH, Perälä M, Räinä SS, Karppinen J, Hakala M, Palm T, Kröger H, Kaitila I, et al. Identification of a novel common genetic risk factor for lumbar disk disease. JAMA. 2001;285(14):1843–9.

Phatarakijnirund V, Mumm S, McAlister WH, Novack DV, Wenkert D, Clements KL, Whyte MP. Congenital insensitivity to pain: fracturing without apparent skeletal pathobiology caused by an autosomal dominant, second mutation in SCN11A encoding voltage-gated sodium channel 1.9. Bone. 2016;84:289–98.

Rausch SM, Gonzalez BD, Clark MM, Patten C, Felten S, Liu H, Li Y, Sloan J, Yang P. SNPs in PTGS2 and LTA predict pain and quality of life in long term lung cancer survivors. Lung Cancer. 2012;77(1):217–23.

Reimann F, Cox JJ, Belfer I, Diatchenko L, Zaykin DV, McHale DP, Drenth JP, Dai F, Wheeler J, Sanders F, et al. Pain perception is altered by a nucleotide polymorphism in SCN9A. Proc Natl Acad Sci. 2010;107(11):5148–53.

Reyes-Gibby CC, Spitz MR, Yennurajalingam S, Swartz M, Gu J, Wu X, Bruera E, Shete S. Role of inflammation gene polymorphisms on pain severity in lung cancer patients. Cancer Epidemiol Biomark Prev. 2009;18(10):2636–42.

Reyes-Gibby CC, Swartz MD, Yu X, Wu X, Yennurajalingam S, Anderson KO, Spitz MR, Shete S. Symptom clusters of pain, depressed mood, and fatigue in lung cancer: assessing the role of cytokine genes. Support Care Cancer. 2013;21(11):3117–25.

Reyes-Gibby CC, Wang J, Silvas MRT, Yu R, Yeung S-CJ, Shete S. MAPK1/ERK2 as novel target genes for pain in head and neck cancer patients. BMC Genet. 2016;17(1):40.

Rinaldi F, Bassi MT, Todeschini A, Rota S, Arnoldi A, Padovani A, Filosto M. A novel mutation in motor domain of KIF5A associated with an HSP/axonal neuropathy phenotype. J Clin Neuromuscul Dis. 2015;16(3):153–8.

Rivière J-B, Ramalingam S, Lavastre V, Shekarabi M, Holbert S, Lafontaine J, Srour M, Merner N, Rochefort D, Hince P, et al. KIF1A, an axonal transporter of synaptic vesicles, is mutated in hereditary sensory and autonomic neuropathy type 2. Am J Hum Genet. 2011;89(2):219–30.

Shatzky S, Moses S, Levy J, Pinsk V, Hershkovitz E, Herzog L, Shorer Z, Luder A, Parvari R. Congenital insensitivity to pain with anhidrosis (CIPA) in Israeli-Bedouins: genetic heterogeneity, novel mutations in the TRKA/NGF receptor gene, clinical findings, and results of nerve conduction studies. Am J Med Genet. 2000;92(5):353–60.

Shekarabi M, Girard N, Rivière J-B, Dion P, Houle M, Toulouse A, Lafrenière RG, Vercauteren F, Hince P, Laganiere J, et al. Mutations in the nervous system–specific HSN2 exon of WNK1 cause hereditary sensory neuropathy type II. J Clin Invest. 2008;118(7):2496.

Sorge RE, Trang T, Dorfman R, Smith SB, Beggs S, Ritchie J, Austin J-S, Zaykin DV, Vander Meulen H, Costigan M, et al. Genetically determined P2X7 receptor pore formation regulates variability in chronic pain sensitivity. Nat Med. 2012;18(4):595–9.

Stephens K, Cooper BA, West C, Paul SM, Baggott CR, Merriman JD, Dhruva A, Kober KM, Langford DJ, Leutwyler H, et al. Associations between cytokine gene variations and severe persistent breast pain in women following breast cancer surgery. J Pain. 2014;15(2):169–80.

Sumiyama D, Kikkawa EF, Kita Y, Shinagawa H, Mabuchi T, Ozawa A, Inoko H. HLA alleles are associated with postherpetic neuralgia but not with herpes zoster. Tokai J Exp Clin Med. 2008;33(4):150–3.

Tanaka BS, Zhao P, Dib-Hajj FB, Morisset V, Tate S, Waxman SG, Dib-Hajj SD. A gain-of-function mutation in Nav1. 6 in a case of trigeminal neuralgia. Mol Med. 2016;22:338.

Tesfaye S, Selvarajah D. Advances in the epidemiology, pathogenesis and management of diabetic peripheral neuropathy. Diabetes Metab Res Rev. 2012;28(S1):8–14.

Tétreault M, Gonzalez M, Dicaire M-J, Allard P, Gehring K, Leblanc D, Leclerc N, Schondorf R, Mathieu J, Zuchner S, et al. Adult-onset painful axonal polyneuropathy caused by a dominant NAGLU mutation. Brain. 2015;138(6):1477–83.

Ursu D, Ebert P, Langron E, Ruble C, Munsie L, Zou W, Fijal B, Qian Y-W, McNearney TA, Mogg A, et al. Gain and loss of function of P2X7 receptors: mechanisms, pharmacology and relevance to diabetic neuropathic pain. Mol Pain. 2014;10:37–48.

Van Hecke O, Kamerman PR, Attal N, Baron R, Bjornsdottir G, Bennett DL, Bennett MI, Bouhassira D, Diatchenko L, Freeman R, et al. Neuropathic pain phenotyping by international consensus (neuroppic) for genetic studies: a NeuPSIG systematic review, Delphi survey, and expert panel recommendations. Pain. 2015;156(11):2337.

Wang X-s, Song H-b, Chen S, Zhang W, Liu J-q, Huang C, Wang H-r, Chen Y, Chu Q. Association of single nucleotide polymorphisms of ABCB1, OPRM1 and COMT with pain perception in cancer patients. J Huazhong Univ Sci Technolog Med Sci. 2015;35(5):752–8.

Warner SC, van Meurs JB, Schiphof D, Bierma-Zeinstra SM, Hofman A, Uitterlinden AG, Richardson H, Jenkins W, Doherty M, Valdes AM. Genome-wide association scan of neuropathic pain symptoms post total joint replacement highlights a variant in the protein-kinase C gene. Eur J Hum Genet. 2017;25(4):446–51.

Yuan J, Matsuura E, Higuchi Y, Hashiguchi A, Nakamura T, Nozuma S, Sakiyama Y, Yoshimura A, Izumo S, Takashima H. Hereditary sensory and autonomic neuropathy type IID caused by an SCN9A mutation. Neurology. 2013;80(18):1641–9.

Zorina-Lichtenwalter K, Parisien M, Diatchenko L. Genetic studies of human neuropathic pain conditions: a review. Pain. 2018;159(3):583.

Spinal Cord Injury-Related Pain and Genomics

13

Angela R. Starkweather and Susan G. Dorsey

Contents

13.1   Introduction

Pain is the most frequently reported comorbidity with spinal cord injury (SCI) and can severely compromise quality of life and functional outcomes (Burke et al. 2018). In addition to the wide prevalence of SCI-related pain, it is notoriously difficult to treat and is often refractory to common analgesics and available therapeutics used to treat pain (Gibbs et al. 2019). In a systematic review of 42 studies that detailed the epidemiology of pain following SCI, the prevalence ranged from 26 to 96% (Dijkers et al. 2009). A reason for this wide variability in documenting SCI-related pain could be due to the challenges associated with classifying the various types of pain described after SCI as evidenced by over 23 different pain taxonomies published in the SCI literature (Bryce et al. 2012).

A. R. Starkweather (✉)
University of Connecticut School of Nursing, Storrs, CT, USA
e-mail: Angela.Starkweather@uconn.edu

S. G. Dorsey
Department of Pain and Translational Symptom Science, University of Maryland School of Nursing, Baltimore, MD, USA
e-mail: SDorsey@umaryland.edu

© Springer Nature Switzerland AG 2020
S. G. Dorsey, A. R. Starkweather (eds.), *Genomics of Pain and Co-Morbid Symptoms*, https://doi.org/10.1007/978-3-030-21657-3_13

In both paraplegic and tetraplegic populations, neuropathic pain, or central pain, is most frequently reported (van Gorp et al. 2014). At-level pain is often reported at the time of injury, while below-level pain can develop months later following injury and may be preceded by sensory hypersensitivity (Hari et al. 2009; Zeilig et al. 2012). The combination of neuropathic pain and spasticity can also have a late onset and become chronic (Finnerup 2017). Multiple studies suggest that residual function of the spinothalamic tract is associated with the development of neuropathic pain (Widerstrom-Noga et al. 2016; Finnerup et al. 2003; Cruz-Almeida et al. 2012; Defrin et al. 2001; Felix and Widerstrom-Noga 2009). At least two studies have reported that sensitivity to cold pain may identify patients at risk for developing neuropathic pain (Siddall et al. 2003; Finnerup et al. 2014); however, other predictors of SCI-related pain remain obscure and the mechanisms surrounding the development of pain following SCI have yet to be identified (Moshourab et al. 2015).

13.2 Classification of SCI-Related Pain

In 2002, the International Association for the Study of Pain Taxonomy of Pain after Spinal Cord Injury was published using a 3-tiered system for classifying pain types (Siddall et al. 2002). Since that time, the International Spinal Cord Injury Pain Classification was developed as a mechanism-based schema to assist clinicians in identifying the most appropriate available treatment options and save costs for differential workup or inappropriate treatment (Bryce et al. 2012). The four main classification types are nociceptive, neuropathic, other, or unknown (Table 13.1). There are several subtypes for nociceptive and neuropathic pain.

In a study evaluating the International Spinal Cord Injury Pain classification among patients referred to a multidisciplinary pain treatment program, it was found that a majority of patients (79%) had neuropathic pain, while 61% had nociceptive pain and 8% had unknown pain (Mahnig et al. 2016). In this sample, more than 50% of patients had more than one pain subtype, with musculoskeletal pain being the most common (58%), followed by at-level neuropathic pain (53%) and below-level neuropathic pain (42%). Pain severity as measured by the numeric rating scale was highest in patients with neuropathic pain (mean rating of 8.1 ± 1.8) followed by nociceptive pain (7.4 ± 1.8). There was no significant difference in pain severity according to age, time since injury, and time since onset of pain; however, patients with higher levels of pain also had higher levels of depression as measured by the Hospital Anxiety and Depression Scale (HADS). Higher pain intensity was also associated with low mental health but not physical health, as measured by the SF-12 mental and physical component summaries.

In a large survey study of individuals with spinal cord injury ($n = 643$), respondents with high pain intensity had significantly lower quality of life than individuals with moderate or no pain, and those with neuropathic pain had significantly lower quality of life than those with nociceptive pain or no pain (Burke et al. 2018). However, pain interference explained a greater degree of the variance in reported

Table 13.1 Types of SCI-related pain

Pain type	Subtype	Description
Nociceptive		
	Musculoskeletal	(e.g., glenohumeral arthritis, lateral epicondylitis, comminuted femur fracture, quadratus lumborum muscle spasm)
	Visceral	(e.g., myocardial infarction, abdominal pain due to bowel impaction, cholecystitis)
	Other	(e.g., autonomic dysreflexia headache, migraine headache, surgical skin incision)
Neuropathic		
	At-level	(e.g., spinal cord compression, nerve root compression, cauda equina compression) caused by a lesion or disease affecting the spinal cord or nerve roots, and located segmentally anywhere within the dermatome of the neurological level of injury and up to three dermatomes below this level
	Below-level	(e.g., spinal cord ischemia, spinal cord compression) caused by a lesion or disease of the spinal cord and pain located more than three dermatomes below the dermatome of the neurological level of injury
Other		
		e.g., fibromyalgia, complex regional pain syndrome type I, interstitial cystitis, irritable bowel syndrome
Unknown		
		Pain that cannot be assigned with any certainty to above categories

Adapted from Bryce TN, Biering-Sorensen F, Finnerup NB, Cardenas DD, Defrin R, Lundeberg T et al. International spinal cord injury pain classification: part I. Background and description. Spinal Cord 2012; 50: 413–417

quality of life, more than intensity or type of pain. Another study on the characteristics of neuropathic pain in patients with SCI reported evidence of diurnal variation in the severity of pain over time, with higher severity of pain in the night hours than other times of the day (Celik et al. 2017).

To facilitate consistent classification and reporting of pain following SCI, an international team of experts representing the International Association for the Study of Pain, International Spinal Cord Society, and American Spinal Injury Association developed the International Spinal Cord Injury Basic Pain Data Set (Widerström-Noga et al. 2014). This data set includes core questions related to the characteristics of pain and its impact on physical, social, and emotional function.

13.3 Mechanisms of Spinal Cord Injury and Intersections with Pain Mechanisms

The primary injury of spinal cord injury entails the structural and vascular tissue injury at the moment of impact that results in cell death and necrosis of tissue as well as shearing of neurons and blood vessels. The primary injury initiates a cascade of cellular responses, some of which contribute to secondary injury. These include:

- Activation of phospholipases from breakdown of cellular membranes
- Release of nitrogen and oxygen species promotes catecholamine oxidation, lipid peroxidation, and oxidative stress, which interrupts ionic and metabolic homeostasis (including malfunction of Ca^{2+} and Na^+/K^+-ATPase ionic pumps and glucose transporters)
- Excitatory mediators (glutamate, aspartate, norepinephrine, and serotonin) increase neuronal depolarization and cause a massive influx of calcium ions, which increased inflammatory mediators, production of reactive oxygen species, and activation of protease systems, and mitochondrial dysfunction (Yezierski 2005; Liu et al. 1990; Tsai et al. 2008)

These immediate responses cause inflammation, cellular apoptosis, and tissue necrosis, as well as glial activation, and several pathways have been noted to influence neuropathic pain in animal models of SCI.

Several mechanisms have been posited to contribute to SCI-related pain, including mechanisms involved in other chronic pain conditions, such as glial activation, changes in synaptic plasticity, changes in cell-signaling pathways, and loss of inhibitory mechanisms (Moshourab et al. 2015). However, some unique features specific to SCI have been identified in animal models of SCI-induced neuropathic pain. In rats with neuropathic pain and allodynia following spinal cord contusion, increased basal and stimulus-evoked levels of c-fos gene expression have been reported in the spinal dorsal horn immediately above the level of injury (Siddall et al. 1999) as well as activation of nuclear factor kappa-beta (NF-kappaB) (Bethea et al. 1998), which serves as a transcription factor of at least 150 genes involved in inflammation, cellular proliferation, and apoptosis. Neuronal hyperexcitability and allodynia may also be influenced by p-38a mitogen-activated protein kinase (MAPK) (Gwak et al. 2009).

Intracellular signaling pathways involved in long-term potentiation and memory have been shown to be involved in the development of neuropathic pain following SCI. Activated MAPKs, including pERK ½, and p-p38 MAPK but not pJNK are upregulated in SCI-injured rats that develop neuropathic pain as compared to rats that do not develop neuropathic pain (Crown et al. 2006; Kasuya et al. 2018). Consistent with these findings, inhibition of spinal c-Jun-NH2-terminal kinase (JNK) improves locomotor activity in SCI-injured rats but does not influence pain (Martini et al. 2016). Spinal neurons expressing neurokinin-1 receptor in the superficial laminae of the spinal cord appear to be involved in the development of SCI-related pain (Hains et al. 2003) and SCI has been shown to trigger changes in sodium channel expression of Nav1.3 in dorsal horn nociceptive neurons, causing hyperexcitability (Yezierski et al. 2004). Hypofunction of GABAergic inhibitory tone brought on by hyperexcitable neurons and glial activation following SCI disrupts the balance of chloride ions, glutamate, and GABA distribution in the spinal dorsal horn, resulting in secondary amplification of ascending signals (Gwak and Hulsebosch 2011). A recent study measured metabolite concentrations obtained by magnetic resonance spectroscopy in the anterior cingulate cortex of participants with SCI who had severe, high-impact neuropathic pain or moderate, low-impact neuropathic pain, or SCI without neuropathic pain and levels were compared with able-bodied healthy controls (Widerström-Noga et al. 2013). Participants with severe, high-impact neuropathic pain had higher levels of myoinositol, creatine, and choline and lower levels of N-acetyl aspartate and glutamate–glutamine ratios than the moderate, low-impact neuropathic pain group. The glutamate–glutamine ratio significantly discriminated high-impact neuropathic pain from the able-bodied healthy controls and participants with SCI without neuropathic pain. Overall, the results suggest that lower glutamate metabolism and proliferation of glia and glial activation contribute to maintaining severe neuropathic pain following SCI.

13.4   Biomarkers of SCI-Related Pain

Neurochemical biomarkers of spinal cord injury have primarily been focused on predicting injury classification, or locomotor and functional recovery (Kwon et al. 2019). An important finding from the biomarker literature on traumatic brain injury (TBI) has been that biomarkers identified in the cerebrospinal fluid of patients with TBI are expressed in circulation (from whole blood, serum, and/or plasma), providing impetus to identify blood-based biomarkers for prediction of outcomes after SCI. In a recent study that assessed routine blood measures (metabolic panel, complete blood count, liver enzymes, and C-reactive protein) and clinical data from patients with SCI, it was found that analytes relating to liver function and acute inflammation significantly increased prediction of sensory function and pain severity at 3 months post-injury (Brown et al. 2019). In male patients with SCI during the early subacute phase, it was found that patients with pain had higher levels of chemokines, CXCL10, and CCL2, compared to males without pain (Mordillo-Mateos et al. 2019). In addition, CCL2 concentrations were positively associated with pain intensity and increased over time, suggesting a potential link with pain maintenance. Another study assessed SNPs in TRPA1 gene among patients with chronic and complete SCI and neuropathic pain and those without neuropathic pain (Rodriguez et al. 2017). They found a greater prevalence of TT homozygous genotype in the rs13255063 variant among patients with SCI and neuropathic pain and greater prevalence of GG homozygous genotype and G allele in rs11988795 among patients with SCI and no pain. These results suggest that the TT genotype in rs13255063 may be a risk factor of neuropathic pain after SCI, while the GG genotype in rs11988795 and G allele could be protective against neuropathic pain after SCI.

Somatosensory phenotypes may also hold value in predicting neuropathic pain after SCI. A review of the literature found several neuropathic pain phenotypes, including mixed thermal/mechanical sensory loss, dynamic mechanical allodynia, thermal allodynia, mechanical hyperalgesia, decreased pain modulation, residual temperature and pain sensation, and reduced electroencephalogram peak frequency (6–12 Hz range) in response to heat stimuli (Widerstrom-Noga 2017). As described earlier, magnetic resonance spectroscopy imaging (MRSI) has been combined with sensory testing to identify specific biomarkers that predict chronic pain after SCI. Lower glutamate–glutamine ratio was observed in the thalamus of SCI patients with severe neuropathic pain compared with less severe or no neuropathic pain and with pain-free controls and significantly associated with greater thermal sensitivity below the level of injury (Widerstrom-Noga et al. 2015).

In addition to MRSI, diffusion tensor imaging (DTI) can identify abnormalities in tissue structures of the brain by tracking the anisotropic (directionally dependent) diffusion of water in nerves and white matter of the brain and spinal cord. Mean diffusivity measures the average molecular motion independent of tissue directionality and is influenced by cellular size and integrity. Thus, a reduction in mean diffusivity may reflect cellular or tissue barriers such as cell proliferation or neuronal sprouting. Two studies have examined DTI in SCI patients with neuropathic pain, finding significantly increased mean diffusivity in premotor, orbitofrontal, dorsolateral prefrontal, and posterior parietal cortices, and the anterior insula compared with SCI patients without neuropathic pain (Gustin et al. 2010; Yoon et al. 2013). Decreased mean diffusivity was identified in the corticospinal and thalamocortical tracts in the internal capsule suggesting abnormalities of brain structures involved in pain modulation.

Although genetic studies in animal models of SCI-related pain have been performed, only one study of genetic expression profiles has been published in humans to date. Using the public Gene Expression Omnibus database, GSE69901, samples of peripheral blood mononuclear cells (PBMCs) derived from patients with complete SCI with a level of injury above T5, 12 patients with intractable neuropathic pain and 13 control patients without pain were analyzed to identify differentially expressed genes (He et al. 2017). Functional enrichment analysis, protein–protein interaction (PPI) network, and a transcriptional regulation network were constructed. Compared to control patients without pain, a total of 70 upregulated and 61 downregulated differentially expressed genes were identified in PBMCs from patients with neuropathic pain, which are significantly involved in focal adhesion, T cell receptor signaling, and mitochondrial function. In the PPI network, glycogen synthase kinase 3 beta (*GSK3B*) was identified as a hub protein while ornithine decarboxylase 1 (*ODC1*) and ornithine aminotransferase (*OAT*) were regulated by additional transcription factors and all three genes were significantly enriched in the function of mitochondrial membrane and DNA binding. Due to heterogeneity in the type of SCI in humans and variation in the timing and characterization of pain, it is going to be imperative for researchers to collaborate across institutional and possibly international boundaries in order to determine whether a unique fingerprint of gene expression profiles exists for each pain type. Well-phenotyped samples with multiple time points could greatly augment the understanding of how pain develops and is maintained over time following SCI.

13.5 Conclusions

The importance of identifying genomic or other somatosensory, biochemical, and imaging biomarkers to identify mechanisms of the development and maintenance of pain after spinal cord injury is a critical frontier in this patient population, as few treatments exist that are highly effective (Hatch et al. 2018; Boldt et al. 2014). Efforts to tailor treatment based on specific chronic pain phenotypes of SCI-related pain will continue to evolve as greater knowledge is accumulated on specific treatment targets and phenotypic response to pharmacologic and non-pharmacologic therapies. Due to the low incidence of spinal cord injury, collaborations across institutions with biobanking availability are becoming a powerful tool for augmenting research, along with the use of standardized measures for categorizing pain types and collecting phenotype measures across studies. Based on the limited research in this patient population, understanding the genomic contributions of risk for developing chronic pain after SCI is still in its infancy and replication studies will be crucial for translation of findings to the bedside.

References

Bethea JR, Castro M, Keane RW, Lee TT, Dietrich WD, Yezierski RP. Traumatic spinal cord injury induces nuclear factor-kappaB activation. J Neurosci. 1998;18:3251–60.

Boldt I, Eriks-Hoogland I, Brinkhof MWG, de Bie R, Joggi D, von Elm E. Non-pharmacological interventions for chronic pain in people with spinal cord injury. Cochrane Database Syst Rev. 2014;11:CD009177. https://doi.org/10.1002/14651858.CD009177.pub2.

Brown SJ, Harrington GMB, Hulme CH, et al. A preliminary cohort study assessing routine blood analyte levels and neurological outcome after spinal cord injury. J Neurotrauma. 2019;36:1–15.

Bryce TN, Biering-Sorensen F, Finnerup NB, Cardenas DD, Defrin R, Lundeberg T, Norrbrink C, Richards JS, Siddall P, Stripling T, Treede RD, Waxman SG, Widerstrom-Noga E, Yezierski RP, Dijkers M. International spinal cord injury pain classification: part I. Background and description. Spinal Cord. 2012;50:413–7.

Burke D, Lennon O, Fullen BM. Quality of life after spinal cord injury: the impact of pain. Eur J Pain. 2018;22:1662–72.

Celik EC, Erhan B, Lakse E. The clinical characteristics of neuropathic pain in patients with spinal cord injury. Spinal Cord. 2017;50:585–9.

Crown ED, Ye Z, Johnson KM, Xu G-Y, McAdoo DJ, Hulsebosch CE. Increases in the activated forms of ERK 1/2, p38 MAPK, and CREB are correlated with the expression of at-level mechanical allodynia following spinal cord injury. Exp Neurol. 2006;199:397–407.

Cruz-Almeida Y, Felix ER, Martinez-Arizala A, Widerström-Noga EG. Decreased spinothalamic and dorsal column–medial lemniscus-mediated function is associated with neuropathic pain after spinal cord injury. J Neurotrauma. 2012;29:2706–15.

Defrin R, Ohry A, Blumen N, Urca G. Characterization of chronic pain and somatosensory function in spinal cord injury subjects. Pain. 2001;89:253–63.

Dijkers M, Bryce T, Zanca J. Prevalence of chronic pain after traumatic spinal cord injury: a systematic review. J Rehabil Res Dev. 2009;46:13–29.

Felix ER, Widerstrom-Noga EG. Reliability and validity of quantitative sensory testing in persons with spinal cord injury and neuropathic pain. J Rehabil Res Dev. 2009;46:69–83.

Finnerup NB. Neuropathic pain and spasticity: intricate consequences of spinal cord injury. Spinal Cord. 2017;55:1046–50.

Finnerup NB, Johannesen IL, Fuglsang-Frederiksen A, Bach FW, Jensen TS. Sensory function in spinal cord injury patients with and without central pain. Brain. 2003;126:57–70.

Finnerup NB, Norrbrink C, Trok K, Piehl F, Johannesen IL, Sorensen JC, Jensen TS, Werhagen L. Phenotypes and predictors of pain following traumatic spinal cord injury: a prospective study. J Pain. 2014;15:40–8.

Gibbs K, Beaufort A, Stein A, Leung TM, Sison C, Bloom O. Assessment of pain symptoms and quality of life using the International Spinal Cord Injury Data Sets in persons with chronic spinal cord injury. Spinal Cord Ser Cases. 2019;5:32.

Gustin SM, Wrigley PJ, Siddall PJ, Henderson LA. Brain anatomy changes associated with persistent neuropathic pain following spinal cord injury. Cereb Cortex. 2010;20:1409–19.

Gwak YS, Hulsebosch CE. GABA and central neuropathic pain following spinal cord injury. Neuropharmacology. 2011;60:799–808.

Gwak YS, Unabia GC, Hulsebosch CE. Activation of p-38α MAPK contributes to neuronal hyperexcitability in caudal regions remote from spinal cord injury. Exp Neurol. 2009;220:154–61.

Hains BC, Klein JP, Saab CY, Craner MJ, Black JA, Waxman SG. Upregulation of sodium channel Nav1.3 and functional involvement in neuronal hyperexcitability associated with central neuropathic pain after spinal cord injury. J Neurosci. 2003;23:8881–92.

Hari AR, Wydenkeller S, Dokladal P, Halder P. Enhanced recovery of human spinothalamic function is associated with central neuropathic pain after SCI. Exp Neurol. 2009;216:428–30.

Hatch MN, Cushing TR, Carlson GD, Chang EY. Neuropathic pain and SCI: identification and treatment strategies in the 21st century. J Neurol Sci. 2018;384:75–83.

He X, Fan L, Wu Z, He J, Cheng B. Gene expression profiles reveal key pathways and genes associated with neuropathic pain in patients with spinal cord injury. Mol Med Rep. 2017;15(4):2120–8.

Kasuya Y, Umezawa H, Hatano M. Stress-activated protein kinases in spinal cord injury: focus on roles of p38. Int J Mol Sci. 2018;19(3):E867.

Kwon BK, Bloom O, Wanner I, Curt A, Schwab JM, Fawcett J, Wang KK. Neurochemical biomarkers in spinal cord injury. Spinal Cord. 2019;7(10):819–31. https://doi.org/10.1038/s41393-019-0319-8.

Liu DX, Valadez V, Sorkin LS, McAdoo DJ. Norepinephrine and serotonin release upon impact injury to rat spinal cord. J Neurotrauma. 1990;7:219–27.

Mahnig S, Landmann G, Stockinger L, Opsommer E. Pain assessment according to the International Spinal Cord Injury Pain classification in patients with spinal cord injury referred to a multidisciplinary pain center. Spinal Cord. 2016;54:809–15.

Martini AC, Forner S, Koepp J, Rae GA. Inhibition of spinal c-Jun-NH2-terminal kinase (JNK) improves locomotor activity of spinal cord injured rats. Neurosci Lett. 2016;621:54–61.

Mordillo-Mateos L, Sanchez-Ramos A, Coperchini F, Bustos-Guadamillas I, Alonso-Bonilla C, Vargas-Baquero E, Rodriguez-Carrion I, Rotondi M, Oliviero A. Development of chronic pain in males with traumatic spinal cord injury: role of circulating levels of the chemokines CCL2 and CXCL10 in subacute stage. Spinal Cord. 2019;57(11):953–9. https://doi.org/10.1038/s41393-019-0311-3.

Moshourab RA, Schafer M, Al-Chaer ED. Chronic pain in neurotrauma: implications on spinal cord and traumatic brain injury, chapter 11. In: Kobeissy FH, editor. Brain neurotrauma: molecular, neuropsychological, and rehabilitation aspects. Boca Raton, FL: CRC Press/Taylor & Francis; 2015.

Rodriguez VS, Aguilar IC, Villa LC, Saenz de Tejada SF. TRPA1 polymorphisms in chronic and complete spinal cord injury patients with neuropathic pain: a pilot study. Spinal Cord Ser Cases. 2017;3:17089.

Siddall PJ, Xu CL, Floyd N, Keay KA. C-fos expression in the spinal cord of rats exhibiting allodynia following contusive spinal cord injury. Brain Res. 1999;851:281–6.

Siddall PJ, Yezierski RP, Loeser JD. Spinal cord injury pain: assessment, mechanisms, management. In: Yezierski RP, Burchiel KJ, editors. Progress in pain research and management, vol. 23. Seattle, WA: IASP Press; 2002. p. 9–24.

Siddall PJ, McClelland JM, Rutkowski SB, Cousins MJ. A longitudinal study of the prevalence and characteristics of pain in the first 5 years following spinal cord injury. Pain. 2003;103:249–57.

Tsai MC, Wei CP, Lee DY, Tseng YT, Tsai MD, Shih YL, Lee YH, Chang SF, Leu SJ. Inflammatory mediators of cerebrospinal fluid from patients with spinal cord injury. Surg Neurol. 2008;70:S19–24.

van Gorp S, Kessels AG, Joosten EA, van Kleef M, Patijn J. Pain prevalence and its determinants after spinal cord injury: a systematic review. Eur J Pain. 2014;19:5–24.

Widerstrom-Noga E. Neuropathic pain and spinal cord injury: phenotypes and pharmacological management. Drugs. 2017;77:967–84.

Widerström-Noga E, Pattany PM, Cruz-Almeida Y, Felix ER, Perez S, Cardenas DD, Martinez-Arizala A. Metabolite concentrations in the anterior cingulate cortex predict high neuropathic pain impact after spinal cord injury. Pain. 2013;154:204–12.

Widerström-Noga E, Biering-Sørensen F, Bryce TN, Cardenas DD, Finnerup NB, Jensen MP, Richards JS, Siddall PJ. The international spinal cord injury pain basic data set (version 2.0). Spinal Cord. 2014;52:282–6.

Widerstrom-Noga E, Cruz-Almeida Y, Felix ER, Pattany PM. Somatosensory phenotype is associated with thalamic metabolites and pain intensity after spinal cord injury. Pain. 2015;156(1):166–74.

Widerstrom-Noga E, Felix ER, Adcock JP, Escalona M, Tibbett J. Multidimensional neuropathic pain phenotypes after spinal cord injury. J Neurotrauma. 2016;11:482–92.

Yezierski RP. Spinal cord injury: a model of central neuropathic pain. Neurosignals. 2005;14:182–93.

Yezierski RP, Yu CG, Mantyh PW, Vierck CJ, Lappi DA. Spinal neurons involved in the generation of at-level pain following spinal injury in the rat. Neurosci Lett. 2004;361:232–6.

Yoon EJ, Kim YK, Shin HI, Lee Y, Kim SE. Cortical and white matter alterations in patients with neuropathic pain after spinal cord injury. Brain Res. 2013;1540:64–73.

Zeilig G, Enosh S, Rubin-Asher D, Lehr B, Defrin R. The nature and course of sensory changes following spinal cord injury: predictive properties and implications on the mechanism of central pain. Brain. 2012;135:418–30.

Sex, Race, and Genomics of Pain

14

Xiaomei Cong, Zewen Tan, and Tessa Weidig

Contents

14.1 Introduction

Differences of sex, gender, and race/ethnicity exist in pain physiology, genomics, psychosocial-cultures, and pain behavior, leading to differential clinical pain assessment and diagnosis, and use of pain management and treatment strategies. In general, the prevalence rates of acute and chronic pain are documented higher in women compared to men, and pain-related diseases and health issues more commonly

X. Cong (✉)
Center for Advancement in Managing Pain, University of Connecticut School of Nursing, Storrs, CT, USA

Biobehavioral Research Laboratory, University of Connecticut School of Nursing, Storrs, CT, USA
e-mail: xiaomei.cong@uconn.edu

Z. Tan
University of Connecticut, Molecular and Cell Biology, Storrs, CT, USA
e-mail: zewen.tan@uconn.edu

T. Weidig
University of Connecticut School of Nursing, Storrs, CT, USA

Connecticut Pediatrics at Community Health Center, Hartford, CT, USA
e-mail: tessa.weidig@uconn.edu

© Springer Nature Switzerland AG 2020
S. G. Dorsey, A. R. Starkweather (eds.), *Genomics of Pain and Co-Morbid Symptoms*, https://doi.org/10.1007/978-3-030-21657-3_14

reported among women (Aloisi 2017; Pieretti et al. 2016). Evidence also exists for ethnic/racial group differences in clinical pain symptoms and experimental pain conditions, such as African Americans demonstrating greater clinical pain severity and experimental pain perception compared to non-Hispanic Whites (Patanwala et al. 2019; Rahim-Williams et al. 2012). Such inter-individual variability of pain experience is still not completely understood, stemming from a combination of complex environmental and underlying genetic factors and must be examined more closely. Identification of sex- and ethnic-specific pain genetic attributes and pathways may inform clinical precision pain management, from pain assessment to personalized pharmacological and non-pharmacological pain treatments.

14.2   Sex, Gender, and Pain Genomics

In discussing the relationship between sex, gender, and pain, we must acknowledge the complexities in defining what sex and gender are. It is known that there is a difference between what we have traditionally considered sex, and the perceived experience of a person's gender. The ways we chose to present ourselves, how we perceive our own gender, and how our gender was assigned to us at birth, come together to create a rich and very contextual human experience of identity and its relation to pain perceptions. Sex is a significant factor in differential response to experimental painful stimulation, in pain attitude such as reporting pain and pain coping behavior, in symptoms and signs of painful disorders, and in response to pain treatment.

Both acute and chronic pain conditions have diverse prevalence among the sexes. Women may have more than twice the prevalence in painful disorders when compared to men. For instance, women suffer more in migraine, fibromyalgia, temporomandibular joint disorders, rheumatoid arthritis, and irritable bowel syndrome (Belfer 2017). In general, women demonstrate higher pain sensitivity and lower pain thresholds in response to painful stimulations than men (Sorge and Totsch 2017). Most recent evidence shows that putative mechanisms underlying sex differences in pain are related to effects of sex-specific genetics, anatomical development, and hormone levels, contributing for pain perception, behaviors, analgesia and treatment and management.

14.2.1 Catechol-o-Methyltransferase (COMT) Gene and Sex Differences in Pain

COMT is one of the enzymes involved in catecholamine metabolism that degrades catecholamine neurotransmitters (such as dopamine, epinephrine, and norepinephrine), influencing concentrations of these neurotransmitters in the brain and affecting cognitive function and mood. COMT is playing a pivotal role in pain perception through the regulation of catecholamine concentrations in pain transmission pathways and is one of the most studied genes in pain genetics research. *COMT* has a number of single nucleotide polymorphisms (SNPs) found among the general population. Common *COMT* SNPs associated with pain are the rs4680 or Val[158]Met, rs6269, rs4633, and rs4818 haplotypes, which can reduce COMT enzymatic activity by three- to fourfold (Diatchenko et al. 2005; Zubieta et al. 2003). Lower enzyme activity is associated with increased catecholamine levels and causes hyperalgesia or higher sensitivity to painful stimuli through stimulation of β2-adrenergic receptors. These *COMT* haplotypes are strongly correlated with pain phenotypes; especially, these functional polymorphic alleles in *COMT* affect pain perception and behaviors much stronger in female than male (Belfer et al. 2013; Diatchenko et al. 2005).

Significant sex differences in the effect of *COMT* genetic variants on pain have been found in both animal models and humans (Belfer et al. 2013; Meloto et al. 2016; Mladenovic et al. 2018). In response to capsaicin-induced pain and thermal pain, the effects of *Comt/COMT* genetic variations leading to lower COMT activity have been found to be much stronger on pain phenotypes in females compared to males in both mice and human species (Belfer et al. 2013). Large variation in estrogen and androgen levels between sexes may be the key factor influencing the COMT activity because estrogens can downregulate *COMT* expression (Xie et al. 1999). In inbred strains of mice, lower *Comt* RNA expression was found in female mice (Segall et al. 2010). In humans, COMT activity in erythrocytes, liver, and postmortem prefrontal cortex was found lower in females than males (Chen et al. 2004; Xie et al. 1999). This evidence of low COMT activity in female indicates its association with the high sensitivity to noxious stimuli.

Diatchenko and associates also found that the least frequent *COMT* haplotype (ACCG, in 10.5% of the population) is associated with the highest pain responsiveness, named high pain sensitivity (HPS), whereas, the most frequently occurring haplotype (ATCA, 48.7%) is associated with average pain sensitivity (named APS) and the second most frequent (GCGG, 36.5%) haplotype is associated with the lowest pain sensitivity (named LPS) (Diatchenko et al. 2005). Both HPS and APS haplotypes can reduce COMT enzymatic activity compared to LPS, and subsequently, increased epinephrine levels activating beta adrenergic receptors leads to exacerbated pain perception (Meloto et al. 2016). Low stress scenarios (QST testing and cognitive puzzle anxiety measured) and high stress scenarios (post-motor vehicle collision) were analyzed for detecting relationships between *COMT* haplotypes, pain sensitivity, stress, and sex (Meloto et al. 2016). Associations between pain sensitivity and HPS haplotypes were only detectable in females in the low stress cohort. In the high stress cohort (post-motor vehicle collision), a significant interaction between HPS haplotype and stress levels was found in males, but not in females. The explanation may be that female *COMT* expression is downregulated by estrogen and women have a higher baseline COMT-dependent pain sensitivity, and therefore *COMT* modulated pain sensitivity can be increased quickly at the same haplotypes for females as opposed to males. Likewise pain sensitivity could reach its plateau faster in females and thus, the study could not detect the association between the HPS haplotype and pain sensitivity at lower levels of epinephrine in females compared to males (Meloto et al. 2016).

COMT SNPs and pain by sex have also been investigated in the temporomandibular disorder (TMD) population (Mladenovic et al. 2018). *COMT* rs4680 and rs6269 SNPs were considered as with Meloto et al. (2016), along with rs165774. Pain threshold in response to electrical pain stimuli and cold pain were evaluated in both males and females with TMD. Females had pain thresholds at a significantly lower level than males and had near significantly higher pain responses to cold. Of the *COMT* SNPs only rs165774 demonstrated a significant relationship with pain, with the heterozygous AG allele demonstrating significantly higher thresholds for electric pain stimulation. Other studies such as Belfer et al. (2013) found more significant relationships between thermal pain, *COMT*, and sex. It was found that though there was no association between haplotypes and thermal pain threshold, HPS types had significantly higher reported pain for thermal stimulation than the APS and LPS types in a sample of 18–45 year old females with TMD. These findings indicate that sex and *COMT* variant contribute for individual variations in pain sensitivity with TMD chronic painful condition and demonstrate the mechanisms of *COMT* gene regulation of pain processes in different sexes.

14.2.2 Opioid Receptor Genes and Sex Differences in Pain

The opioid receptor mu 1 (*OPRM1*) gene encodes mu (μ) opioid receptor in humans. Opioid receptors are an important component of the endogenous opioid system that regulate pain perception, reward feelings, and addictive behaviors. *OPRM1* SNP A118G (rs1799971), a functional variant, has been widely studied in relation to pain and analgesic responses. The association of the A118G SNP of *OPRM1* with experimental pain sensitivity was examined and showed that the rare allele was associated with higher pressure pain thresholds, and in response to heat pain, minor allele was associated with lower pain ratings in men, but higher among women (Fillingim et al. 2005). An interaction between sex and A118G genotype regarding pain intensity was also found in patients with lower back pain and sciatica after lumbar disc herniation. This minor allele increased by two times the pain intensity in women, but had a protective effect in men 12 months after the disc herniation (Olsen et al. 2012). A more recent study suggests consistent findings of the sex-dependent effect of A118G genotype on pain recovery in patients with low back pain and lumbar radicular pain, specifically, *OPRM1* rs179971 A > G in men was associated with better long-term pain recovery (Bjorland et al. 2017).

Polymorphisms in other opioid receptor genes, such as *OPRD* and *OPRK* may also contribute to the pain sensitivity by sex. In a study of healthy participants, *OPRD* SNP rs2234918 was associated with a sex-specific difference in thermal pain sensitivity (Kim et al. 2004). However, the *OPRD* haplotype was not shown to have a consistent sex-specific pain effect in a similar study by the same researchers (Kim et al. 2006). A recent study showed that *OPRK* rs6473799 and gender contributed to 34% of pain variability in thermal skin painful stimulation (Sato et al. 2013). Furthermore, *OPRD1* demonstrates sex-specific regulation on morphine analgesia in experimental pain tests. As such, variability in morphine analgesia to rectal thermal stimulation in males was significantly associated with *OPRD1* SNPs rs2234918 and rs533123 (Nielsen et al. 2017).

14.2.3 GTP Cyclohydrolase (GCH1) Gene and Sex Differences in Pain

Recent research also shows evidence of polymorphisms in the GTP cyclohydrolase (*GCH1*) gene contributing to pain pathophysiology and sex-specific pain sensitivity. The *GCH1* provides instructions for making GTP cyclohydrolase, an enzyme involved in tetrahydrobiopterin (BH4) synthesis, a cofactor for nitric oxide synthases and aromatic hydroxylases and a key modulator of inflammatory pain. In a sickle cell anemia and pain study, polymorphisms in *GCH1* gene were observed to be associated with pain crises limited to females, but not among males (Belfer et al. 2014). Thus, *GCH1* haplotypes may be protective for peripheral neuropathic pain in men (Belfer 2017). *GCH1* variant demonstrates a marker for intrinsic regulator of pain sensitivity and traits of pain chronicity.

The *AVPR1A* gene involved in encoding the vasopressin-1A receptor (V1AR) has been found to be responsible for inflammatory pain sensitivity. The V1AR is predominated in the central nervous system and recognized to be male-specific and genetically determined in human affiliation and social communication behaviors (Mogil et al. 2011). A significant association of the *AVPR1A* SNP (rs10877969) with capsaicin pain was found, but only in male subjects reporting stress at the time of testing, not in females. Mice models also confirmed the male-specific interaction of V1AR and stress, suggesting that vasopressin may activate endogenous analgesia mechanisms during pain process and analgesic efficacy may depend on the emotional state of the recipient (Mogil et al. 2011).

14.2.4 Genome-Wide Association Studies (GWAS) and Sex Differences in Pain

Along with the advancement of genomic technology, more recent genome-wide association studies (GWAS) provide evidence for sex-specific contribution of genetic factors to pain in humans. In a large scale study of diabetic neuropathic pain, Meng et al. (2015) found significant sex-specific loci for women (in the Chr1p35.1; ZSCAN20-TLR12P) and for males (in the Chr8p23.1; HMGB1P46). Sex-specific narrow-sense heritability was also demonstrated in this study to be higher in men than women (Meng et al. 2015). Another recent GWAS study by Smith et al. (2019) in TMD orofacial pain identified 3 distinct loci in the sex-segregated analyses, and thus, an SNP on chromosome 3 (rs13078961) was found to be significantly associated with TMD in males only. This male-specific genetic effect may contribute to the lower TMD pain in men (Smith et al. 2019). These GWAS study findings indicate the need for further research pain disorders with sex differences.

14.2.5 Hormonal Effects on Pain Changes in Transgenders

Most of the modern literature fails to recognize whether or not individuals who may not identify with their assigned gender were included in the research. Though it is always possible that individuals may have been included who chose not disclose this information, a disservice is done in not attempting to voluntarily recognize the inclusion of minority populations to the conversation. As with countless other minorities before, individuals who identify outside of the binary gender they were assigned at birth (most commonly described as transgender or trans) are often forgotten from medical discourse on gender variants in health, especially in pain disorders. At risk of medicalizing gender experience unnecessarily, people who identify outside of their assigned gender can have a variety of chromosomal, physiological, psychological, and hormonal considerations which must be included in the equation for pain research. Pain research in transsexual population by Aloisi et al. (2007) showed that sex hormones affected the pain occurrence and associated symptoms when cross-sex hormone treatments provided for sex reassignment for at least 1 year. About 30% of the male to female (MtF) transsexuals developed chronic pain and reported increased pain sensitivity after hormonal therapy, while about 50% female to male (FtM) showed a significant reduced chronic pain (e.g., headaches, breast, and musculoskeletal pain) already present before, with reduced pain sensitivity following the hormonal treatments (Aloisi et al. 2007). This remarkable study suggests that marked changes in sex hormones affect the pain prevalence and symptoms, but the underlying mechanisms including peripheral or central actions of sex steroids and genomic contribution are largely unknown. Further pain research must include individuals outside of the cisgender population in future investigations.

14.3   Race, Ethnicity, and Pain Genomics

Increasing numbers of studies on racial/ethnical differences in pain and pain management have shown that minorities in the USA experience higher prevalence and greater levels of pain than non-Hispanic Whites, whether it be in clinical settings, in chronic pain, or in experimental pain (Mossey 2011; Wyatt 2013). These differences can contribute to important clinical outcomes such as differing treatment decisions, increased morbidity and mortality, and lowered quality of life. Before further discussion, it is important to know the difference between race and ethnicity, two terms commonly used interchangeably. Race refers to physical characteristics, such as skin, eye, and hair color, all based on genetics. Ethnicity refers more to cultural and sociological factors such as regional heritage and customs. There is consistent evidence that shows both acute and chronic pain are more prevalent in non-Hispanic Blacks and Hispanics compared to non-Hispanic Whites in the emergency service (Beaudoin et al. 2018) and in the primary care settings (Reyes-Gibby et al. 2007; Staton et al. 2007). Here, the discussion focuses mainly on racial/ethnical and genetic factors contributing to pain sensitivity, occurrence, and clinical characteristics.

The *COMT* SNPs not only contribute to sex differences in pain, but are significantly associated with ethnicity. In a study of fibromyalgia (FM), females and African Americans were more likely to have a diagnosis of FM, compared to men and Caucasians, respectively, and African Americans were found to be 11.3 times more likely to have a *COMT* HPS haplotype, regardless of the fibromyalgia diagnosis (Lee et al. 2018). Meanwhile, the study shows that the minor alleles of *COMT* SNPs rs4680, rs4818, rs4633, and rs6269 were overrepresented in the FM and had ethnic differences compared with general populations (Lee et al. 2018). This study suggests that African Americans had a higher risk of developing a FM diagnosis than Caucasians or Hispanics, and African-American women had the highest risk overall. Another study focusing on the role of *COMT* SNPs in migraine with a cohort of carefully clinical characterized Caucasian migraineurs recruited reported that *COMT* variations were not associated with migraine susceptibility and clinical phenotypes (De Marchis et al. 2015). Together with other study findings (Corominas et al. 2009; Todt et al. 2009), it seems no correlations exist between the *COMT* SNP rs4680 and migraine in Caucasians population, but a putative correlation was found in Korean and Turkish migraine population (Emin Erdal et al. 2001; Park et al. 2007). Further large scale studies are needed to investigate the correlation of differences in *COMT* variant frequency and race/ethnicity in different populations.

Ethnic-dependent association of *OPRM1* genotype with pain sensitivity has also been demonstrated in recent studies. Ethnic differences in *OPRM1* A118G SNP with experimental pain responses were found in the study by Hastie et al. (2012). Fewer African Americans had the rare allele of *OPRM1* compared to non-Hispanic Whites and the association of G allele effects with decreased pain sensitivity was only shown among Whites. The mechanism is still unclear for this dichotomy but may be related to ethnic differences in the haplotypic structure (Hastie et al. 2012). *OPRM1* SNPs were also found to be associated with amounts of self-administered morphine by patients and among different ethnic groups in opioid analgesia. Post-hysterectomy patients carrying the GG allele reported higher pain scores and used the most amount of morphine compared to patients carrying the AA allele, and specifically significant for the Chinese and Asian Indians ethnic groups (Sia et al. 2013). A meta-analysis supports Sia et al.'s study that also shows *OPRM1* A118G polymorphism was significantly associated with requirement for postoperative opioids in Asian patients, but not in Caucasians (Hwang et al. 2014). The exact mechanisms for the ethnic differences in *OPRM1* genotype, pain, and analgesia remain unclear. Additional research is warranted and will provide valuable information regarding the inter-individual variability in pain response and needs of analgesic doses to achieve personalized pain management.

Few studies have been conducted to investigate the association of polymorphisms in *GCH1* with ethnic-specific pain differences. The *GCH1* pain-protective haplotype was not found having a significant effect on pain patterns or severity in Caucasian patients with chronic pancreatitis or recurrent acute pancreatitis (Lazarev et al. 2008). In African-American patients with sickle cell disease, *GCH1* SNPs were associated with the variability of acute and chronic pain crisis and two of the five *GCH1* SNPs (rs8007267 and rs3783641) showed significant association with pain in these sick cell disease patients (Sadhu et al. 2018). These findings indicate potential contribution of *GCH1* genetic variations to the variability of pain in populations with different ethnicities and may guide clinical pain assessment and treatment.

Evidence supporting racial/ethnic differences in pain from GWAS research is still limited. A large scale GWAS meta-analysis of chronic back pain (CBP) in European ancestry population reported three loci associated with CBP present for over 3–6 months (Suri et al. 2018). Data were used from the Heart and Aging Research in Genomic Epidemiology consortium and from the UK Biobank. Significant genome-wide associations were found for the genetic loci (intronic variant rs12310519 in *SOX5*; intergenic variant rs7833174 in *CCDC26/GSDMC*, and intronic variant rs4384683 in *DCC*) with CBP in European ancestry (Suri et al. 2018). Another discovery GWAS of TMD pain disorder was conducted using the U.S. Hispanic Community Health Study/Study of Latinos data, with the replication cohort from the USA, and Germany, Finland, and Brazil (Sanders et al. 2017). A locus near the sarcoglycan alpha (*SGCA*, rs4794106) associated with TMD was suggestive in the discovery analysis and two additional genome-wide significant loci were identified in females in the sex-stratified analysis in this Hispanic population. Even though few GWAS studies on race, ethnicity, and pain are available, these findings provide valuable insights into understanding the genomic and biological mechanisms in chronic pain among different ethnic groups.

14.4　Conclusion

The relationships between sex, gender, and race/ethnicity with pain are remarkably complex. Individuals may vary dramatically in response to experimental pain stimulations and pain-related health conditions, and these variations could be more pronounced when considering sex and race/ethnicity as the influential factors. Healthcare providers are most likely to underestimate pain in certain sexual and/or ethnic group patients (e.g., female, transgender, and African American) compared to other patients when these factors are not considered. The underlying mechanisms involved in these complex relationships are still largely unknown. Genetic and genomic findings provide significant insights into these differences in pain sensitivity, tolerance, and recovery among sexes and races. Evidence supports gene variations involved in pain processing pathways, such as polymorphisms in *COMT*, opioid receptor genes (*OPRM1, OPRD and OPRK*), *GCH1*, and *AVPR1A* genes, and contributing for sex- and ethnic-specific pain variability. Even though GWAS studies are limited, most recent findings suggest the associations of certain genetic loci with pain and sex and ethnicity. The interaction of pain characteristics with ethnicity, sex, and genetic traits can be used for clinical personalized diagnose of pain and pain-related diseases and treatment of patients with acute and chronic pain conditions.

References

Aloisi AM. Why we still need to speak about sex differences and sex hormones in pain. Pain Ther. 2017;6(2):111–4. https://doi.org/10.1007/s40122-017-0084-3.

Aloisi AM, Bachiocco V, Costantino A, Stefani R, Ceccarelli I, Bertaccini A, Meriggiola MC. Cross-sex hormone administration changes pain in transsexual women and men. Pain. 2007;132(Suppl 1):S60–7. https://doi.org/10.1016/j.pain.2007.02.006.

Beaudoin FL, Gutman R, Zhai W, Merchant RC, Clark MA, Bollen KA, McLean SA. Racial differences in presentations and predictors of acute pain after motor vehicle collision. Pain. 2018;159(6):1056–63. https://doi.org/10.1097/j.pain.0000000000001186.

Belfer I. Pain in women. Agri. 2017;29(2):51–4. https://doi.org/10.5505/agri.2017.87369.

Belfer I, Segall SK, Lariviere WR, Smith SB, Dai F, Slade GD, Diatchenko L. Pain modality- and sex-specific effects of COMT genetic functional variants. Pain. 2013;154(8):1368–76. https://doi.org/10.1016/j.pain.2013.04.028.

Belfer I, Youngblood V, Darbari DS, Wang Z, Diaw L, Freeman L, Taylor JGT. A GCH1 haplotype confers sex-specific susceptibility to pain crises and altered endothelial function in adults with sickle cell anemia. Am J Hematol. 2014;89(2):187–93. https://doi.org/10.1002/ajh.23613.

Bjorland S, Roe C, Moen A, Schistad E, Mahmood A, Gjerstad J. Genetic predictors of recovery in low back and lumbar radicular pain. Pain. 2017;158(8):1456–60. https://doi.org/10.1097/j.pain.0000000000000934.

Chen J, Lipska BK, Halim N, Ma QD, Matsumoto M, Melhem S, Weinberger DR. Functional analysis of genetic variation in catechol-O-methyltransferase (COMT): effects on mRNA, protein, and enzyme activity in postmortem human brain. Am J Hum Genet. 2004;75(5):807–21. https://doi.org/10.1086/425589.

Corominas R, Ribases M, Camina M, Cuenca-Leon E, Pardo J, Boronat S, Macaya A. Two-stage case-control association study of dopamine-related genes and migraine. BMC Med Genet. 2009;10:95. https://doi.org/10.1186/1471-2350-10-95.

De Marchis ML, Barbanti P, Palmirotta R, Egeo G, Aurilia C, Fofi L, Guadagni F. Look beyond catechol-O-Methyltransferase genotype for catecolamines derangement in migraine: the BioBIM rs4818 and rs4680 polymorphisms study. J Headache Pain. 2015;16:520. https://doi.org/10.1186/s10194-015-0520-x.

Diatchenko L, Slade GD, Nackley AG, Bhalang K, Sigurdsson A, Belfer I, Maixner W. Genetic basis for individual variations in pain perception and the development of a chronic pain condition. Hum Mol Genet. 2005;14(1):135–43. https://doi.org/10.1093/hmg/ddi013.

Emin Erdal M, Herken H, Yilmaz M, Bayazit YA. Significance of the catechol-O-methyltransferase gene polymorphism in migraine. Brain Res Mol Brain Res. 2001;94(1–2):193–6. https://doi.org/10.1016/s0169-328x(01)00219-4.

Fillingim RB, Kaplan L, Staud R, Ness TJ, Glover TL, Campbell CM, Wallace MR. The A118G single nucleotide polymorphism of the mu-opioid receptor gene (OPRM1) is associated with pressure pain sensitivity in humans. J Pain. 2005;6(3):159–67. https://doi.org/10.1016/j.jpain.2004.11.008.

Hastie BA, Riley JL 3rd, Kaplan L, Herrera DG, Campbell CM, Virtusio K, Fillingim RB. Ethnicity interacts with the OPRM1 gene in experimental pain sensitivity. Pain. 2012;153(8):1610–9. https://doi.org/10.1016/j.pain.2012.03.022.

Hwang IC, Park JY, Myung SK, Ahn HY, Fukuda K, Liao Q. OPRM1 A118G gene variant and postoperative opioid requirement: a systematic review and meta-analysis. Anesthesiology. 2014;121(4):825–34. https://doi.org/10.1097/ALN.0000000000000405.

Kim H, Neubert JK, San Miguel A, Xu K, Krishnaraju RK, Iadarola MJ, Dionne RA. Genetic influence on variability in human acute experimental pain sensitivity associated with gender, ethnicity and psychological temperament. Pain. 2004;109(3):488–96. https://doi.org/10.1016/j.pain.2004.02.027.

Kim H, Mittal DP, Iadarola MJ, Dionne RA. Genetic predictors for acute experimental cold and heat pain sensitivity in humans. J Med Genet. 2006;43(8):e40. https://doi.org/10.1136/jmg.2005.036079.

Lazarev M, Lamb J, Barmada MM, Dai F, Anderson MA, Max MB, Whitcomb DC. Does the pain-protective GTP cyclohydrolase haplotype significantly alter the pattern or severity of pain in humans with chronic pancreatitis? Mol Pain. 2008;4:58. https://doi.org/10.1186/1744-8069-4-58.

Lee C, Liptan G, Kantorovich S, Sharma M, Brenton A. Association of Catechol-O-methyltransferase single nucleotide polymorphisms, ethnicity, and sex in a large cohort of fibromyalgia patients. BMC Rheumatol. 2018;2:38. https://doi.org/10.1186/s41927-018-0045-4.

Meloto CB, Bortsov AV, Bair E, Helgeson E, Ostrom C, Smith SB, Diatchenko L. Modification of COMT-dependent pain sensitivity by psychological stress and sex. Pain. 2016;157(4):858–67. https://doi.org/10.1097/j.pain.0000000000000449.

Meng W, Deshmukh HA, Donnelly LA, Smith BH, Wellcome Trust Case Control Consortium 2 (WTCCC2), Surrogate markers for Micro- and Macro-vascular hard endpoints for Innovative diabetes Tools (SUMMIT) study group. A genome-wide association study provides evidence of sex-specific involvement of Chr1p35.1 (ZSCAN20-TLR12P) and Chr8p23.1 (HMGB1P46) with diabetic neuropathic pain. EBioMedicine. 2015;2(10):1386–93. https://doi.org/10.1016/j.ebiom.2015.08.001.

Mladenovic I, Krunic J, Supic G, Kozomara R, Bokonjic D, Stojanovic N, Magic Z. Pulp sensitivity: influence of sex, psychosocial variables, COMT gene, and chronic facial pain. J Endod. 2018;44(5):717–721 e711. https://doi.org/10.1016/j.joen.2018.02.002.

Mogil JS, Sorge RE, LaCroix-Fralish ML, Smith SB, Fortin A, Sotocinal SG, Fillingim RB. Pain sensitivity and vasopressin analgesia are mediated by a gene-sex-environment interaction. Nat Neurosci. 2011;14(12):1569–73. https://doi.org/10.1038/nn.2941.

Mossey JM. Defining racial and ethnic disparities in pain management. Clin Orthop Relat Res. 2011;469(7):1859–70. https://doi.org/10.1007/s11999-011-1770-9.

Nielsen LM, Christrup LL, Sato H, Drewes AM, Olesen AE. Genetic influences of OPRM1, OPRD1 and COMT on morphine analgesia in a multi-modal, multi-tissue human experimental pain model. Basic Clin Pharmacol Toxicol. 2017;121(1):6–12. https://doi.org/10.1111/bcpt.12757.

Olsen MB, Jacobsen LM, Schistad EI, Pedersen LM, Rygh LJ, Roe C, Gjerstad J. Pain intensity the first year after lumbar disc herniation is associated with the A118G polymorphism in the opioid receptor mu 1 gene: evidence of a sex and genotype interaction. J Neurosci. 2012;32(29):9831–4. https://doi.org/10.1523/JNEUROSCI.1742-12.2012.

Park JW, Lee KS, Kim JS, Kim YI, Shin HE. Genetic contribution of catechol-O-methyltransferase polymorphism in patients with migraine without Aura. J Clin Neurol. 2007;3(1):24–30. https://doi.org/10.3988/jcn.2007.3.1.24.

Patanwala AE, Norwood C, Steiner H, Morrison D, Li M, Walsh K, Karnes JH. Psychological and genetic predictors of pain tolerance. Clin Transl Sci. 2019;12(2):189–95. https://doi.org/10.1111/cts.12605.

Pieretti S, Di Giannuario A, Di Giovannandrea R, Marzoli F, Piccaro G, Minosi P, Aloisi AM. Gender differences in pain and its relief. Ann Ist Super Sanita. 2016;52(2):184–9. https://doi.org/10.4415/ANN_16_02_09.

Rahim-Williams B, Riley JL 3rd, Williams AK, Fillingim RB. A quantitative review of ethnic group differences in experimental pain response: do biology, psychology, and culture matter? Pain Med. 2012;13(4):522–40. https://doi.org/10.1111/j.1526-4637.2012.01336.x.

Reyes-Gibby CC, Aday LA, Todd KH, Cleeland CS, Anderson KO. Pain in aging community-dwelling adults in the United States: non-Hispanic whites, non-Hispanic blacks, and Hispanics. J Pain. 2007;8(1):75–84. https://doi.org/10.1016/j.jpain.2006.06.002.

Sadhu N, Jhun EH, Yao Y, He Y, Molokie RE, Wilkie DJ, Wang ZJ. Genetic variants of GCH1 associate with chronic and acute crisis pain in African Americans with sickle cell disease. Exp Hematol. 2018;66:42–9. https://doi.org/10.1016/j.exphem.2018.07.004.

Sanders AE, Jain D, Sofer T, Kerr KF, Laurie CC, Shaffer JR, Smith SB. GWAS identifies new loci for painful temporomandibular disorder: Hispanic community health study/study of Latinos. J Dent Res. 2017;96(3):277–84. https://doi.org/10.1177/0022034516686562.

Sato H, Droney J, Ross J, Olesen AE, Staahl C, Andresen T, Drewes AM. Gender, variation in opioid receptor genes and sensitivity to experimental pain. Mol Pain. 2013;9:20. https://doi.org/10.1186/1744-8069-9-20.

Segall SK, Nackley AG, Diatchenko L, Lariviere WR, Lu X, Marron JS, Wiltshire T. Comt1 genotype and expression predicts anxiety and nociceptive sensitivity in inbred strains of mice. Genes Brain Behav. 2010;9(8):933–46. https://doi.org/10.1111/j.1601-183X.2010.00633.x.

Sia AT, Lim Y, Lim EC, Ocampo CE, Lim WY, Cheong P, Tan EC. Influence of mu-opioid receptor variant on morphine use and self-rated pain following abdominal hysterectomy. J Pain. 2013;14(10):1045–52. https://doi.org/10.1016/j.jpain.2013.03.008.

Smith SB, Parisien M, Bair E, Belfer I, Chabot-Dore AJ, Gris P, Diatchenko L. Genome-wide association reveals contribution of MRAS to painful temporomandibular disorder in males. Pain. 2019;160(3):579–91. https://doi.org/10.1097/j.pain.0000000000001438.

Sorge RE, Totsch SK. Sex differences in pain. J Neurosci Res. 2017;95(6):1271–81. https://doi.org/10.1002/jnr.23841.

Staton LJ, Panda M, Chen I, Genao I, Kurz J, Pasanen M, Cykert S. When race matters: disagreement in pain perception between patients and their physicians in primary care. J Natl Med Assoc. 2007;99(5):532–8.

Suri P, Palmer MR, Tsepilov YA, Freidin MB, Boer CG, Yau MS, Williams FMK. Genome-wide meta-analysis of 158,000 individuals of European ancestry identifies three loci associated with chronic back pain. PLoS Genet. 2018;14(9):e1007601. https://doi.org/10.1371/journal.pgen.1007601.

Todt U, Netzer C, Toliat M, Heinze A, Goebel I, Nurnberg P, Kubisch C. New genetic evidence for involvement of the dopamine system in migraine with aura. Hum Genet. 2009;125(3):265–79. https://doi.org/10.1007/s00439-009-0623-z.

Wyatt R. Pain and ethnicity. Virtual Mentor. 2013;15(5):449–54. https://doi.org/10.1001/virtualmentor.2013.15.5.pfor1-1305.

Xie T, Ho SL, Ramsden D. Characterization and implications of estrogenic down-regulation of human catechol-O-methyltransferase gene transcription. Mol Pharmacol. 1999;56(1):31–8. https://doi.org/10.1124/mol.56.1.31.

Zubieta JK, Heitzeg MM, Smith YR, Bueller JA, Xu K, Xu Y, Goldman D. COMT val158met genotype affects mu-opioid neurotransmitter responses to a pain stressor. Science. 2003;299(5610):1240–3. https://doi.org/10.1126/science.1078546.

Placebo Hypoalgesic Effects and Genomics

15

Luana Colloca and Nandini Raghuraman

Contents

15.1 Introduction

It is well established that placebo-induced hypoalgesic effects are due to patient expectations, prior experience, and patient–provider interactions (Colloca and Benedetti 2005; Colloca 2018) making it one of the most well-known mechanisms to alter clinical outcomes.

L. Colloca (✉)
Department of Pain and Translational Symptom Science, School of Nursing, University of Maryland, Baltimore, MD, USA

Departments of Anesthesiology and Psychiatry, School of Medicine, University of Maryland, Baltimore, MD, USA

Center to Advance Chronic Pain Research, University of Maryland, Baltimore, MD, USA
e-mail: colloca@umaryland.edu

N. Raghuraman
Department of Pain and Translational Symptom Science, School of Nursing, University of Maryland, Baltimore, MD, USA

Program in Epidemiology and Human Genetics, University of Maryland School of Medicine, Baltimore, MD, USA

© Springer Nature Switzerland AG 2020
S. G. Dorsey, A. R. Starkweather (eds.), *Genomics of Pain and Co-Morbid Symptoms*, https://doi.org/10.1007/978-3-030-21657-3_15

In recent times placebo responses have had even a larger significant impact on randomized clinical trials (Tuttle et al. 2015), making it inconvenient to validate the efficacy of a drug under investigation. In order to separate drug and placebo effects from potential confounders (Ernst and Resch 1995), a no-treatment arm (Vase et al. 2015) and possibly a measurement of expectations (Colloca 2017) should be included.

Placebo effects have the potential to influence any condition or treatment (Colloca and Barsky 2020). This phenomenon has been particularly well-investigated in the areas of experimental and clinical pain including different chronic pain population like idiopathic pain, where pain arises spontaneously and neuropathic pain (Vase et al. 2014; Petersen et al. 2014, 2012), low back pain (Hashmi et al. 2012; Carvalho et al. 2016), knee osteoarthritis (Tetreault et al. 2016), irritable bowel syndrome (Vase et al. 2003, 2005; Kaptchuk et al. 2008), and migraine (Kam-Hansen et al. 2014).

Various neuroimaging and pharmacological studies have shown that placebo hypoalgesia engages multiple neurobiological mechanisms including the release of endogenous opioids, endocannabinoids, oxytocin, vasopressin, and dopamine, with effects that depend on the target system and illness.

15.2 Placebo- and Drug-Related Effects

Placebo effects affect any pharmacological and non-pharmacological interventions (Colloca 2019). In some cases like integrative treatments (Vickers et al. 2012) (e.g., homeopathy (Mathie et al. 2017)) the overall therapeutic effect can often be exclusively driven by expectancy and placebo effects (Colloca et al. 2013). In fact, expectancy and placebo effects can account for 50% of the effectiveness of pain treatments such as morphine (Colloca et al. 2004). For example, given a treatment in full view as compared to a hidden administration can boost the overall positive outcome across distinct pain treatments such as opioid and non-opioid based treatments (e.g., buprenorphine, tramadol, ketorolac, and metamizole) (Colloca et al. 2004). Moreover, research has indicated that features such as price (higher price larger placebo effects) (Kam-Hansen et al. 2014; Waber et al. 2008), labeling (generic vs brand) (Faasse et al. 2016), and mode of administration (sham acupuncture vs oral placebo) (Meissner et al. 2013) influence the occurrence and magnitude of placebo effects.

In experimental research settings, the duration of painful stimulations, for example, phasic or tonic pain and its modality either thermal or chemical can also influence the occurrence and magnitude of placebo hypoalgesic effects (Vase et al. 2009).

15.3 Neurobiological Studies

In the last four decades, pain researchers have been able to link the placebo effect to the endogenous opioid system. A series of neuro-pharmacological studies using opioid antagonist naloxone (Levine et al. 1978; Eippert et al. 2009a, b; Benedetti

1996; Amanzio and Benedetti 1999) and in vivo receptor binding of μ-opioid receptors (Wager et al. 2007; Zubieta et al. 2005) have elucidated that the opioid system is engaged in the formation of placebo analgesia. The mentioned findings corroborated Levine et al. pioneering study in 1978, where it was demonstrated that placebo hypoalgesic effects could be blocked by administering the opioid receptor antagonist naloxone (Levine et al. 1978).

A morphine-like effect at the level of pain endurance is antagonized by naloxone (10 mg/kg) when a placebo is given after consecutive administrations of morphine either 1 day (Amanzio and Benedetti 1999) or 1 week (Benedetti et al. 2007) apart. The μ-opioid receptors (MOPRs) are widely distributed, with elevated concentrations in the thalamus and periaqueductal gray (PAG) (Oroszi and Goldman 2004), and are pivotal for pain reduction by therapeutic opioids thereby making the importance of the opioid system for placebo effects less astounding. Endogenous opioids are those naturally released in the brain, and exogenous opioids are drugs synthetically manufactured to activate the opioid receptors in distinct locations (Stoeber et al. 2018). Understanding the receptor binding mechanisms of both endogenous and exogenous opioids will be critical for drug development for pain, and prevention of side effects.

Other systems, namely the cannabinoid and dopamine, have been pharmacologically explored by employing antagonists to reverse the behavioral effects of placebo. For example, a pharmacological conditioning with ketorolac, a non-opioid non-steroidal anti-inflammatory drug (NSAID) induced placebo effects while replaced by a placebo given after several administration of NSAIDs (Amanzio and Benedetti 1999).

Also, the CB1 receptor antagonist rimonabant reversed ketorolac-like effects indicating that the endogenous cannabinoid system can be "trained" to form cannabinoid-like placebo effects (Benedetti et al. 2011, 2013).

As far as the involvement of the dopamine system is concerned, a PET imaging study using a carbon-11 labeled radio-ligand raclopride has demonstrated the relationship between the ventral basal ganglia, including the nucleus accumbens and placebo-induced pain reduction (Scott et al. 2008).

Pharmacological approaches in healthy participants with dopamine antagonists and agonists did not show any significant influence on placebo analgesia in healthy participants (Wrobel et al. 2014), and also in neuropathic chronic pain patients (Skyt et al. 2017), whereby the dopamine antagonist haloperidol did not reverse placebo analgesia. Although these results highlight that dopamine antagonists and agonists may not be vital for generating placebo-induced pain reduction, they do moderate expectations of pain relief (Skyt et al. 2017) and recalled efficacy of placebo (Jarcho et al. 2016).

Conversely, studies with Parkinsonian patients using in vivo single neural activity recording and pharmacological conditioning, for example, using apomorphine (Benedetti et al. 2004, 2009; Mercado et al. 2006) and also in PET studies (Lidstone et al. 2010; de la Fuente-Fernandez et al. 2001) displayed dopamine-like effects suggesting that in Parkinson's disease placebo effects are mediated by the dopamine system when conditioning with dopamine agonist is performed.

15.4 Enhancement of Placebo Analgesic Effects Through Vasopressin and Oxytocin Agonist Agents

In order to improve pain management and to cope, researchers have been recently able to show the possibility to increase placebo hypoalgesic effects pharmacologically. They were able to derive that intranasal administration of both oxytocin and vasopressin agonists possessed the ability to enhance expectancy-induced placebo analgesia (Kessner et al. 2013; Colloca et al. 2016). The distribution of these two hormones in the CNS is characterized by sex-specific effects (Bielsky et al. 2005; Donaldson and Young 2008; Ebstein et al. 2009; Heinrichs and Domes 2008; Heinrichs et al. 2009; Young and Wang 2004) across different species and the regulation of social behaviors (Donaldson and Young 2008; Kogan et al. 2011) is also based on dimorphic effects. In male animals, at the level of the septum, anterior hypothalamus and central gray vasopressin promoted aggression while in female animals, via actions in the septum and ventral pallidum it promotes affiliative behaviors (Bielsky et al. 2005). In humans, vasopressin plays a role in appeasing behaviors (Feng et al. 2015) as well as interpersonal communication (Thompson et al. 2004, 2006), advocating a "tend-and-befriend" behavioral patterns in women toward other women, and in men "fight-or-flight" towards other men (Thompson et al. 2006). Intranasal administration of synthetic vasopressin in humans has shown to reach the central nervous system through the nasal mucous membranes, and achieve a steady state within 30–50 min, and its concentration measured by radioimmunoassay remained stable for 80 min in the cerebrospinal fluid (Born et al. 2002). Colloca et al. (2016) performed a randomized, double-blinded, parallel design trial to test the effects of arginine vasopressin a non-selective agonist of the Avp1a and Avp1b receptors against no treatment, oxytocin, and saline. These effects were explored using a model of verbally induced placebo effects with suggestions of pain reduction along with a sham intervention. A relatively low dosage of oxytocin (24IU) was used compared to the dosage (40IU) used by Keller et al., where the increase of placebo effects was reported in men (Kessner et al. 2013). Contrarily, in the above study vasopressin enhanced the verbally induced placebo analgesic effect in women but not in men and those women with lower acute cortisol levels and lower dispositional anxiety had larger placebo effects. As these results are consistent with animal and human studies (Goodson and Thompson 2010; Ferris et al. 1997; Gobrogge et al. 2009), it has increased the prospect of using vasopressin agonists as potential pharmacological therapeutic targets. The probable link between psychophysiological changes and behavior could pave the way to understand drug and placebo responsiveness (Meissner et al. 2011). In the socially monogamous coppery titi monkey, receptor autography studies demonstrated Avp1a receptors' crucial involvement in social behaviors. The Avp1a receptor distribution expressed at the level of the cortex (cingulate, insular, and occipital), central amygdala, nucleus accumbens, caudate, putamen, endopiriform nucleus, and hippocampus (Freeman et al. 2014). Activation of brain reward and salience circuits has also been shown in men and women (Thompson et al. 2006). Vasopressin most likely shaped the enhancement of expectancy-induced analgesic responses, emphasizing the role of "response to meaning" in forming placebo effects (Moerman and Jonas 2002).

15.5 Brain Imaging Phenotypes of Pain-Related Placebo Responsiveness

Importantly, predicting placebo effects in patients suffering from chronic pain can open up new avenues for a better pain management. Some recent studies illuminate this concept by looking at neural biomarkers of placebo hypoalgesic responses. Brain activity changes and enhanced functional coupling have been reported to consolidate the nature of placebo hypoalgesic effects (Zunhammer et al. 2018). Areas of the brain in which changes have been documented include primarily but are not limited to, the dorsolateral prefrontal cortex (DLPFC), the anterior cingulate cortex (ACC), and subcortical regions including the hypothalamus, amygdala, and the periaqueductal gray (PAG) (Eippert et al. 2009b; Bingel et al. 2006; Craggs et al. 2008; Wager et al. 2004). The DLPFC initiates the placebo hypoalgesic effects as demonstrated by different groups and approaches (Krummenacher et al. 2010; Lui et al. 2010). Changes in the rACC that is connected with the PAG correlate with placebo hypoalgesic effects (Eippert et al. 2009b; Bingel et al. 2006). Placebo hypoalgesic effects can decrease brain activity in areas such as the thalamus (Th), insula (INS), and the somatosensory cortex (Eippert et al. 2009b; Bingel et al. 2006; Wager et al. 2004; Lui et al. 2010), showing relationship with nociception and placebo effects. Interestingly, placebo hypoalgesic effects can modulate the activity at the level of the spinal cord with changes that occur in the dorsal horn ipsilateral to the area of painful stimulation (Eippert et al. 2009a). Using functional magnetic resonance imaging of the spinal cord, Eippert et al. showed in humans that placebo hypoalgesia results in a reduction of activity in the spinal cord indicating a potential top-down mechanism suppressing pain signaling from the periphery to subcortical and cortical brain regions (Eippert et al. 2009a). The modulation at the level of the spinal cord is bidirectional. A negative conditioning induced behavioral nocebo hyperalgesic effect along with a strong activation in the spinal cord corresponding to stimulated area and ipsilateral C5/C6 dermatomes (Geuter and Buchel 2013).

Several attempts have recently made to look at brain structure (e.g., gray matter density) and *functional* connectivity as predictors of individual placebo responsiveness status. Using voxel-based morphometry, gray matter density in the DLPFC, INS, and nucleus accumbens (NAc) has been correlated with larger placebo hypoalgesia (Schweinhardt et al. 2009). The structural differences in NAc and DLPFC were associated with novelty seeking and behavioral activation—dopamine psychologically—related factors (Schweinhardt et al. 2009). Kong et al. have also demonstrated that pretest resting-state functional connectivity correlated with individual expectations placebo hypoalgesia (Kong et al. 2013). The baseline resting-state *functional* connectivity at the level of the right fronto-parietal regions with the rostral ACC was positively associated with the magnitude of expectations of pain reduction, and the resting-state *functional* connectivity between the somatosensory areas and the cerebellum was correlated with individual placebo hypoalgesic effects (Kong et al. 2013). For example, the *functional* connectivity between the dorsomedial PFC and bilateral INSs cortices was associated with the magnitude of placebo hypoalgesia and the likelihood to have better outcomes in patients suffering from low back pain. Those patients who responded to the 2 week placebo pills had a lower functional connectivity between dorsomedial PFC and INSs as compared to those who did not benefit from taking placebo pills and showed no trend to recover (Hashmi et al. 2012). The functional connectivity between dorsomedial PFC-INS and DLPFC-midcingulate cortex predicted the clinical low back pain recovery with very high accuracy (about 90%) (Hashmi et al. 2012).

Stein et al. measured local white matter anisotropy and *structural* connectivity between a priori cortical and subcortical regions of interest, and indicated that individual placebo hypoalgesic effects were predicted and associated with a higher fractional anisotropy in the rostral ACC and in left DLPFC, and with a stronger fiber connection of these two regions with the periaqueductal gray (PAG) (Stein et al. 2012).

Recently, Apkarian and his team demonstrated that the individual placebo hypoalgesic effects depend on brain structure and function. Namely, volume asymmetry in the limbic regions, thickness of the sensorimotor cortices and functional coupling of prefrontal regions, anterior cingulate, and periaqueductal gray are predictive of placebo responsiveness. Importantly, these neural signatures remained unchanged before, during and after washout periods. Psychologically speaking, interoceptive awareness and openness predicted the magnitude of placebo responsiveness. These results will help detect placebo responders in RCTs in chronic pain patients, and might suggest that the long-term placebo effects observed clinically, are predictable at least in part with brain imaging approaches (Vachon-Presseau et al. 2018).

15.6 Genomics of Placebo Hypoalgesic Effects

One of the goals of the genomics of placebo effects is to identify specific DNA variations that increase or decrease the susceptibility to respond to treatments and placebo procedures (Hall and Kaptchuk 2013). Recent studies suggest that genomics may help identify those patients and healthy participants who are likely to respond to placebos and placebo effects (for reviews, see refs. (Colagiuri et al. 2015; Hall et al. 2015a)). Not all individuals respond to placebos: some do not respond at all, some a little bit, and some perceive complete remission of their symptoms such as pain. Given that pain responses vary across individuals (Mogil 2009, 2012) and are more than 50% heritable according to twin studies (Nielsen et al. 2008; Hartvigsen et al. 2009; Norbury et al. 2007), it is plausible that specific genetic variations play a role in responsiveness (Hall et al. 2015a). In our recent review of the literature, we reported (Colagiuri et al. 2015) three single nucleotide polymorphisms (SNPs) that have been associated with placebo effects in the arena of pain when well-controlled laboratory procedures are adopted performed in healthy participants and patients (e.g., inclusion of the no-intervention arm also called natural history (Colloca 2017)). These three SNPs are located in the opioid receptor mu subunit (*OPRM1* rs1799971) gene, the fatty acid amide hydrolase (*FAAH* rs324420) gene, and the catechol-O-methyltransferase (*COMT* rs4680) gene, respectively (See Fig. 15.1 and Table 15.1).

Pharmacological and imaging studies using the opioid antagonist naloxone have shown that the endogenous opioid system plays an important role in driving pain-related placebo effects (Eippert et al. 2009b; Amanzio and Benedetti 1999; Pecina and Zubieta 2015a, b; Fields 2004; Levine and Gordon 1984). Zubieta et al. first demonstrated that the opioid receptors and the opioid receptors mu subunit gene *(OPRM1)* A118G variant (Asn40Asp, rs1799971) that is thought to be expressed at lower levels (Zhang et al. 2005) play a role in pain sensitivity and regulation (Zubieta et al. 2001, 2002, 2003, 2005).

Given that brain responses to pain and other stressors are regulated by interactions among neurochemical systems, Zubieta et al. explored the functional influence of *COMT* on the responses to sustained experimental pain in humans. They found that individuals homozygous for the met158 allele of COMT polymorphism (val158met) showed diminished brain mu-opioid system responses to pain compared with heterozygotes, higher sensory and affective pain self-reports and a highly negative internal affective state (Zubieta et al. 2003). The COMT val158met polymorphism influences the human brain regional mu-opioid response to pain, behavioral inter-individual differences in pain and other stressful stimuli.

The rs4680 polymorphism in *COMT* which encodes a valine to methionine amino acid substitution at codon 158 (val158met) and reduces the enzymatic activity of the protein product of this gene by three- to fourfold (Lotta et al. 1995) has been also associated with better placebo-induced outcomes in patients with irritable bowel syndrome (Hall et al. 2012) and brain imaging placebo analgesia in healthy subjects (Yu et al. 2014). These two studies indicate that the met/met carriers for the *COMT* gene are those displaying larger clinical (Hall et al. 2012) and experimental (Yu et al. 2014) pain reduction. However, the reward system is likely to attribute a

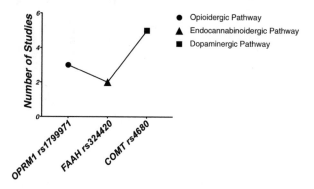

Fig. 15.1 Published findings (2012–2019) linking placebo hypoalgesia and genetic variants

Table 15.1 Published studies (2012–2019) linking placebo hypoalgesia and genetic variants

Gene	SNP	Sample size (n)	Type of pain	Placebo manipulation	Findings	References
COMT	rs4680	104	Venipuncture stimulation	Verbal suggestions	COMT val158met polymorphism predict placebo hypoalgesia IBS patients	Hall et al. (2012)
		233	Contact heat stimulation	Verbal suggestions	COMT val158met Met-Allele was associated with negative processing of fear of medical pain but no relationship with placebo hypoalgesia prediction	Forsberg et al. (2018)
		160	Electrical and contact heat stimulation	Verbal suggestions and conditioning	Three-way interaction among OPRM1_AA, COMT met/met, FAAH Pro/Pro, provides the most robust estimation of placebo responsiveness	
		296	Heat pain stimulation	Verbal suggestion	Potential OPRM1 Asn/Asn and COMT Met/Met and Val/Met combination indicator of Placebo Hypoalgesia	Aslaksen et al. (2018)
		48	Heat pain stimulation	Conditioning	Linear correlation between number of Met-alleles in COMT polymorphism and pain suppression	Yu et al. (2014)
OPRM1	rs1799971	50	Sustained pain induced by the infusion of 5% hypertonic saline into left masseter muscle		The OPRM1 A118G polymorphism contributes towards variations in neurotransmission and potentially influence Placebo hypoalgesia	Pecina et al. (2014)
		296	Heat pain stimulation	Verbal suggestions	Potential OPRM1 Asn/Asn and COMT Met/Met and Val/Met combination indicator of placebo hypoalgesia	Aslaksen et al. (2018)
		160	Electrical and contact heat stimulation	Verbal suggestions and conditioning	Three-way interaction among OPRM1 AA, COMT met/met, FAAH Pro/Pro, provides the most robust estimation of placebo responsiveness	
FAAH	rs324420	42	Sustained pain induced by the infusion of 5% hypertonic saline into left masseter muscle		FAAH Pro129/Pro129 homozygotes reported higher placebo analgesia and more positive affective states immediately and 24 h after placebo administration	Pecina et al. (2014)
		160	Electrical and contact heat stimulation	Verbal suggestions and conditioning	Three-way interaction among OPRM1 AA, COMT met/met, FAAH Pro/Pro, provides the most robust estimation of placebo responsiveness	

perception of benefit from a specific treatment and predict clinical outcomes (e.g., pain reduction, improved quality of life, patient satisfaction) differently in pain patients (Hall et al. 2012, 2015b; Diatchenko et al. 2005) than healthy participants (Yu et al. 2014).

Moreover, it is likely that the interaction between *OPRM1 rs1799971* and *COMT rs4680* has an impact on placebo hypoalgesia. A recent study indicates that participants being *COMT* Met/Met or Val/Met and *OPRM1* A/A carriers compared to those who are the Val/Val—G carriers reported level of placebo-induced pain reduction that was 4–6 times higher (Aslaksen et al. 2018).

The cannabinoid system is another system that has been implicated in placebo hypoalgesia using the CB1/CB2 receptor antagonist rimonabant (Benedetti et al. 2013, 2011). The functional variant in *FAAH* gene and rs324420 polymorphism encodes a Pro129Thr missense substitution that leads to distinct endocannabinoid levels (Cajanus et al. 2016). Genetic variants in *FAAH* rs324420 affect placebo hypoalgesia behaviorally and at the level of PET responses (Pecina et al. 2014). We performed recently for the first time three-way gene-to-gene interaction study using the three identified SNPs and verbal suggestions and conditioning as by placebo procedures to determine the potential influence of combined genetic variants and psychological approach on placebo analgesic effects. Controlling for age, sex, and race we found that the interaction among *OPRM1* rs1799971, *COMT* rs4680, *FAAH*, rs324420 and placebo procedure was predictive of distinct patterns of placebo responsiveness. Reinforcing verbal suggestion (and presumably expectancies) via learning generated significant larger behavioral effect sizes. The model with *OPRM1* rs1799971 by *COMT* rs4680 by *FAAH* rs324420 by placebo procedure predicted placebo responsiveness of an accuracy of 0.773.

Previous studies with genotypes, brain imaging, and placebo hypoalgesia have reported significance independently for three distinct SNPs (rs1799971; rs4680; rs324420) located in the *OPRM1*, *COMT*, and *FAAH* genes, respectively. Put more simply, those participants with a genetic variant in one of these three genes tend to experience larger placebo-induced pain reduction (Hall and Kaptchuk 2013; Hall et al. 2015a). We demonstrated that adding a 2-way interaction and even 3-way interaction improved the prediction of placebo effects marginally rather the gene-to-gene interact combined with the placebo procedure provided the stronger estimation of placebo effects. We found a main effect of *OPRM1 rs1799971* that is significantly linked to placebo reduction induced by verbal suggestion and learning. The *OPRM1* rs1799971 polymorphism has been associated with pain reduction in dysmenorrhea women (Wei et al. 2017) and hypnotizability and pain (Presciuttini et al. 2018). The role of the COMT SNP in placebo effects is less consistent with both positive findings detecting an association (Forsberg et al. 2018; Aslaksen et al. 2017) as well as negative findings (Hall et al. 2012). Molecular (but not behavioral) findings provided evidence of μ-opioid-mediated placebo related changes in brain regions associated with pain, reward-motivated learning and memory processing are modulated by *FAAH* gene, C385A (rs324420) (Pecina et al. 2014). We provided a novel demonstration that the three-way genetic interaction—*OPRM1* rs1799971 by *COMT* rs4680 by *FAAH* rs324420—produced distinct profiles of placebo responsiveness with a comparable effect size. Distinct placebo profiles were driven by this three-way genetic interaction. Namely, individuals who are carriers for *OPRM1* rs1799971 AA and *FAAH* rs324420 Pro/Pro and individuals carrying *COMT* rs4680 met/met together with *FAAH* rs324420 Pro/Pro showed significant placebo analgesic effects. Moreover, participants who are *COMT* met/val carriers showed significant placebo effects independently of *OPRM1* rs1799971 and *FAAH* rs324420 allele combinations. Additionally, G val/val Thr carriers showed significant hypoalgesia. The magnitude of placebo hypoalgesia in the identified placebo responder profiles was comparable in magnitude.

Overall, it exists a tangible difficulty to control for type I error with type II error—a key issue in association studies of placebo analgesia is the small sample size. Larger genetic studies are needed when contribution of genetic variants to placebo-induced endogenous pain modulation is explored. Yet, the use of specific SNP combinations (e.g., *COMT, OPRM1,* and *FAAH* SNPs gene interactions) is inexpensive and such information may be integrated into clinical practice for assessing the individual's propensity to benefit from the endogenous modulatory systems and placebo responsiveness.

15.7   Future Directions

Future directions should expand upon the investigation of the genetics of placebo hypoalgesic effects to approaches such as gene expression profiling and Genome Wide Association Studies (GWAS).

In addition to the candidate-gene approach, the genome-wide approach has recently been used to investigate the determinants of pain disorders. The genome-wide gene expression microarrays are in nature different from the candidate-gene hypothesis-driven approaches. The genome-wide approach employs high-throughput technologies to analyze biomarkers across the entire human genome in order to find associations with observable traits and individuals' features. Genome-wide gene expression microarray studies are widely utilized to analyze hundreds of thousands of biomarkers by high-throughput technologies using RNA derived from peripheral blood cells. Gene expression profiling has been used to attempt to isolating biomarkers of pain disorders such as acute low back pain (Starkweather et al. 2016), visceral pain (Gupta et al. 2017), fibromyalgia (Jones et al. 2016; Lukkahatai et al. 2015), osteoarthritis (Attur et al. 2011), as well as migraine and headache (Eising et al. 2017; Gerring et al. 2018; Perry et al. 2016).

Future studies should be conducted to explore whether and how genome-wide gene expression could help predict placebo responders from nonresponders in well-controlled laboratory settings and randomized clinical trials. Gene expression profiles in those with and without placebo-induced pain relief phenotypes can be promising to identify individual biomarkers of proneness to respond to placebo pills and placebo manipulations including pharmacological conditioning, verbal suggestions, observationally induced placebo hypoalgesia. The possibility that gene-related biomarkers could identify those who are likely to respond to placebo effects and may activate in turn, the pain modulatory systems across various pain populations represents a new emerging possible avenue of research.

15.8 Conclusions

Research on genomics of placebo hypoalgesic effects is still nascent but there is a potential to inform and guide more effective, personalized, and mechanistic-based therapeutic strategies. Moreover, genomics-related information may help separate drug from placebo effects (Wang et al. 2017) in randomized clinical trials and predict individual responses to treatments and placebos would have a terrific impact on drug development and pain medicine. Future research that combines sophisticated genetic approaches with careful phenotyping of patients will help translate science of placebo into better pain management plans.

Acknowledgments This research is supported by NIDCR (R01 DE025946).

Competing Interests None.

References

Amanzio M, Benedetti F. Neuropharmacological dissection of placebo analgesia: expectation-activated opioid systems versus conditioning-activated specific subsystems. J Neurosci. 1999;19(1):484–94.

Aslaksen PM, Forsberg JT, Gjerstad J. The mu-opioid receptor gene OPRM1 as a genetic marker for placebo analgesia. bioRxiv. 2017:139345.

Aslaksen PM, Forsberg JT, Gjerstad J. The opioid receptor mu 1 (OPRM1) rs1799971 and catechol-O-methyltransferase (COMT) rs4680 as genetic markers for placebo analgesia. Pain. 2018;159(12):2585–92.

Attur M, Belitskaya-Levy I, Oh C, Krasnokutsky S, Greenberg J, Samuels J, et al. Increased interleukin-1beta gene expression in peripheral blood leukocytes is associated with increased pain and predicts risk for progression of symptomatic knee osteoarthritis. Arthritis Rheum. 2011;63(7):1908–17.

Benedetti F. The opposite effects of the opiate antagonist naloxone and the cholecystokinin antagonist proglumide on placebo analgesia. Pain. 1996;64(3):535–43.

Benedetti F, Colloca L, Torre E, Lanotte M, Melcarne A, Pesare M, et al. Placebo-responsive Parkinson patients show decreased activity in single neurons of subthalamic nucleus. Nat Neurosci. 2004;7(6):587–8.

Benedetti F, Pollo A, Colloca L. Opioid-mediated placebo responses boost pain endurance and physical performance: is it doping in sport competitions? J Neurosci. 2007;27(44):11934–9.

Benedetti F, Lanotte M, Colloca L, Ducati A, Zibetti M, Lopiano L. Electrophysiological properties of thalamic, subthalamic and nigral neurons during the anti-parkinsonian placebo response. J Physiol. 2009;587(Pt 15):3869–83.

Benedetti F, Amanzio M, Rosato R, Blanchard C. Nonopioid placebo analgesia is mediated by CB1 cannabinoid receptors. Nat Med. 2011;17(10):1228–30.

Benedetti F, Thoen W, Blanchard C, Vighetti S, Arduino C. Pain as a reward: changing the meaning of pain from negative to positive co-activates opioid and cannabinoid systems. Pain. 2013;154(3):361–7.

Bielsky IF, Hu SB, Young LJ. Sexual dimorphism in the vasopressin system: lack of an altered behavioral phenotype in female V1a receptor knockout mice. Behav Brain Res. 2005;164(1):132–6.

Bingel U, Lorenz J, Schoell E, Weiller C, Buchel C. Mechanisms of placebo analgesia: rACC recruitment of a subcortical antinociceptive network. Pain. 2006;120(1–2):8–15.

Born J, Lange T, Kern W, McGregor GP, Bickel U, Fehm HL. Sniffing neuropeptides: a transnasal approach to the human brain. Nat Neurosci. 2002;5(6):514–6.

Cajanus K, Holmstrom EJ, Wessman M, Anttila V, Kaunisto MA, Kalso E. Effect of endocannabinoid degradation on pain: role of FAAH polymorphisms in experimental and postoperative pain in women treated for breast cancer. Pain. 2016;157(2):361–9.

Carvalho C, Caetano JM, Cunha L, Rebouta P, Kaptchuk TJ, Kirsch I. Open-label placebo treatment in chronic low back pain: a randomized controlled trial. Pain. 2016;157(12):2766–72.

Colagiuri B, Schenk LA, Kessler MD, Dorsey SG, Colloca L. The placebo effect: from concepts to genes. Neuroscience. 2015;307:171–90.

Colloca L. Treatment of pediatric migraine. N Engl J Med. 2017;376(14):1387–8.

Colloca L. In: Colloca L, editor. Neurobiology of the placebo effect part I. 1st ed. Cambridge: Elsevier/Academic Press; 2018. 322 p.

Colloca L. The placebo effect in pain therapies. Annu Rev Pharmacol Toxicol. 2019;59:191–211.

Colloca L, Barsky AJ. Placebo and Nocebo Effects. N Engl J Med. 2020;382(6):554–61.

Colloca L, Benedetti F. Placebos and painkillers: is mind as real as matter? Nat Rev Neurosci. 2005;6(7):545–52.

Colloca L, Lopiano L, Lanotte M, Benedetti F. Overt versus covert treatment for pain, anxiety, and Parkinson's disease. Lancet Neurol. 2004;3(11):679–84.

Colloca L, Klinger R, Flor H, Bingel U. Placebo analgesia: psychological and neurobiological mechanisms. Pain. 2013;154(4):511–4.

Colloca L, Pine DS, Ernst M, Miller FG, Grillon C. Vasopressin boosts placebo analgesic effects in women: a randomized trial. Biol Psychiatry. 2016;79(10):794–802.

Craggs JG, Price DD, Perlstein WM, Verne GN, Robinson ME. The dynamic mechanisms of placebo induced analgesia: evidence of sustained and transient regional involvement. Pain. 2008;139(3):660–9.

de la Fuente-Fernandez R, Ruth TJ, Sossi V, Schulzer M, Calne DB, Stoessl AJ. Expectation and dopamine release: mechanism of the placebo effect in Parkinson's disease. Science. 2001;293(5532):1164–6.

Diatchenko L, Slade GD, Nackley AG, Bhalang K, Sigurdsson A, Belfer I, et al. Genetic basis for individual variations in pain perception and the development of a chronic pain condition. Hum Mol Genet. 2005;14(1):135–43.

Donaldson ZR, Young LJ. Oxytocin, vasopressin, and the neurogenetics of sociality. Science. 2008;322(5903):900–4.

Ebstein RP, Israel S, Lerer E, Uzefovsky F, Shalev I, Gritsenko I, et al. Arginine vasopressin and oxytocin modulate human social behavior. Ann N Y Acad Sci. 2009;1167:87–102.

Eippert F, Finsterbusch J, Bingel U, Buchel C. Direct evidence for spinal cord involvement in placebo analgesia. Science. 2009a;326(5951):404.

Eippert F, Bingel U, Schoell ED, Yacubian J, Klinger R, Lorenz J, et al. Activation of the opioidergic descending pain control system underlies placebo analgesia. Neuron. 2009b;63(4):533–43.

Eising E, Pelzer N, Vijfhuizen LS, Vries B, Ferrari MD, Hoen PA, et al. Identifying a gene expression signature of cluster headache in blood. Sci Rep. 2017;7:40218.

Ernst E, Resch KL. Concept of true and perceived placebo effects. BMJ. 1995;311(7004):551–3.

Faasse K, Martin LR, Grey A, Gamble G, Petrie KJ. Impact of brand or generic labeling on medication effectiveness and side effects. Health Psychol. 2016;35(2):187–90.

Feng C, Hackett PD, DeMarco AC, Chen X, Stair S, Haroon E, et al. Oxytocin and vasopressin effects on the neural response to social cooperation are modulated by sex in humans. Brain Imaging Behav. 2015;9(4):754–64.

Ferris CF, Melloni RH Jr, Koppel G, Perry KW, Fuller RW, Delville Y. Vasopressin/serotonin interactions in the anterior hypothalamus control aggressive behavior in golden hamsters. J Neurosci. 1997;17(11):4331–40.

Fields H. State-dependent opioid control of pain. Nat Rev Neurosci. 2004;5(7):565–75.

Forsberg JT, Gjerstad J, Flaten MA, Aslaksen PM. Influence of catechol-o-methyltransferase Val158Met on fear of pain and placebo analgesia. Pain. 2018;159:168–74.

Freeman SM, Walum H, Inoue K, Smith AL, Goodman MM, Bales KL, et al. Neuroanatomical distribution of oxytocin and vasopressin 1a receptors in the socially monogamous coppery titi monkey (Callicebus cupreus). Neuroscience. 2014;273:12–23.

Gerring ZF, Powell JE, Montgomery GW, Nyholt DR. Genome-wide analysis of blood gene expression in migraine implicates immune-inflammatory pathways. Cephalalgia. 2018;38(2):292–303.

Geuter S, Buchel C. Facilitation of pain in the human spinal cord by nocebo treatment. J Neurosci. 2013;33(34):13784–90.

Gobrogge KL, Liu Y, Young LJ, Wang Z. Anterior hypothalamic vasopressin regulates pair-bonding and drug-induced aggression in a monogamous rodent. Proc Natl Acad Sci U S A. 2009;106(45):19144–9.

Goodson JL, Thompson RR. Nonapeptide mechanisms of social cognition, behavior and species-specific social systems. Curr Opin Neurobiol. 2010;20(6):784–94.

Gupta A, Cole S, Labus JS, Joshi S, Nguyen TJ, Kilpatrick LA, et al. Gene expression profiles in peripheral blood mononuclear cells correlate with salience network activity in chronic visceral pain: a pilot study. Neurogastroenterol Motil. 2017;29(6):e13027.

Hall KT, Kaptchuk TJ. Genetic biomarkers of placebo response: what could it mean for future trial design? Clin Investig. 2013;3(4):311–4.

Hall KT, Lembo AJ, Kirsch I, Ziogas DC, Douaiher J, Jensen KB, et al. Catechol-O-Methyltransferase val158met polymorphism predicts placebo effect in irritable bowel syndrome. PLoS One. 2012;7(10):e48135.

Hall KT, Loscalzo J, Kaptchuk TJ. Genetics and the placebo effect: the placebome. Trends Mol Med. 2015a;21(5):285–94.

Hall KT, Tolkin BR, Chinn GM, Kirsch I, Kelley JM, Lembo AJ, et al. Conscientiousness is modified by genetic variation in catechol-o-methyltransferase to reduce symptom complaints in IBS patients. Brain Behavior. 2015b;5(1):39–44.

Hartvigsen J, Nielsen J, Kyvik KO, Fejer R, Vach W, Iachine I, et al. Heritability of spinal pain and consequences of spinal pain: a comprehensive genetic epidemiologic analysis using a population-based sample of 15,328 twins ages 20-71 years. Arthritis Rheum. 2009;61(10):1343–51.

Hashmi JA, Baria AT, Baliki MN, Huang L, Schnitzer TJ, Apkarian AV. Brain networks predicting placebo analgesia in a clinical trial for chronic back pain. Pain. 2012;153(12):2393–402.

Heinrichs M, Domes G. Neuropeptides and social behaviour: effects of oxytocin and vasopressin in humans. Prog Brain Res. 2008;170:337–50.

Heinrichs M, von Dawans B, Domes G. Oxytocin, vasopressin, and human social behavior. Front Neuroendocrinol. 2009;30(4):548–57.

Jarcho JM, Feier NA, Labus JS, Naliboff B, Smith SR, Hong JY, et al. Placebo analgesia: self-report measures and preliminary evidence of cortical dopamine release associated with placebo response. Neuroimage Clin. 2016;10:107–14.

Jones KD, Gelbart T, Whisenant TC, Waalen J, Mondala TS, Ikle DN, et al. Genome-wide expression profiling in the peripheral blood of patients with fibromyalgia. Clin Exp Rheumatol. 2016;34(2 Suppl 96):S89–98.

Kam-Hansen S, Jakubowski M, Kelley JM, Kirsch I, Hoaglin DC, Kaptchuk TJ, et al. Altered placebo and drug labeling changes the outcome of episodic migraine attacks. Sci Transl Med. 2014;6(218):218ra5.

Kaptchuk TJ, Kelley JM, Conboy LA, Davis RB, Kerr CE, Jacobson EE, et al. Components of placebo effect: randomised controlled trial in patients with irritable bowel syndrome. BMJ. 2008;336(7651):999–1003.

Kessner S, Sprenger C, Wrobel N, Wiech K, Bingel U. Effect of oxytocin on placebo analgesia: a randomized study. JAMA. 2013;310(16):1733–5.

Kogan A, Saslow LR, Impett EA, Oveis C, Keltner D, Saturn SR. Thin-slicing study of the oxytocin receptor (OXTR) gene and the evaluation and expression of the prosocial disposition. Proc Natl Acad Sci U S A. 2011;108(48):19189–92.

Kong J, Jensen K, Loiotile R, Cheetham A, Wey HY, Tan Y, et al. Functional connectivity of the frontoparietal network predicts cognitive modulation of pain. Pain. 2013;154(3):459–67.

Krummenacher P, Candia V, Folkers G, Schedlowski M, Schonbachler G. Prefrontal cortex modulates placebo analgesia. Pain. 2010;148(3):368–74.

Levine JD, Gordon NC. Influence of the method of drug administration on analgesic response. Nature. 1984;312(5996):755–6.

Levine JD, Gordon NC, Fields HL. The mechanism of placebo analgesia. Lancet. 1978;2(8091):654–7.

Lidstone SC, Schulzer M, Dinelle K, Mak E, Sossi V, Ruth TJ, et al. Effects of expectation on placebo-induced dopamine release in Parkinson disease. Arch Gen Psychiatry. 2010;67(8):857–65.

Lotta T, Vidgren J, Tilgmann C, Ulmanen I, Melen K, Julkunen I, et al. Kinetics of human soluble and membrane-bound catechol O-methyltransferase: a revised mechanism and description of the thermolabile variant of the enzyme. Biochemistry. 1995;34(13):4202–10.

Lui F, Colloca L, Duzzi D, Anchisi D, Benedetti F, Porro CA. Neural bases of conditioned placebo analgesia. Pain. 2010;151(3):816–24.

Lukkahatai N, Walitt B, Espina A, Wang D, Saligan LN. Comparing genomic profiles of women with and without fibromyalgia. Biol Res Nurs. 2015;17(4):373–83.

Mathie RT, Ramparsad N, Legg LA, Clausen J, Moss S, Davidson JR, et al. Randomised, double-blind, placebo-controlled trials of non-individualised homeopathic treatment: systematic review and meta-analysis. Syst Rev. 2017;6(1):63.

Meissner K, Bingel U, Colloca L, Wager TD, Watson A, Flaten MA. The placebo effect: advances from different methodological approaches. J Neurosci. 2011;31(45):16117–24.

Meissner K, Fassler M, Rucker G, Kleijnen J, Hrobjartsson A, Schneider A, et al. Differential effectiveness of placebo treatments: a systematic review of migraine prophylaxis. JAMA Intern Med. 2013;173(21):1941–51.

Mercado R, Constantoyannis C, Mandat T, Kumar A, Schulzer M, Stoessl AJ, et al. Expectation and the placebo effect in Parkinson's disease patients with subthalamic nucleus deep brain stimulation. Mov Disord. 2006;21(9):1457–61.

Moerman DE, Jonas WB. Deconstructing the placebo effect and finding the meaning response. Ann Intern Med. 2002;136(6):471–6.

Mogil JS. Are we getting anywhere in human pain genetics? Pain. 2009;146(3):231–2.

Mogil JS. Pain genetics: past, present and future. Trends Genet. 2012;28(6):258–66.

Nielsen CS, Stubhaug A, Price DD, Vassend O, Czajkowski N, Harris JR. Individual differences in pain sensitivity: genetic and environmental contributions. Pain. 2008;136(1–2):21–9.

Norbury TA, MacGregor AJ, Urwin J, Spector TD, McMahon SB. Heritability of responses to painful stimuli in women: a classical twin study. Brain. 2007;130(Pt 11):3041–9.

Oroszi G, Goldman D. Alcoholism: genes and mechanisms. Pharmacogenomics. 2004;5(8):1037–48.

Pecina M, Zubieta JK. Over a decade of neuroimaging studies of placebo analgesia in humans: what is next? Mol Psychiatry. 2015a;20(4):415.

Pecina M, Zubieta JK. Molecular mechanisms of placebo responses in humans. Mol Psychiatry. 2015b;20(4):416–23.

Pecina M, Martinez-Jauand M, Hodgkinson C, Stohler CS, Goldman D, Zubieta JK. FAAH selectively influences placebo effects. Mol Psychiatry. 2014;19(3):385–91.

Perry CJ, Blake P, Buettner C, Papavassiliou E, Schain AJ, Bhasin MK, et al. Upregulation of inflammatory gene transcripts in periosteum of chronic migraineurs: implications for extracranial origin of headache. Ann Neurol. 2016;79(6):1000–13.

Petersen GL, Finnerup NB, Norskov KN, Grosen K, Pilegaard HK, Benedetti F, et al. Placebo manipulations reduce hyperalgesia in neuropathic pain. Pain. 2012;153(6):1292–300.

Petersen GL, Finnerup NB, Grosen K, Pilegaard HK, Tracey I, Benedetti F, et al. Expectations and positive emotional feelings accompany reductions in ongoing and evoked neuropathic pain following placebo interventions. Pain. 2014;155(12):2687–98.

Presciuttini S, Curcio M, Sciarrino R, Scatena F, Jensen MP, Santarcangelo EL. Polymorphism of opioid receptors mu1 in highly hypnotizable subjects. Int J Clin Exp Hypn. 2018;66(1):106–18.

Schweinhardt P, Seminowicz DA, Jaeger E, Duncan GH, Bushnell MC. The anatomy of the mesolimbic reward system: a link between personality and the placebo analgesic response. J Neurosci. 2009;29(15):4882–7.

Scott DJ, Stohler CS, Egnatuk CM, Wang H, Koeppe RA, Zubieta JK. Placebo and nocebo effects are defined by opposite opioid and dopaminergic responses. Arch Gen Psychiatry. 2008;65(2):220–31.

Skyt I, Moslemi K, Baastrup C, Grosen K, Benedetti F, Petersen GL, et al. Dopaminergic tone does not influence pain levels during placebo interventions in patients with chronic neuropathic pain. Pain. 2017;159(2):261–72.

Starkweather AR, Heineman A, Storey S, Rubia G, Lyon DE, Greenspan J, et al. Methods to measure peripheral and central sensitization using quantitative sensory testing: a focus on individuals with low back pain. Appl Nurs Res. 2016;29:237–41.

Stein N, Sprenger C, Scholz J, Wiech K, Bingel U. White matter integrity of the descending pain modulatory system is associated with interindividual differences in placebo analgesia. Pain. 2012;153(11):2210–7.

Stoeber M, Jullie D, Lobingier BT, Laeremans T, Steyaert J, Schiller PW, et al. A genetically encoded biosensor reveals location bias of opioid drug action. Neuron. 2018;98(5):963–76.

Tetreault P, Mansour A, Vachon-Presseau E, Schnitzer TJ, Apkarian AV, Baliki MN. Brain connectivity predicts placebo response across chronic pain clinical trials. PLoS Biol. 2016;14(10):e1002570.

Thompson R, Gupta S, Miller K, Mills S, Orr S. The effects of vasopressin on human facial responses related to social communication. Psychoneuroendocrinology. 2004;29(1):35–48.

Thompson RR, George K, Walton JC, Orr SP, Benson J. Sex-specific influences of vasopressin on human social communication. Proc Natl Acad Sci U S A. 2006;103(20):7889–94.

Tuttle AH, Tohyama S, Ramsay T, Kimmelman J, Schweinhardt P, Bennett GJ, et al. Increasing placebo responses over time in U.S. clinical trials of neuropathic pain. Pain. 2015;156(12):2616–26.

Vachon-Presseau E, Berger SE, Abdullah TB, Huang L, Cecchi GA, Griffith JW, et al. Brain and psychological determinants of placebo pill response in chronic pain patients. Nat Commun. 2018;9(1):3397.

Vase L, Robinson ME, Verne GN, Price DD. The contributions of suggestion, desire, and expectation to placebo effects in irritable bowel syndrome patients. An empirical investigation. Pain. 2003;105(1–2):17–25.

Vase L, Robinson ME, Verne GN, Price DD. Increased placebo analgesia over time in irritable bowel syndrome (IBS) patients is associated with desire and expectation but not endogenous opioid mechanisms. Pain. 2005;115(3):338–47.

Vase L, Petersen GL, Riley JL 3rd, Price DD. Factors contributing to large analgesic effects in placebo mechanism studies conducted between 2002 and 2007. Pain. 2009;145(1–2):36–44.

Vase L, Petersen GL, Lund K. Placebo effects in idiopathic and neuropathic pain conditions. Handb Exp Pharmacol. 2014;225:121–36.

Vase L, Amanzio M, Price DD. Nocebo vs. placebo: the challenges of trial design in analgesia research. Clin Pharmacol Ther. 2015;97(2):143–50.

Vickers AJ, Cronin AM, Maschino AC, Lewith G, MacPherson H, Foster NE, et al. Acupuncture for chronic pain: individual patient data meta-analysis. Arch Intern Med. 2012;172(19):1444–53.

Waber RL, Shiv B, Carmon Z, Ariely D. Commercial features of placebo and therapeutic efficacy. JAMA. 2008;299(9):1016–7.

Wager TD, Rilling JK, Smith EE, Sokolik A, Casey KL, Davidson RJ, et al. Placebo-induced changes in FMRI in the anticipation and experience of pain. Science. 2004;303(5661):1162–7.

Wager TD, Scott DJ, Zubieta JK. Placebo effects on human mu-opioid activity during pain. Proc Natl Acad Sci U S A. 2007;104(26):11056–61.

Wang RS, Hall KT, Giulianini F, Passow D, Kaptchuk TJ, Loscalzo J. Network analysis of the genomic basis of the placebo effect. JCI Insight. 2017;2(11):93911.

Wei SY, Chen LF, Lin MW, Li WC, Low I, Yang CJ, et al. The OPRM1 A118G polymorphism modulates the descending pain modulatory system for individual pain experience in young women with primary dysmenorrhea. Sci Rep. 2017;7:39906.

Wrobel N, Wiech K, Forkmann K, Ritter C, Bingel U. Haloperidol blocks dorsal striatum activity but not analgesia in a placebo paradigm. Cortex. 2014;57:60–73.

Young LJ, Wang Z. The neurobiology of pair bonding. Nat Neurosci. 2004;7(10):1048–54.

Yu R, Gollub RL, Vangel M, Kaptchuk T, Smoller JW, Kong J. Placebo analgesia and reward processing: integrating genetics, personality, and intrinsic brain activity. Hum Brain Mapp. 2014;35(9):4583–93.

Zhang Y, Wang D, Johnson AD, Papp AC, Sadee W. Allelic expression imbalance of human mu opioid receptor (OPRM1) caused by variant A118G. J Biol Chem. 2005;280(38):32618–24.

Zubieta JK, Smith YR, Bueller JA, Xu Y, Kilbourn MR, Jewett DM, et al. Regional mu opioid receptor regulation of sensory and affective dimensions of pain. Science. 2001;293(5528):311–5.

Zubieta JK, Smith YR, Bueller JA, Xu Y, Kilbourn MR, Jewett DM, et al. Mu-opioid receptor-mediated antinociceptive responses differ in men and women. J Neurosci. 2002;22(12):5100–7.

Zubieta JK, Heitzeg MM, Smith YR, Bueller JA, Xu K, Xu Y, et al. COMT val158met genotype affects mu-opioid neurotransmitter responses to a pain stressor. Science. 2003;299(5610):1240–3.

Zubieta JK, Bueller JA, Jackson LR, Scott DJ, Xu Y, Koeppe RA, et al. Placebo effects mediated by endogenous opioid activity on mu-opioid receptors. J Neurosci. 2005;25(34):7754–62.

Zunhammer M, Bingel U, Wager TD, Placebo Imaging Consortium. Placebo effects on the neurologic pain signature: a meta-analysis of individual participant functional magnetic resonance imaging data. JAMA Neurol. 2018;75(11):1321–30.

Pain Analgesic Developments in the Genomic Era

16

Aaron Jesuthasan, Daniel Bullock, Rafael González-Cano, and Michael Costigan

Contents

16.1 Introduction

Attempts to understand, categorize, and manage pain have continued for millennia. To properly understand pain therapeutic progress, it is necessary to examine the evolution of both pain pathways and analgesia throughout documented history. Opioids, for example, have been widely used as analgesics due to their relatively strong efficacy across multiple conditions, though not without serious adverse effects including sedation, dizziness, nausea, constipation, tolerance, respiratory depression, and physical dependence (Schug et al. 1992). Indeed, the abuse liability and dangers of illicit opioid use are starkly illustrated by the ongoing opioid crisis (Barlas 2017). In addition, certain chronic pain conditions such as neuropathic pain

Aaron Jesuthasan and Daniel Bullock contributed equally.

A. Jesuthasan · D. Bullock · R. González-Cano · M. Costigan (✉)
Anesthesia Department, Boston Children's Hospital, Harvard Medical School,
Boston, MA, USA
e-mail: a.jesuthasan@nhs.net; daniel.bullock@childrens.harvard.edu; rgcano@ugr.es;
Michael.Costigan@childrens.harvard.edu

can be refractory to the analgesic effects of opioids. Worse, after long-term use, they can themselves cause pain (Afsharimani et al. 2015). Given the heterogeneity and diversity of pain-associated conditions, it seems unlikely for a single therapeutic to combat all chronic pain presentations.

Thus, research into the mechanisms underlying pain has expanded in recent decades, leading to better insight of the phenotypic intricacies of differing pain types and the molecular cascades that drive them. Understanding these molecular mechanisms will allow for the development of reliable diagnostic methods and therapeutic strategies for each unique pain condition. Ideally, novel tailored analgesics should be non-addictive and devoid of serious side effects. There remains much to learn about the debilitating conditions that comprise the overarching disease of chronic pain. Here, we provide a summary of the differing types of pain and describe associated chronic ailments. We then review specific neural signaling pathways implicated in normal and pathological conditions and discuss the evolution of analgesics with a focus on recent advances. We conclude by discussing modern genetic identification and editing techniques, which we believe will aid the production of effective analgesics.

16.2   Pain Subtypes

While many would consider pain to be a single entity, mechanistically it can be separated into four major subtypes: nociceptive (transient protective pain), inflammatory (pain due to peripheral tissue inflammation), nociplastic (chronic hypersensitivity with unclear mechanisms), and neuropathic (pain due to frank nerve damage) (Costigan et al. 2009). In some cases, these pain subtypes can be separated molecularly. For instance, ablating specific nociceptor nerve fibers expressing the tetrodotoxin (TTX)-resistant sodium channel Nav1.8 eliminates inflammatory pain but leaves neuropathic hypersensitivity intact, suggesting different sensory fibers contribute to each subtype (Abrahamsen et al. 2008). While the net effect of these different mechanisms is pain, their distinction is important to ensure treatments are appropriately targeted towards their individual pathways (Table 16.1).

Nociceptive pain is a reaction to acute and high intensity stimuli, evolved from the primitive nervous system as a protective mechanism (Woolf 2010). It functions to prevent tissue damage by initiating the withdrawal reflex, activating unpleasant sensations, producing emotional distress, and triggering avoidance behaviors to hinder future exposure to the harmful stimulus. This functional protective pain usually resolves following removal of the stimulus, although it may continue upon repeated or prolonged exposure. Furthermore, repetitive nociceptive stimulation alters all levels of pain processing. This leads to chronic pain, usually characterized as pain lasting or recurring for greater than 3–6 months (Treede et al. 2015). In the dorsal root ganglion (DRG) and sensory nerves, peripheral sensitization, ectopic neuronal activity, transcriptional changes, presynaptic modulation, and acquisition of new sensory modalities facilitate this transition (Berta et al. 2017). Altered gene and protein expression also occur in the spinal cord, resulting in central sensitization

Table 16.1 A physiological description of the nociceptive, inflammatory, nociplastic, and neuropathic subtypes of pain, with example stimuli

Pain subtype	Definition	Peripheral processing	Central processing	Acute example	Chronic example	Molecular mechanisms
Nociceptive	Reflexive reaction to noxious stimuli	Transient neural signaling; sensitization when chronic stimulus	Central sensitization when chronic stimulus	Stubbed foot; pin prick; twisted ankle	Bee sting; burn; capsaicin	Posttranslational
Inflammatory	Immune-modulated response to tissue injury or infection	Inflammatory mediators from blood and resident immune cells in injured tissue increase sensitization; enhanced presynaptic transmission	Immune cell recruitment and activation; central sensitization; central disinhibition	Transient infection	Rheumatoid arthritis; inflammatory bowel disease	Largely translational changes; some transient transcription (i.e., cFos); peripheral changes in chemokines/cytokines
Nociplastic	Pain without apparent nerve or tissue damage	Potentially impaired sensory processing; involvement of the autonomic nervous system	Amplified central signaling	Myofascial pain	Temporomandibular disorders; irritable bowel syndrome; fibromyalgia	Largely unknown—likely translational; altered chemokines/cytokines
Neuropathic	Direct lesions to peripheral and central nervous system	Drastic changes to injured neurons; immune cell recruitment into damaged nerves; altered satellite cells; hypersensitivity and maladaptive plasticity	Drastic changes to injured neurons; immune cell recruitment into damaged tissue; altered satellite cells; neuronal/glia cell death; hypersensitivity and maladaptive plasticity; central sensitization; central disinhibition	Transient low grade nerve damage after tissue damage (i.e., procedural pain)	Mechanical trauma; viral infection; chemical damage (i.e., anti-chemotherapeutic) cancer; stroke; multiple sclerosis	Transcription in injured neurons and immune cells (recruited and activated)

and increased signaling in projection pathways. Further, imaging studies demonstrate additional changes in pain-associated areas of the brain leading to pain hypersensitivity (Walitt et al. 2016; Tracey et al. 2019). The acute warning function of physical nociception is lost as a consequence of these pathophysiological changes (Basbaum et al. 2009), and here the aim is to prevent these changes by resetting the system to normality.

Inflammatory pain is triggered by tissue injury or infection, leading to immune cell recruitment. Several resident and migrating innate immune cells, including mast cells, neutrophils, and macrophages are activated and recruited to the injury site to release pro-inflammatory mediators (Lopes et al. 2017). Inflammatory factors are also released from the blood and any locally injured cells. These mediators include adenosine triphosphate, glutamate, protons, potassium, bradykinin, prostaglandin E2, proteases, cytokines, nerve growth factor, brain-derived neurotropic factor, glial cell-derived neurotropic factor, and insulin-like growth factor (Amaya et al. 2013). Together, these chemicals alter receptor expression in the sensory nerves and DRG to elevate pain sensitivity and promote tissue recovery and prevent further tissue injury (Lin et al. 2011). Changes at the level of the spinal dorsal horn facilitate pain hypersensitivity through enhancement of postsynaptic transmission, loss of local inhibitory input, and activation of microglia and astrocytes (Xu and Yaksh 2011). Hyperalgesia (amplification of an already painful stimulus) and allodynia (pain from a previously innocuous stimulus) consequently develop. Inflammatory pain usually subsides with resolution of the triggering injury, though it may persist in conditions characterized by chronic inflammation, such as rheumatoid arthritis or inflammatory bowel disease, often leading to organ dysfunction and potentially death (Lee 2013; Coates et al. 2013).

Dysfunctional or nociplastic pain occurs without the presence of a noxious stimulus, inflammatory process, or damage to the nervous system (Costigan et al. 2009). It has been defined as "pain arising from altered nociception despite no clear evidence of actual or threatened tissue damage causing the activation of peripheral nociceptors, or evidence for disease or lesion of the somatosensory system causing the pain" (IASP, 2017). This new definition addresses the underlying pathophysiology of the pain in greater depth (Treede 2018). Fibromyalgia, irritable bowel syndrome, and tension-type headache are conditions exemplifying this subtype (Nagakura 2015). Although the exact mechanisms that lead to nociplastic pain are unclear; amplified nociceptive signaling of the central nervous system, a disruption in the balance of central excitation and inhibition, and impaired sensory processing are considered key to disease etiology (Costigan et al. 2009).

Finally, neuropathic pain is the result of frank damage to the nervous system and is subdivided into peripheral and central components. Peripheral neuropathic pain results from lesions to the peripheral nervous system induced by several conditions, including metabolic diseases, infection, cancer, and mechanical trauma (Colloca et al. 2017). Central neuropathic pain involves lesions to the central nervous system, commonly caused by spinal cord injury, stroke, and multiple sclerosis (Watson and Sandroni 2016). The initial lesions involved in both forms trigger a cascade of changes leading to maladaptive plasticity within the nervous system and inevitably neuropathic hypersensitivity (Costigan et al. 2009). The maladaptive plasticity of neuropathic pain is now thought to be the predominant feature of this subtype, suggesting a mechanistic target for the development of novel treatments (Li et al. 2016).

Beyond the seriously debilitating effects of ongoing pain, the presence of multiple comorbidities including mental health and sleep conditions further complicate patient problems (Davis et al. 2011). It remains unclear if a causal relationship underpins each of these associations, however, there is strong evidence to suggest anxiety and depression are etiologically related to neuropathic pain (de Heer et al. 2014). Multiple mechanisms have been suggested for these links, including damaged noradrenergic and serotonergic systems leading to impaired descending inhibitory inputs to pain pathways (Kleiber et al. 2005). In addition, excessive glutamatergic, reduced GABAergic, and impaired dopaminergic activity have all been highlighted as central drivers of chronic pain and these co-morbid diseases (Han and Pae 2015). Whatever the nature of the link between chronic pain and mental health, recent research by our group suggests that for neuropathic pain and anxiety at least, this comorbidity is driven by ongoing painful input over the medium- to long-term rather than prior anxiety being a driver toward neuropathic hypersensitivity (Sieberg et al. 2018).

16.3   Molecular Mechanisms of Nociception and Chronic Pain

The experience of pain can be split into three main components: the physiological painful sensation, an emotional response, and an associated negative memory. An understanding of pain physiology is therefore necessary for proper management of symptoms (Fig. 16.1). Pain signaling commences with a sufficiently severe peripheral stimulus, which triggers action potentials (AP) within nociceptive sensory A-delta (Aδ) and C-fibers, sub-groups of primary afferent neurons. Aδ fibers act as high-threshold mechanoreceptors as well as noxious and non-noxious thermoreceptors. These lightly myelinated fibers are responsible for immediate, well-defined painful sensations through the rapid transmission of signals from the peripheral to central nervous system (CNS) (Khalid and Tubbs 2017). C-fibers also transmit pain sensation and are able to respond to multiple noxious stimuli (Sneddon 2018). These unmyelinated, slowly transmitting axons transduce the diffuse, prolonged "burning pain" that is present in inflammation. Additionally, myelinated, rapid-transmitting A-beta (Aβ) fibers normally act as cutaneous mechanoreceptors, but can start to elicit painful responses following injury (Zhu et al. 2012). Although polymodal nociceptive fibers are not particularly discriminative between individual types of stimuli (thermal, mechanical, or chemical), they express receptors on their peripheral nerve terminals that are specialized to identify distinct noxious substances, cytokines, chemokines, and other signaling molecules. These receptors transduce the initial stimulus into an electrical signal, which is subsequently transmitted along the nociceptive fibers to central synapses in the dorsal spinal cord or brainstem. Sensory neurons are pseudo-unipolar, meaning they consist of a single axon beginning in the periphery and terminating in the CNS. These axons send out

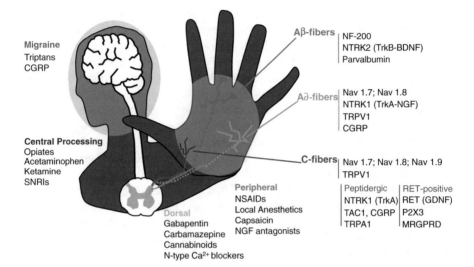

Fig. 16.1 A graphical abstract of common pain therapeutics and target sites in the human nervous system. Analgesic compounds targeting cranial, central, dorsal, and peripheral neurological sites are indicated on the left and bottom. Examples of prominent sensory fibers, subdivided into A-beta (Aβ), A-delta (Aδ), and C-fibers, are indicated on the right side

an additional branch to their cell bodies, which reside in the ganglia of the nerves. The upper and lower body are served by cranial and dorsal root ganglia, respectively. Excitatory and inhibitory interneurons within the dorsal horn further influence the electrical signals before they are relayed for perception to higher centers including the thalamus, periaqueductal gray, primary somatosensory cortex, insula, and other cortical areas (Xie et al. 2009).

Peripherally expressed transduction receptors include acid-sensing ion channels (ASICs), purinoceptors (the ionotropic P2RXs and G protein-coupled P2RYs), transient receptor potential channels (TRPs), Piezo channels, serotonin receptors, other G-protein coupled receptors (GPCRs) and chemokine, cytokine and growth factor receptors (Julius and Basbaum 2001). Activation of ionotropic transducers enables passage of cations and other signaling molecules into the neuron leading to its depolarization, termed a receptor potential. For GPCRs and other non-ionotropic receptors, signaling occurs via multiple well-studied intracellular cascades (Kannampalli and Sengupta 2015). A subset of GPCRs includes the peripheral cannabinoid receptor CB_2, which can notably reduce inflammatory, mechanical, and thermal nociception (Fine and Rosenfeld 2013). ASICs and purinoceptors, being activated by changes in extracellular pH and adenosine triphosphate levels, respectively, are considered particularly important in augmenting ongoing inflammatory hypersensitivity (Wemmie et al. 2013; Liang et al. 2010). Piezo channels, particularly Piezo2, discriminate light pressure (touch), vibration, and proprioception under normal circumstances (Szczot et al. 2017). However, Piezo2 signaling can be enhanced in the presence of inflammatory mediators via protein kinase

A- and C-dependent mechanisms, to facilitate noxious mechanotransduction (Dubin et al. 2012). The TRP channels, primarily TRPA1, TRPM3, TRPV1, TRPM2, and TRPM8 are calcium-permeable thermoreceptors, but can respond to multiple stimuli. TRPA1 is responsible for transducing many noxious stimuli, TRPM3 encodes responses to heat, and the capsaicin receptor TRPV1 is a key heat receptor in the higher to marginally noxious range (Gonzalez-Ramirez et al. 2017). Additionally, TRPM2 is a sensor of reactive oxygen species, and TRPM8 is responsible for transducing the response to ambient and noxious cold (Basso and Altier 2017).

A variety of modulators may influence sensory transduction to affect the threshold of signal onset and the level of response (AP frequency). For example, tissue damage leads to recruitment of inflammatory and non-inflammatory cells, including mast cells and macrophages. Mast cells release amines (serotonin and histamine) and arachidonic acid metabolites, while macrophages release cytokines (IL1, IL6, TNFα) and growth factors (nerve growth factor (NGF) and leukemia inhibitory factor). Bradykinin, other peptides, and multiple chemokines may also be released from recruited immune cells (Kidd and Urban 2001). These factors act to amplify sensory transduction, which usually manifests as hypersensitivity and sustained pain. Some of these inflammatory products may directly depolarize nociceptive fibers by acting on the aforementioned receptors, though most sensitize the nerve terminal through posttranslational modification of the receptors and channels. Additionally, many of these molecules induce transcriptional or translational changes in sensory neurons by activating kinase cascades to further enhance neuronal excitability (Miller et al. 2009; Kays et al. 2018).

Once stimuli are transduced into receptor potentials, several voltage-gated ion channels are activated, including voltage-gated sodium (VGSCs) and potassium channels (VGKCs). These influence AP generation and propagation to central synapses, often within the dorsal horn. Several VGSCs are thought to be involved, including TTX-sensitive Nav1.1, Nav1.6, and Nav1.7, and TTX-insensitive Nav1.8 and Nav1.9 channels (Basbaum et al. 2009). Nav1.7 and Nav1.8 are likely the predominant VGSCs responsible for AP generation in Aδ and C-fibers, with Nav1.9 acting to lower the threshold for AP generation (Dib-Hajj et al. 2010). Dysfunction of Nav1.7 as well as Nav1.8 and Nav1.9 is associated with several pain disorders (Emery et al. 2016).

VGKCs, on the other hand, promote potassium ion efflux to negatively regulate AP thresholds and firing rates (Tsantoulas and McMahon 2014). Furthermore, leak and calcium-activated potassium channels are also present to counteract depolarization. Certain potassium channel subunits and their genetic variation within individuals are thought to be integral to one's susceptibility to develop pain. The $K_v1.1$, K_v7, K_{2P}, and BK_{CA} subunits are shown to be particularly important, whereas genetic analyses implicate the *KCNS1*, *GIRKs*, and *TRESK* genes in various aspects of pain transmission (Tsantoulas 2015). Voltage-gated calcium channels (VGCCs), located at the nerve terminals, activate after membrane depolarization to promote intracellular calcium ion transport. Calcium ions entering through these channels and from intracellular stores act as second messengers for a range of processes, including neurotransmitter release into the synaptic cleft to augment pain transmission (Cain and Snutch 2011).

The dorsal horn is the first site of synaptic transfer in the nociceptive pathway, where terminating A delta and C fibers release glutamate to bind several postsynaptic receptors. These include N-methyl-D-aspartate receptors (NMDARs), α-amino-3-hydroxy-5-methyl-4-isoxazolepropionic acid (AMPA), and kainate receptors. NMDA receptors are ion channels, usually blocked at resting membrane potential by a magnesium ion sitting within their pores. Following glutamatergic activation of AMPA receptors, sodium, potassium, and calcium ions travel into the dorsal horn cell to elicit depolarization and fast excitatory postsynaptic potentials, which signal information regarding the peripheral noxious stimuli. More intense nociceptor activation prompts further peptide and neuromodulator release to generate sustained depolarization and slow synaptic potentials.

Furthermore, with damage, C-fibers can fire spontaneously, causing an augmented release of neurotransmitters and neuromodulators from the presynaptic terminal, which act on various receptors to initiate several signal transduction pathways. This leads to depolarization-induced removal of the magnesium ion from the NMDAR pore to allow glutamate binding, calcium ion influx, and subsequent cell depolarization. NMDAR phosphorylation also occurs to facilitate long-lasting changes in the excitability of the dorsal horn neurons. Posttranslational modification of NMDARs causes further excitability changes and enables relief from the magnesium ion block without the need for cell depolarization. Consequently, subthreshold inputs may then excite postsynaptic neurons. Brief C-fiber inputs additionally lead to rapid excitability changes, manifesting as a cumulative excitability increase lasting several hours, termed central sensitization (Costigan and Woolf 2000). This is attributed to posttranslational changes in the dorsal horn, which allows affected neurons to recall previous inputs and remain sensitive to them hours later. This phenomenon is thought to underlie the processing of allodynia (Latremoliere and Woolf 2009).

16.4   The Origin of Pain

Nociceptive properties, including sensation to noxious as well as innocuous stimuli, can be traced across species and time to better understand conserved pathways and channels. Early radial animals including Cnidaria and Ctenophora demonstrate a net-shaped neural system capable of somatosensation. As coelomate bilateria evolved, they developed a central and peripheral nervous system with diverse nociceptive cell types (Sneddon 2018; Smith and Lewin 2009). While an exhaustive examination of nociceptive evolution in all non-mammal classes is impractical, many milestones have been observed with available data. Invertebrates including *Hirudo medicinalis* (leeches) have cells capable of mechanical, acid, and thermal sensation, including capsaicin sensitivity, though with a higher activation threshold than mammals. Other species, such as *Aplysia californica* (sea slugs), possess ventrocaudal neurons capable of sensing a range of mechanical pressures, similar to mammalian spinal wide dynamic range neurons (Smith and Lewin 2009).

Nematodes including *Caenorhabditis elegans* also respond to noxious mechanical and thermal stimuli. These and other more primitive species possess opioid receptors, demonstrating decreased nociceptive signaling after opioid administration (Sneddon 2018). Arthropods including *Drosophila melanogaster* are nocifensive against noxious heat and touch, possessing the *painless* gene, a homolog of mammalian TRPA1 (Smith and Lewin 2009), as well as TRPA1 itself (Neely et al. 2010; Kang et al. 2010) and multiple other receptors and cascades that transmit noxious sensation (Neely et al. 2010). It has been recently demonstrated that *Drosophila* feel allodynia following peripheral nerve injury in a process that involves the disruption of central GABA signaling (Khuong et al. 2019). While many homologs of mammalian nociceptors exist across species, environmental conditions can alter their specific properties. Invertebrates including cephalopods demonstrate mechanoreceptor activity, but are not sensitive to temperatures that would induce noxious sensitivity in mammals, likely due to their lack of exposure to these temperature ranges (Smith and Lewin 2009). Lower vertebrates such as Elasmobranchii have a low number of unmyelinated neurons and may not strongly detect noxious stimuli. Others including the *Oncorhynchus mykiss* (rainbow trout) have a balance of unmyelinated to myelinated neurons, with noxious thermal, mechanical, and acid sensitivity which is greater than mammals and responds well to analgesics (Sneddon 2018). They also have limited C-fibers, with Aδ-fibers assuming some of these roles (Smith and Lewin 2009). Reptiles and birds have developed nervous systems more comparable to mammals, capable of noxious and non-noxious responses. As in mammals, μ, δ, and κ opioid receptors are present and function in CNS pain processing (Sneddon 2018).

Specific conserved nociceptive molecules can also be examined across species. Mammalian TRPV4, employed in mechanical hyperalgesia, is an orthologue of OSM-9 in *C. elegans*. While *C. elegans* employs MEC4 for mechanosensation, this role is encoded by Piezo2 in mammals. Modulators of ASICs are found in both mice and *C. elegans* (Smith and Lewin 2009). For noxious heat, sensitivity varies significantly between species, and these varied mechanisms are likely to be due to multiple heat-sensing channels (Sneddon 2018). Research into noxious cold has identified TRPM8 as the predominant cold sensing channel (Blanquart et al. 2019). The ability to respond to noxious pH changes can be partially attributed to ASICs and TRPV1 tracing back to invertebrates, although other channels play a role here. Finally, electrical signaling demonstrates some inconsistencies; for example, *C. elegans* lacks sodium channels, which are prominent in AP generation in higher animals. Studying the evolution and function of identified receptors and ion channels has led to a better understanding of nociception and ongoing pain, revealing several new molecular targets for the development of therapeutics.

For instance, TRP channels were discovered in *Drosophila* in the 1960s through experiments aimed at understanding the perception of light (Cosens and Manning 1969), followed by further elaboration of this channel family using fly cloning methods in the 1980s (Montell and Rubin 1989). Thus, it is hoped that modern analgesics will be able to achieve better efficacy on targeted mechanisms that are shared between experimental animals and patients (Latremoliere and Costigan 2018). However, it is important to remember that divergent signaling systems may, and often do, lead to marked differences between species. Continuous cross-checking between laboratory and patient-based studies is therefore essential to prevent these differences from significantly damaging translation of results.

16.5   Analgesic Development

16.5.1 Empirical Analgesics: Historical Advances

Significant research has been dedicated to the identification and development of analgesic treatments since the 1960s (Kissin 2010) (Table 16.2). For many years, treatment was facilitated through opioids and non-steroidal anti-inflammatory drugs (NSAIDs), although alcohol was also frequently used for this purpose. Now largely forgotten, alcohol use for pain relief strongly points towards GABAergic mechanisms being an effective analgesic target (Davies 2003). The GABAergic system is also thought to be the molecular target of many general anesthetics used today (Forman and Miller 2016). Analgesic research has largely mirrored drug efficacy and general use, although the predominance of work into opioids and NSAIDs has given way somewhat recently to newer therapies aimed at treating less tractable conditions such as neuropathic and nociplastic pain (Fig. 16.2).

Opioids are derived from poppy seed heads and have been used empirically to treat pain for millennia. This led to the extraction of morphine from opium in 1803 (Rosenblum et al. 2008). While the addictive and withdrawal effects of morphine were exposed during the nineteenth century, opiate peptides were crucially identified as endogenous ligands for opioid receptors in the 1970s (Hughes et al. 1975). Investigations into anatomical locations and pharmacological profiles of opioids led to the proposal of several opioid receptor subtypes. This was later proven, as many stereospecific opiate binding sites showed a non-uniform distribution across the central nervous system (Brownstein 1993). This subsequently led to investigation into the structure–activity relationships of peptides for the opioid receptors. Developing agonists that create the analgesic effects of opioid receptor activation without the adverse CNS effects of euphoria, dependence, and peripheral issues including constipation is a very exciting avenue in current molecular pain research (Aldrich and McLaughlin 2012; Noble and Marie 2018).

A variety of morphine analogues such as buprenorphine, and substitutes including fentanyl, have been used for chronic pain relief in recent decades. Buprenorphine binds the μ-opioid receptor with a lower affinity than morphine and has been used extensively to help combat opioid withdrawal (Herring et al. 2019). Fentanyl chiefly binds the μ-opioid receptor and is a more potent analgesic than morphine (Ramos-Matos and Lopez-Ojeda 2019). As with other opioids, the addictive nature of fentanyl continues to be a concern, and it is often illegally distributed in mixture with heroin or extracted from transdermal patches (Kuczynska et al. 2018). Due to the

Table 16.2 Analgesic classes and therapeutic examples are presented, alongside their mechanisms of action and targeted channels

Drug class	Drug examples	Pain type	Target	Mechanism	PubMed reference ID
Opioids	Morphine, buprenorphine, fentanyl; tramadol	Nociceptive, inflammatory, post-operative, neuropathic	OPRM1	Decreased peripheral and central sensitization	30616926; 23509136
NSAIDs	Aspirin, ibuprofen; celecoxib, rofecoxib	Nociceptive, inflammatory, post-operative	PTGS1; PTGS2	Decreased inflammatory action	8454631; 29796239
Local anesthetics	Lidocaine; saxitoxin, neosaxitoxin; procaine	Nociceptive, inflammatory, neuropathic (peripheral)	Voltage-gated sodium channels (VGSCs)	Inhibition of VGSCs to decrease action potential propagation	21041957; 29146176; 25192265
Sodium channel modulators	Tetrodotoxin, saxitoxin, neosaxitoxin; carbamazepine; QX314	Nociceptive, inflammatory	VGSCs	Inhibition of VGSCs for nerve block of action potential propagation	29146176; 21249671; 21457220
Cannabinoids	Dronabinol	Neuropathic, inflammatory, cancer	CNR1, CNR2	Agonist of cannabinoid receptors	17828291
NMDA antagonists	Ketamine	Post-operative	NMDA receptor	Postsynaptic inhibition of receptor	27965560
Calcium channel modulators	Ziconotide	Neuropathic, nociplastic, cancer	N-type voltage-gated calcium channels	VGCC inhibition to decrease neurotransmitter release	21882460; 29440821
Neuromodulators	Tanezumab; capsaicin	Nociceptive; inflammatory	Nerve growth factor (NGF); TRPV1	Binds NGF to prevent receptor activation; overstimulation and desensitization of TRPV1	28967370, 31026504; 21852280
Anticonvulsants	Carbamazepine; gabapentin, pregabalin	Neuropathic, nociplastic	VGSCs; CACNA2D1	Inhibition of VGSC and VGCC to decrease AP propagation	21249671; 21882460
Antidepressants	Amitriptyline; duloxetine	Neuropathic; nociplastic	VGSCs; serotonin transporter (SERT)	Serotonin and noradrenaline reuptake inhibitors	23235657; 30769838
Triptans	Sumatriptan	Migraine	Serotonin receptors	Receptor agonist, reduction of CGRP release	18471110, 23786565
Antimigraine	Fremanezumab, galcanezumab, erenumab; olcegepant	Migraine	CGRP and CGRP receptor	Bind circulating CGRP to prevent receptor activation; inhibits receptor directly for symptomatic relief	30725283; 30242830

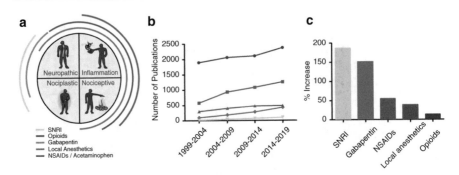

Fig. 16.2 A visual representation of each chronic pain type is illustrated. Effects of therapies including serotonin and norepinephrine reuptake inhibitors (SNRIs), opioids, gabapentin, local anesthetics, non-steroidal anti-inflammatory drugs (NSAIDs), and acetaminophen are indicated (**a**). A timeline of publications discussing each of these therapeutics is presented for mouse and rat studies. The number of publications is indicated on the y-axis; 5-year intervals are indicated on the x-axis (**b**). Percent increase (y-axis) in these publications is indicated in (**c**). For (**b**) and (**c**), data was obtained using PubMed searches of the respective analgesic categories

high risks associated with overdose, careful monitoring of individual pain conditions and methods to decrease illicit access to opiates are imperative (Preuss et al. 2019; Ciccarone 2017). It is suggested tramadol, an agent with mixed opioid, serotonergic and noradrenergic effects, may be better positioned for clinical use, due to possessing a lower risk for addiction. However, this is contingent on the individual's genotype (Young and Juurlink 2013).

As alternatives to opioids, NSAID-related treatments have additionally progressed since the discovery and isolation of salicin from willow bark in the early eighteenth century (Rao et al. 2010). Non-selective cyclooxygenase (COX) inhibition to inhibit prostaglandin (PG) synthesis was later defined as key to the efficacy of several popular NSAID medications, including aspirin and ibuprofen (Meade et al. 1993). The identification of the second COX isoform (COX-2) followed in the early 1990s, allowing the development of modern alternative inhibitors. Celecoxib and rofecoxib are two such drugs that came about from the identification of COX-2, designed by Monsanto and Merck, respectively, to specifically block the active site of COX-2 while sparing COX-1 to provide more targeted therapy (Rao et al. 2010; Hawkey 2005; Beales 2018). These selective compounds escape the gastrointestinal adverse effects commonly exhibited by the traditional NSAIDs. However, normal renal, reproductive, and athero-protective functions rely upon COX-2 action, meaning its selective inhibition may predispose individuals to renal ischemia, complicated pregnancy, heart attacks, and strokes (Zarghi and Arfaei 2011).

Additional non-opioid analgesics introduced in the late 1800s include acetanilide and phenacetin. Concerns of toxicity later identified their metabolite paracetamol (acetaminophen) as a safe and effective antipyretic and analgesic agent, though without the anti-inflammatory effects of NSAIDs (Toussaint et al. 2010; Candido et al. 2017). While NSAID mechanisms are largely understood, the likely multifaceted effects of paracetamol remain unclear. Multiple lines of evidence suggest that at least some of its analgesic effect is through inhibition of PG synthesis through blockage of central COX-dependent pathways (Graham and Scott 2005). It is difficult to isolate a direct target of paracetamol; thus, it is possible for multiple mechanisms to be functioning at several sites, perhaps synergistically. Paracetamol also demonstrates a synergistic effect when used alongside NSAIDs (Toussaint et al. 2010). While therapeutic dosing of paracetamol may be less toxic than opioids or NSAIDs in higher doses, it is minimally effective for osteoarthritis and ineffective for lower back pain (Borenstein et al. 2017). Furthermore, high doses of paracetamol lead to hepatotoxicity and oxidative stress-induced brain toxicity (Ghanem et al. 2016).

Coca leaf extracts have been employed for pain relief for thousands of years in Central and South America, later attributed to their active ingredient cocaine. The numbing effects of cocaine point to its role as a VGSC blocker, and development of cocaine derivatives has resulted in routinely used local anesthetics, such as lidocaine (Burgess and Williams 2010). Electrophysiology, particularly with voltage-clamp techniques, has since enabled further characterization of VGSCs into different subtypes (Bennett et al. 2019). The use of TTX and saxitoxin (STX) provides insight into the individual pore selectivity of these VGSCs. Additionally, neosaxitoxin, a recently developed analogue of STX, shows superior potency to TTX, STX, and other STX analogues for long-lasting nerve blocks (Riquelme et al. 2018). Moreover, human genetic linkage studies and extensive mouse knockout data have implicated Nav1.7 as a particularly important VGSC for the AP generation and signal conveyance in peripheral neurons that are involved in pain production. Much ongoing research has focused on developing selective inhibitors of this channel, with a drug of sufficient efficacy continuing to be a challenge to construct. However, recent advances in research technology will continue to improve our understanding of Nav1.7, hopefully leading to potential analgesic therapies with more efficient targeting (McDermott et al. 2019). Additionally, VGSCs including Nav1.8 and Nav1.9 have now been linked to congenic pain syndromes; hence, inhibitors of these channels also represent a potential avenue for novel analgesic drug development (Burgess and Williams 2010; Bullock et al. 2019).

Triptans were specifically developed through research efforts for treating migraine and, as such, represent a major success in rational analgesic design (Kissin 2010). The launch of the first triptan, sumatriptan, stemmed from the identification of serotonin for migraine cessation in the 1960s and subsequent investigation into the therapeutic potential of serotonin receptors (Humphrey 2008). Sumatriptan has been revolutionary for migraine relief, thought to act via a combination of constricting distended cranial vessels and inhibiting trigeminal afferent neurons (Humphrey 2008). Ongoing molecular studies examining the affinity of sumatriptan to different serotonin receptor subtypes will provide a better understanding of its precise mechanism. Additionally, increases in our understanding of triptans have yielded better routes of administration, formulations, and efficacy while limiting vascular side effects (Macone and Perloff 2017).

Migraine research has also revealed modulators of calcitonin gene-related peptide (CGRP) as a potential drug therapy. CGRP and its receptors are shown to be distributed in sensory neurons of the trigeminovascular system (Benemei et al. 2009), expressed in approximately half of neurons projecting from the trigeminal ganglion (Edvinsson 2019). Upon activation of trigeminal sensory fibers, CGRP acts as a potent vasodilator of cerebral arteries and arterioles. It significantly contributes to the protective trigeminovascular reflex, in which it is released by trigeminal nerves to combat local vasoconstriction and regulate cerebral blood flow, while also mediating pain signal transmission (Hargreaves and Olesen 2019). This mechanism is thought to be central to migraine pathophysiology, as CGRP is selectively released by trigeminal nerves, and elevated levels are additionally found within the cranial circulation during migraine episodes (Benemei et al. 2009).

It is also suggested that prolonged activation of trigeminal pathways by CGRP may be important for the transformation of episodic to chronic migraine (Hargreaves and Olesen 2019).

CGRP therapies include biologic inhibitors of circulating CGRP, such as fremanezumab and galcanezumab, and antibodies that block CGRP receptors, such as erenumab (Tepper 2018). The first CGRP receptor antagonist, olcegepant, is quite effective, but cannot be given orally (Edvinsson 2019). More recently, small molecule CGRP receptor antagonists, such as ubrogepant, atogepant, and rimegepant are currently in final stages of clinical trials and are effective for acute relief of migraine symptoms rather than prophylactic care. Additionally, studies suggest that CGRP modulators may also be applicable to non-headache pain conditions, including those characterized by somatic, visceral, neuropathic, and inflammatory pain (Schou et al. 2017).

Capsaicin, a well-known TRPV1 agonist, was isolated in the early nineteenth century from paprika and cayenne. Its chili plant origin, capsicum, has been understood and documented since Mesoamerican times (Yang and Zheng 2017; Powis et al. 2013), with the use of capsaicin as a pain treatment recognized since at least 1850 (Anand and Bley 2011). More recently, low-dose capsaicin creams (0.025–0.1%) have become widely available, without prescription, in many countries. The use of a TRPV1 agonist as an analgesic in the first instance seems paradoxical; however, it is proposed that capsaicin calms nociceptive input through overstimulation of TRPV1 channels leading to sensory cell exhaustion (Pecze et al. 2013).

16.5.2 Targeted Analgesics: Modern Successes and Challenges

The cloning of the receptors and signaling cascade effectors that encode sensation has sponsored a shift toward developing more targeted analgesics in the last half century. Areas of focus include TRPV1 agents, cannabinoids, N-type calcium channel blockers, NMDA antagonists, gabapentin, repurposed antidepressants, nerve growth factor, fatty acid amide hydrolase inhibition, substance P (SP) receptor antagonists, Piezo2 channel inhibition, tetrahydrobiopterin (BH4) inhibition, purinergic signaling, and catechol-O-methyltransferase (COMT) studies. In common with the examples above, many of these signaling systems have been used empirically for years. However, rational drug design based on these molecular mechanisms has only been possible recently.

Outside of capsaicin cream, research into TRPV1 therapies has produced multiple specific antagonists. However, progress has been hindered due to unexpected on-target effects on centrally expressed TRPV1 receptors, leading to conditions of altered core body temperature including hypothermia (Garami et al. 2018). Another analgesic approach employing TRPV1 has been recently devised using VGSC blocking agents, such as QX314. Such agents can specifically target TRPV1-expressing sensory neurons by using their non-selective pores as a convenient entry port. Such compounds can therefore specifically inhibit VGSCs of nociceptive neurons, while sparing other sensory and motor axons (Roberson et al. 2011). As QX314 remains trapped within the sensory neuron for some time, these approaches can lead to a long-lasting sensory block (Lim et al. 2007).

Cannabinoids, such as dronabinol (Pertwee 2008), were first isolated in the late 1800s from cannabis, which has been used for medicinal purposes for at least 5000 years (Pertwee 2006). Cannabinoids have since demonstrated analgesic effects in several animal models of pain, particularly acting agonistically on CB_1 and CB_2 receptors to alleviate neuropathic, inflammatory, and cancer pain subtypes (Elikkottil et al. 2009). Clinical studies have additionally demonstrated their utility against chronic pain uncontrolled by other medications (Tawfik et al. 2019), and further work may point towards an ability to treat side effects of opioid therapy. However, the potential for abuse and misuse of cannabinoids, with side effects of compromised immunity, potential cardiovascular hyperactivity, and impaired cognition hinder their wider clinical use and more extensive research into analgesic roles (Tawfik et al. 2019). With recent medical and recreational legalization in the USA and other countries, close policy regulation and therapeutic roles will undoubtedly be further explored (Tawfik et al. 2019; Pacula and Smart 2017).

N-type calcium channel blockers stemmed from the identification of peptides found in marine Conus snail venom in 1984, which demonstrated inhibition of hyperalgesia. This was later mechanistically determined to be due to inhibition of a VGCC, leading to development of calcium channel-based analgesics (Vink and Alewood 2012). One example, ziconotide is now routinely used to treat chronic pain from cancerous and non-cancerous causes (Nair et al. 2018). This compound is a very effective analgesic, although studies have demonstrated a limited penetration of this drug across the blood–brain barrier, and thus it is administered intrathecally and only after other drugs are deemed ineffective. Additionally, associated risks of syncope, confusion, and meningitis require novel targeted methods to improve overall efficacy and safety (Nair et al. 2018).

The NMDA antagonist, ketamine, arose in the 1960s after the anesthetic phencyclidine was found to induce severe delirium making it unsuitable for human medical use (Lodge and Mercier 2015). Researchers subsequently sought to develop shorter-acting analogues of phencyclidine with similar anesthetic potential. CI-581, later named ketamine, was found to demonstrate anesthetic effects while inducing profound analgesia. Its clinical utility as a strong analgesic and sedative have now expanded, in part due to minimal adverse respiratory effects. Ketamine is regarded especially useful as an adjunct for controlling post-operative pain more safely than opioid therapies (Li and Vlisides 2016).

Many analgesics stem from drugs initially approved for non-pain related indications. For example, anticonvulsant drugs including carbamazepine have been used for pain management since the 1960s after they first revolutionized the management of epilepsy (Wiffen et al. 2011). While exact analgesic mechanisms most likely vary between individual anticonvulsants, with carbamazepine, for instance, blocking VGSCs, these drugs are shown to be effective for managing neuropathic pain. Gabapentin and pregabalin represent two of the more recent anticonvulsants, adapted for pain management. Gabapentin, initially developed as a GABA analogue to target epilepsy, later demonstrated a potential use in treatment for refractory pain (Chaplan et al. 2010). Preclinical investigations and clinical trials then supported its analgesic use, leading to official approval for treating pain associated with postherpetic neuralgia in 2002. Gabapentin's success warranted investigation into the analgesic efficacy of further GABA analogues, and pregabalin was later identified for therapeutic use in neuropathic pain. Later, the mechanistic target of gabapentin was revealed as the α_2/δ_1 subunit of VGCCs (Field et al. 2006) which is heavily upregulated in damaged sensory neurons (Luo et al. 2001).

Tricyclic antidepressants (TCAs) and serotonin–norepinephrine reuptake inhibitors (SNRIs), such as amitriptyline and duloxetine, respectively, are also effective anti-neuropathic agents (Sansone and Sansone 2008). Amitriptyline was constructed in 1961 to treat depression. It has since been used off-label and, in some cases, is regarded as the first-line treatment for neuropathic pain (Ramachandraih et al. 2011; Moore et al. 2012; Thour and Marwaha 2019), among other pain indications including fibromyalgia (Rico-Villademoros et al. 2015). Similarly, duloxetine was initially developed by Eli Lilly and Company as an antidepressant (Pitsikas 2000), gaining FDA approval in 2004 for use in major depressive disorder, and becoming licensed for used in diabetic neuropathic pain later that year (Yasuda et al. 2011; Wright et al. 2010). Analgesic effects of TCAs are suggested to be independent of their antidepressant action, and, therefore, pain relief can be possible at doses lower than for their antidepressant functions (Sansone and Sansone 2008). However, TCA use has been associated with several side effects, including orthostatic hypotension and cardiovascular complications, making SNRIs a more suitable antidepressant for patient use. In contrast, analgesic SNRI doses resemble those for treatment of depression (Onutu 2015). SNRIs increase noradrenaline levels in the spinal cord by blocking its reuptake, thus leading to noradrenaline accumulation and inhibition of chronic pain transmission through α_2-adrenergic receptors (Hayashida and Obata 2019).

Recent compounds developed specifically for analgesia include drugs targeting nerve growth factor and its receptor, tropomyosin receptor kinase A (TrkA, now NTRK1). NGF is produced following nociceptor activation by noxious stimuli to enhance pain signaling. It binds and complexes with NTRK1, which is endocytosed and transported to cell bodies to modulate the expression of several pain-promoting receptors and channels. Additionally, this complex induces posttranslational modification of TRPV1 leading to increased receptor activity, central sensitization, mast cell activation to produce histamine and additional NGF, and sensitization of adjacent nociceptive neurons to inflammation. Multiple drugs have been designed to hinder NGF effects, including those which eliminate free NGF such as tanezumab (Miller et al. 2017) (NGF-capturing agents), prevent NGF binding to its receptor (NGF receptor antagonists), and inhibit NTRK1 pathways (NTRK1 antagonists) (Shang et al. 2017). Interestingly, there are also congenital diseases resulting in profoundly altered pain thresholds due to mutations of the NTRK1 receptor and its ligands (Franco et al. 2016). In this case, however, the disease is caused by selective death of the neurons in development, which require this neurotrophic signaling cascade to reach maturity (Indo 2018).

Fatty acid amide hydrolase (FAAH), a significant enzyme in the metabolism of endocannabinoids, has received attention as a potential analgesic target. Reversible or irreversible FAAH inhibition leads to accumulation of endocannabinoids including anandamide, which bind CB_1 and CB_2 to decrease pain and anxiety without the impairing effects of exogenous cannabinoids (Ahn et al. 2009a, b; Mallet et al. 2016). Unfortunately, clinical trials have been met with challenges. The inhibitor BIA10-2474 was shelved in 2016 after a high dose caused serious rapid adverse neurological events in five participants and led to the death of another (Mallet et al. 2016). While it was later demonstrated that this was likely due to multiple off-target neurological effects on similar serine hydrolases and not FAAH inhibition (Huang et al. 2019), many other FAAH inhibitors were shelved after this event (Mallet et al. 2016). Clinical trials of the irreversible FAAH inhibitor PF-04457845 failed to reduce pain associated with osteoarthritis (Di Marzo 2012), though it has not yet been examined in models of neuropathic pain. A highly selective irreversible FAAH inhibitor, PF-3845, has demonstrated promise in mouse models of inflammatory pain (Ahn et al. 2009b), though this is currently being further optimized in other pain types (Bhuniya et al. 2019). Despite these obstacles, FAAH inhibition continues to remain relevant. Recently, a 66-year-old woman was reportedly unable to feel pain or anxiety, and had an ability to rapidly heal from injury. She demonstrated co-inheritance of a hypomorphic *FAAH* SNP and a nearby microdeletion which deactivates a pseudogene termed *FAAH-OUT*. Her pain insensitivity identified *FAAH-OUT* as a potential pain target (Habib et al. 2019).

Similarly, the neuropeptide substance P has been a difficult analgesic target. First isolated in 1931 from equine intestine and brain, and later sequenced in 1971, SP is a neurokinin produced by the *TAC1* gene and is part of the tachykinin family (Schank and Heilig 2017; Douglas and Leeman 2011). It chiefly binds the Neurokinin 1 receptor (NK1R) in the spinal cord and peripheral sites, promoting smooth muscle contraction, stress responses, anxiety, addiction, neurogenic inflammation, and pain signaling (Schank and Heilig 2017). NK1R binding induces NF-κB-mediated pro-inflammatory signaling in the CNS and PNS (Douglas and Leeman 2011), and also heightens postsynaptic glutamatergic pain propagation (Schank and Heilig 2017). Chronic pain and inflammatory conditions, including osteoarthritis and rheumatoid arthritis, demonstrate positive correlation with increased serum SP concentration. Additionally, pain intensity is positively correlated with SP concentration and NK1R density in these conditions (Lisowska et al. 2015). This data appears to identify the NK1R as a significant pain target. However, aside from aprepitant, an antagonist marketed for emesis and nausea, clinical SP therapies have been largely unsuccessful since the 1990s (Schank and Heilig 2017). Antagonists such as RP-67580 have been effective in some laboratory studies (Rupniak et al. 1993), but ineffective in others (Wallace et al. 1998). Possible explanations include that these inhibitors may not occupy sufficient NK1 receptors needed to successfully combat stress, pain, and inflammation (Schank and Heilig 2017). Alternatively, recent research supports the theory that the NK1R is not chiefly responsible for these painful and inflammatory responses, but that the Mas-related G-protein-coupled receptor (MrgprX2), existing solely on mast cells, serves this role. Further work should help elucidate whether novel analgesics targeting the MrgprX2 receptor can effectively mitigate SP's contribution to chronic pain and inflammation (Navratilova and Porreca 2019).

A further exciting avenue of therapeutic research seeks to exploit the Piezo2 channels and their role in physiological and noxious mechanotransduction. There is subsequent potential for developing drugs which target and inhibit this channel to combat mechanical allodynia. A recent agent termed D-GsMTx4 is suggested to reversibly and dose-dependently inhibit Piezo2 channel currents in response to mechanical force (Alcaino et al. 2017).

The identification of abnormally elevated peripheral BH4 levels in patients with low back pain, supported by animal models of chronic pain in which the somatosensory system is injured, additionally reveals BH4 synthesis as a novel therapeutic target (Latremoliere and Costigan 2018). BH4 serves as a co-factor for the production of several neurotransmitters and signaling molecules, including serotonin, dopamine, epinephrine, norepinephrine, and nitric oxide. It is likely that BH4 is associated with chronic pain through the excessive production of these chemicals (Costigan et al. 2012), although recent research into the role of BH4 in energy metabolism also opens this mechanism up to its action in chronic pain (Cronin et al. 2018). A proposed approach of targeting BH4 synthesis is inhibiting its rate-limiting enzyme, GTP cyclohydrolase 1 (GCH1). However, this would excessively deplete BH4 levels, thus impairing normal physiological synthesis of its downstream products (Latremoliere and Costigan 2018). A safer proposed avenue would be to inhibit the enzyme responsible for the final stage of the BH4 synthesis pathway, sepiapterin reductase, which does not completely impair physiological BH4 production and would allow sepiapterin to reliably serve as a biomarker of drug efficacy (Cronin et al. 2018).

The purinergic system, first proposed in 1972, has also been a significant focus for pain research (Burnstock 2017). While its P2X and P2Y receptors are found on many sensory neurons, it was the realization that P2X3 and P2X2/3 receptors are expressed predominantly within small nociceptive DRG sensory neurons that led to the system's association with the initiation of pain (Burnstock 2013). This supported known pivotal roles of ATP in the production of sympathetic, vascular-related, and tumor-induced pain. The importance of P2X receptors in neuropathic pain has been further validated by the correlation of their activation with allodynia in mice (Fukuhara et al. 2000). The P2X4, P2X7, and $P2Y_{12}$ receptors, found on spinal cord microglia, were more recently shown to be key to this association (Tsuda et al. 2003, 2010; Jarvis 2010). P2Y receptors are additionally suggested to modulate pain transmission, potentiating pain via TRPV1 channels. Furthermore, purinergic mechanosensory transduction in visceral, cutaneous, and musculoskeletal nociception has been detailed elsewhere (Burnstock 2016).

Reduced activity of COMT, an enzyme responsible for the metabolism of catecholamines, L-dopa, catechol estrogens, ascorbic acid, and dihydroxyindolic intermediates of melanin, correlates with increased pain sensitivity in experimental models and human studies (Kambur and Mannisto 2010; Nackley et al. 2007). This is supported by the pro-nociceptive effects of COMT inhibitors in animal studies, as well as associations between COMT polymorphisms and certain pain conditions, such as fibromyalgia and chronic widespread pain (Tammimaki and Mannisto 2012). Low COMT activity has additionally been shown to increase opioid receptor expression and enhance the effects of opioid therapy. However, the relationship between COMT and pain is complex, with neuropathic pain models paradoxically demonstrating nitecapone, a COMT inhibitor, to be anti-allodynic. This is probably due to an elaborate interplay between adrenergic and dopaminergic pathways at different portions of the nociceptive system, alongside compensatory changes in other neurotransmitters (Kambur and Mannisto 2010). Studies of the receptors involved in COMT signaling have revealed that β_2-and β_3- adrenergic receptors are integral to the elevated pain sensitivity produced by low COMT activity (Nackley et al. 2007). Recent evidence suggests that this association is due to the synthesis of nitric oxide and pro-inflammatory cytokines following stimulation of these receptors, although further studies are warranted before this can translate to potential analgesic targets (Hartung et al. 2014).

16.5.3 Molecular Pain Genetics: New Analgesic Targets

As novel data-driven approaches to genetic pain research continue to develop (Bullock et al. 2019), modern targeted therapies will ideally integrate these resources. Techniques such as genome-wide association studies (GWAS) have been employed to interpret genomic data from large patient populations, with the goal of identifying new pain-associated genes (Manolio 2010). This has been successful in determining potential clinical targets in many conditions, including headache (Meng et al. 2018) and back pain (Lemmela et al. 2016).

These targeted approaches to medicine can also allow for more personalized treatments. An invaluable tool for promoting this has been genomic sequencing, a technique which has rapidly evolved since its inception in the later twentieth century. While sequencing in mammals was initially focused more on characterizing genes associated with clear disease phenotypes, it has since become more rapid and versatile, with next-generation genome sequencing leading to better understandings of therapeutic targets (Koboldt et al. 2013; Esplin et al. 2014). Sequencing has been employed in cases of patients with rare congenital pain conditions including congenital insensitivity to pain (CIP). This condition is characterized by insensitivity to noxious temperature, acute pain, or inflammatory pain. In one study, genomes of several families with autosomal recessive CIP were sequenced and screened, identifying a role of the epigenetic transcriptional regulator *PRDM12*. Results demonstrated mostly homozygous missense mutations in conserved regions of *PRDM12*, which altered its functionality. It was found that *PRDM12* is largely localized in DRG neurons early in neurogenesis. Mechanistically, mutations which can lead to *PRDM12* aggregation lead to a decreased ability to regulate neurogenesis. Thus, an observed alteration in peripheral sensory C-fiber and Aδ-fiber signaling in some patients is likely due to its mutation (Chen et al. 2015).

Recent studies in mice revealed that PRDM12 co-localizes with NTRK1 during development, and that knockout of PRDM12 also prevents the development of NTRK1-expressing cells, without affecting the development of non-nociceptive mechanoreceptors (Bartesaghi et al. 2019). In humans, *PRDM12* has also been found to promote NTRK1 expression. Thus, these genetic techniques have determined a crucial and necessary role for *PRDM12* in initiation of neurogenesis across evolution (Desiderio et al. 2019). It has been demonstrated that, in general, patients presenting with CIP often present with mutations in NTRK1 (Altassan et al. 2017; Wang et al. 2018). This information clearly demonstrates NGF-NTRK1 signaling as a focal point in our understanding of pain disorders and analgesia. However, one familial study of an autosomal dominant pain insensitivity due to a point mutation in ZFHX2 revealed no direct correlation with the NGF-NTRK1 axis, suggesting an unrelated mechanism to this hypoalgesia (Habib et al. 2018).

Finally, methods of genomic targeting such as RNA silencing and clustered regularly interspaced short palindromic repeats (CRISPR)/Cas9 genome editing have evolved significantly over recent years, which will surely prove useful in analgesia. RNA silencing allows temporary modification of gene expression, including through gene knockdown, and has been recently employed in pain studies. Examples include helping to elucidate a pain-promoting role of the NLRP2 inflammasome in mouse DRG (Matsuoka et al. 2019), and knockdown of TRPV1 to alleviate mechanically and thermally associated bone cancer pain in rats (Zhang et al. 2019). Additionally, the more recently developed CRISPR/Cas9 genome editing system allows direct manipulation of a target genome to alter or insert DNA sequences and produces conditions such as gene knockout or knock-in, as we have previously detailed (Bullock et al. 2019). It has been employed in in vitro and in vivo *Drosophila* models to develop libraries capable of targeting thousands of genes to better understand gene correlation to phenotype (Bassett et al. 2015; Viswanatha et al. 2018; Meltzer et al. 2019). In pain studies, the CRISPR/Cas9 tool has been employed in ablation studies of osteoarthritis-associated pain in mice (Zhao et al. 2019) and was used in human cell lines to illustrate that the chronic pain-promoting FAM173B methyltransferase is necessary for ATP production (Malecki et al. 2019). Finally, induced pluripotent stem cell nociceptors have been used as a screening platform to model predicted pathogenic variants, with CRISPR/Cas9 being able to correct these variants. This allows researchers to causally link such variants with phenotypes, as has been demonstrated in recent Nav1.7 studies (McDermott et al. 2019). These novel tools can therefore further our understanding of ion channel variants, while additionally providing opportunities to target pain-associated genes (Sun et al. 2016).

16.6 Conclusion

The techniques used to develop analgesics and improve the utility of existing drugs have progressed significantly over time, aided by major advances in research methods and drug screening technology. However, there remain large gaps in our knowledge, which hinders further analgesic development. Future work in targeted analgesic discovery will likely interrogate specific genes known to be selectively expressed in at least one part of the pain pathway, with targets expressed specifically in the PNS at a premium to limit potential side effects (Costigan and Woolf 2000). Mechanisms that could be targeted include inhibiting peripheral sensitization, ectopic discharges, sympathetically maintained pain, central sensitization, and general pain transmission (Costigan et al. 2009; Baron et al. 2010). Inflammation and anti-inflammatory therapies should clearly remain a major focus in modern analgesic research, with the new appreciation of the close symbiosis of sensory neural and immune actions at the forefront to better understand relevant targets (Reardon et al. 2018; Chavan et al. 2017). One major unmet milestone is defining and developing drugs to act on the mechanisms at play in nociplastic pain, although the high rates of comorbid anxiety with this pain subtype suggest common mechanisms, as is the case with neuropathic pain (Zhuo 2016). Novel scientific methods, including those interrogated by molecular pain genetics and new computing algorithms to deal with complex gene-gene interactions and the large amounts of data to be processed (Capriotti et al. 2019; de Los Campos et al. 2018) should provide a solution to several of these questions, enabling construction of superior analgesics.

Acknowledgments None.

Conflict of Interest The authors declare that this study received funding from Amgen. The funder was not involved in the study design, collection, analysis, interpretation of data, the writing of this article, or the decision to submit it for publication.

Funding

We thank Amgen for funding. RG-C was supported by the Alfonso Martín Escudero Fellowship.

Abbreviations

AMPA	α-amino-3-hydroxy-5-methyl-4-isoxazolepropionic acid
AP	Action potentials
ASICs	Acid-sensing ion channels
Aδ	A-delta
BH4	Tetrahydrobiopterin

CGRP	Calcitonin gene-related peptide
CIP	Congenital insensitivity to pain
CNS	Central nervous system
COMT	Catechol-O-methyltransferase
COX	Cyclooxygenase
CRISPR	Clustered regularly interspaced short palindromic repeats
DRG	Dorsal root ganglion
FAAH	Fatty acid amide hydrolase
GCH1	GTP cyclohydrolase 1
GPCRs	G-protein coupled receptors
GWAS	Genome-wide association studies
MrgprX2	Mas-related G-protein-coupled receptor
NGF	Nerve growth factor
NK1R	Neurokinin 1 receptor
NMDARs	N-methyl-D-aspartate receptors
NSAIDs	Non-steroidal anti-inflammatory drugs
NTRK1	Tropomyosin receptor kinase A
PG	Prostaglandin
SNRIs	Serotonin–norepinephrine reuptake inhibitors
SP	Substance P
STX	Saxitoxin
TCAs	Tricyclic antidepressants
TRPs	Transient receptor potential channels
TTX	Tetrodotoxin
VGCCs	Voltage-gated calcium channels
VGKCs	Voltage-gated potassium channels
VGSCs	Voltage-gated sodium channels

References

Abrahamsen B, Zhao J, Asante CO, Cendan CM, Marsh S, Martinez-Barbera JP, et al. The cell and molecular basis of mechanical, cold, and inflammatory pain. Science. 2008;321(5889):702–5.

Afsharimani B, Kindl K, Good P, Hardy J. Pharmacological options for the management of refractory cancer pain-what is the evidence? Support Care Cancer. 2015;23(5):1473–81.

Ahn K, Johnson DS, Cravatt BF. Fatty acid amide hydrolase as a potential therapeutic target for the treatment of pain and CNS disorders. Expert Opin Drug Discovery. 2009a;4(7): 763–84.

Ahn K, Johnson DS, Mileni M, Beidler D, Long JZ, McKinney MK, et al. Discovery and characterization of a highly selective FAAH inhibitor that reduces inflammatory pain. Chem Biol. 2009b;16(4):411–20.

Alcaino C, Knutson K, Gottlieb PA, Farrugia G, Beyder A. Mechanosensitive ion channel Piezo2 is inhibited by D-GsMTx4. Channels. 2017;11(3):245–53.

Aldrich JV, McLaughlin JP. Opioid peptides: potential for drug development. Drug Discov Today Technol. 2012;9(1):e23–31.

Altassan R, Saud HA, Masoodi TA, Dosssari HA, Khalifa O, Al-Zaidan H, et al. Exome sequencing identifies novel NTRK1 mutations in patients with HSAN-IV phenotype. Am J Med Genet A. 2017;173(4):1009–16.

Amaya F, Izumi Y, Matsuda M, Sasaki M. Tissue injury and related mediators of pain exacerbation. Curr Neuropharmacol. 2013;11(6):592–7.

Anand P, Bley K. Topical capsaicin for pain management: therapeutic potential and mechanisms of action of the new high-concentration capsaicin 8% patch. Br J Anaesth. 2011;107(4):490–502.

Barlas S. U.S. and states ramp up response to opioid crisis: regulatory, legislative, and legal tools brought to bear. P T. 2017;42(9):569–92.

Baron R, Binder A, Wasner G. Neuropathic pain: diagnosis, pathophysiological mechanisms, and treatment. Lancet Neurol. 2010;9(8):807–19.

Bartesaghi L, Wang Y, Fontanet P, Wanderoy S, Berger F, Wu H, et al. PRDM12 is required for initiation of the nociceptive neuron lineage during neurogenesis. Cell Rep. 2019;26(13):3484–92. e4.

Basbaum AI, Bautista DM, Scherrer G, Julius D. Cellular and molecular mechanisms of pain. Cell. 2009;139(2):267–84.

Bassett AR, Kong L, Liu JL. A genome-wide CRISPR library for high-throughput genetic screening in Drosophila cells. J Genet Genomics. 2015;42(6):301–9.

Basso L, Altier C. Transient receptor potential channels in neuropathic pain. Curr Opin Pharmacol. 2017;32:9–15.

Beales ILP. Time to reappraise the therapeutic place of celecoxib. Ther Adv Chronic Dis. 2018;9(5):107–10.

Benemei S, Nicoletti P, Capone JG, Geppetti P. CGRP receptors in the control of pain and inflammation. Curr Opin Pharmacol. 2009;9(1):9–14.

Bennett DL, Clark AJ, Huang J, Waxman SG, Dib-Hajj SD. The role of voltage-gated sodium channels in pain signaling. Physiol Rev. 2019;99(2):1079–151. https://doi.org/10.1152/physrev.00052.2017.

Berta T, Qadri Y, Tan PH, Ji RR. Targeting dorsal root ganglia and primary sensory neurons for the treatment of chronic pain. Expert Opin Ther Targets. 2017;21(7):695–703.

Bhuniya D, Kharul RK, Hajare A, Shaikh N, Bhosale S, Balwe S, et al. Discovery and evaluation of novel FAAH inhibitors in neuropathic pain model. Bioorg Med Chem Lett. 2019;29(2):238–43.

Blanquart S, Borowiec AS, Delcourt P, Figeac M, Emerling CA, Meseguer AS, et al. Evolution of the human cold/menthol receptor, TRPM8. Mol Phylogenet Evol. 2019;136:104–18.

Borenstein DG, Hassett AL, Pisetsky D. Pain management in rheumatology research, training, and practice. Clin Exp Rheumatol. 2017;35(Suppl 107):S2–7.

Brownstein MJ. A brief history of opiates, opioid peptides, and opioid receptors. Proc Natl Acad Sci U S A. 1993;90(12):5391–3.

Bullock D, Jesuthasan A, Gonzalez-Cano R, Costigan M. Reading and writing: the evolution of molecular pain genetics. Pain. 2019;160:2177.

Burgess G, Williams D. The discovery and development of analgesics: new mechanisms, new modalities. J Clin Invest. 2010;120(11):3753–9.

Burnstock G. Purinergic mechanisms and pain--an update. Eur J Pharmacol. 2013;716(1–3):24–40.

Burnstock G. Purinergic mechanisms and pain. Adv Pharmacol. 2016;75:91–137.

Burnstock G. Introduction to the special issue on purinergic receptors. Adv Exp Med Biol. 2017;1051:1–6.

Cain SM, Snutch TP. Voltage-gated calcium channels and disease. Biofactors. 2011;37(3):197–205.

Candido KD, Perozo OJ, Knezevic NN. Pharmacology of acetaminophen, nonsteroidal antiinflammatory drugs, and steroid medications: implications for anesthesia or unique associated risks. Anesthesiol Clin. 2017;35(2):e145–e62.

Capriotti E, Ozturk K, Carter H. Integrating molecular networks with genetic variant interpretation for precision medicine. Wiley Interdiscip Rev Syst Biol Med. 2019;11(3):e1443.

Chaplan SR, Eckert IW, Carruthers NI. Drug discovery and development for pain. In: Kruger L, Light AR, editors. Translational pain research: from mouse to man. Frontiers in neuroscience. Boca Raton: CRC Press; 2010.

Chavan SS, Pavlov VA, Tracey KJ. Mechanisms and therapeutic relevance of neuro-immune communication. Immunity. 2017;46(6):927–42.

Chen YC, Auer-Grumbach M, Matsukawa S, Zitzelsberger M, Themistocleous AC, Strom TM, et al. Transcriptional regulator PRDM12 is essential for human pain perception. Nat Genet. 2015;47(7):803–8.

Ciccarone D. Fentanyl in the US heroin supply: a rapidly changing risk environment. Int J Drug Policy. 2017;46:107–11.

Coates MD, Lahoti M, Binion DG, Szigethy EM, Regueiro MD, Bielefeldt K. Abdominal pain in ulcerative colitis. Inflamm Bowel Dis. 2013;19(10):2207–14.

Colloca L, Ludman T, Bouhassira D, Baron R, Dickenson AH, Yarnitsky D, et al. Neuropathic pain. Nat Rev Dis Primers. 2017;3:17002.

Cosens DJ, Manning A. Abnormal electroretinogram from a Drosophila mutant. Nature. 1969;224(5216):285–7.

Costigan M, Woolf CJ. Pain: molecular mechanisms. J Pain. 2000;1(3 Suppl):35–44.

Costigan M, Scholz J, Woolf CJ. Neuropathic pain: a maladaptive response of the nervous system to damage. Annu Rev Neurosci. 2009;32:1–32.

Costigan M, Latremoliere A, Woolf CJ. Analgesia by inhibiting tetrahydrobiopterin synthesis. Curr Opin Pharmacol. 2012;12(1):92–9.

Cronin SJF, Seehus C, Weidinger A, Talbot S, Reissig S, Seifert M, et al. The metabolite BH4 controls T cell proliferation in autoimmunity and cancer. Nature. 2018;563(7732):564–8.

Davies M. The role of GABAA receptors in mediating the effects of alcohol in the central nervous system. J Psychiatry Neurosci. 2003;28(4):263–74.

Davis JA, Robinson RL, Le TK, Xie J. Incidence and impact of pain conditions and comorbid illnesses. J Pain Res. 2011;4:331–45.

de Heer EW, Gerrits MM, Beekman AT, Dekker J, van Marwijk HW, de Waal MW, et al. The association of depression and anxiety with pain: a study from NESDA. PLoS One. 2014;9(10):e106907.

de Los Campos G, Vazquez AI, Hsu S, Lello L. Complex-trait prediction in the era of big data. Trends Genet. 2018;34(10):746–54.

Desiderio S, Vermeiren S, Van Campenhout C, Kricha S, Malki E, Richts S, et al. Prdm12 directs nociceptive sensory neuron development by regulating the expression of the NGF receptor TrkA. Cell Rep. 2019;26(13):3522–36. e5.

Di Marzo V. Inhibitors of endocannabinoid breakdown for pain: not so FA(AH)cile, after all. Pain. 2012;153(9):1785–6.

Dib-Hajj SD, Cummins TR, Black JA, Waxman SG. Sodium channels in normal and pathological pain. Annu Rev Neurosci. 2010;33:325–47.

Douglas SD, Leeman SE. Neurokinin-1 receptor: functional significance in the immune system in reference to selected infections and inflammation. Ann N Y Acad Sci. 2011;1217:83–95.

Dubin AE, Schmidt M, Mathur J, Petrus MJ, Xiao B, Coste B, et al. Inflammatory signals enhance piezo2-mediated mechanosensitive currents. Cell Rep. 2012;2(3):511–7.

Edvinsson L. Role of CGRP in migraine. Handb Exp Pharmacol. 2019;255:121.

Elikkottil J, Gupta P, Gupta K. The analgesic potential of cannabinoids. J Opioid Manag. 2009;5(6):341–57.

Emery EC, Luiz AP, Wood JN. Nav1.7 and other voltage-gated sodium channels as drug targets for pain relief. Expert Opin Ther Targets. 2016;20(8):975–83.

Esplin ED, Oei L, Snyder MP. Personalized sequencing and the future of medicine: discovery, diagnosis and defeat of disease. Pharmacogenomics. 2014;15(14):1771–90.

Field MJ, Cox PJ, Stott E, Melrose H, Offord J, Su TZ, et al. Identification of the alpha2-delta-1 subunit of voltage-dependent calcium channels as a molecular target for pain mediating the analgesic actions of pregabalin. Proc Natl Acad Sci U S A. 2006;103(46):17537–42.

Fine PG, Rosenfeld MJ. The endocannabinoid system, cannabinoids, and pain. Rambam Maimonides Med J. 2013;4(4):e0022.

Forman SA, Miller KW. Mapping general anesthetic sites in heteromeric gamma-aminobutyric acid type a receptors reveals a potential for targeting receptor subtypes. Anesth Analg. 2016;123(5):1263–73.

Franco ML, Melero C, Sarasola E, Acebo P, Luque A, Calatayud-Baselga I, et al. Mutations in TrkA causing congenital insensitivity to pain with anhidrosis (CIPA) induce misfolding, aggregation, and mutation-dependent neurodegeneration by dysfunction of the autophagic flux. J Biol Chem. 2016;291(41):21363–74.

Fukuhara N, Imai Y, Sakakibara A, Morita K, Kitayama S, Tanne K, et al. Regulation of the development of allodynia by intrathecally administered P2 purinoceptor agonists and antagonists in mice. Neurosci Lett. 2000;292(1):25–8.

Garami A, Pakai E, McDonald HA, Reilly RM, Gomtsyan A, Corrigan JJ, et al. TRPV1 antagonists that cause hypothermia, instead of hyperthermia, in rodents: compounds' pharmacological profiles, in vivo targets, thermoeffectors recruited and implications for drug development. Acta Physiol. 2018;223(3):e13038.

Ghanem CI, Perez MJ, Manautou JE, Mottino AD. Acetaminophen from liver to brain: new insights into drug pharmacological action and toxicity. Pharmacol Res. 2016;109:119–31.

Gonzalez-Ramirez R, Chen Y, Liedtke WB, Morales-Lazaro SL. TRP channels and pain. In: TLR E, editor. Neurobiology of TRP channels. Frontiers in neuroscience. 2nd ed. Boca Raton: CRC Press; 2017. p. 125–47.

Graham GG, Scott KF. Mechanism of action of paracetamol. Am J Ther. 2005;12(1):46–55.

Habib AM, Matsuyama A, Okorokov AL, Santana-Varela S, Bras JT, Aloisi AM, et al. A novel human pain insensitivity disorder caused by a point mutation in ZFHX2. Brain. 2018;141(2):365–76.

Habib AM, Okorokov AL, Hill MN, Bras JT, Lee MC, Li S, et al. Microdeletion in a FAAH pseudogene identified in a patient with high anandamide concentrations and pain insensitivity. Br J Anaesth. 2019;123(2):e249–e53.

Han C, Pae CU. Pain and depression: a neurobiological perspective of their relationship. Psychiatry Investig. 2015;12(1):1–8.

Hargreaves R, Olesen J. Calcitonin gene-related peptide modulators - the history and renaissance of a new migraine drug class. Headache. 2019;59(6):951–70.

Hartung JE, Ciszek BP, Nackley AG. beta2- and beta3-adrenergic receptors drive COMT-dependent pain by increasing production of nitric oxide and cytokines. Pain. 2014;155(7):1346–55.

Hawkey CJ. COX-2 chronology. Gut. 2005;54(11):1509–14.

Hayashida KI, Obata H. Strategies to treat chronic pain and strengthen impaired descending noradrenergic inhibitory system. Int J Mol Sci. 2019;20(4):822.

Herring AA, Perrone J, Nelson LS. Managing opioid withdrawal in the emergency department with buprenorphine. Ann Emerg Med. 2019;73(5):481–7.

Huang Z, Ogasawara D, Seneviratne UI, Cognetta AB 3rd, Am Ende CW, Nason DM, et al. Global portrait of protein targets of metabolites of the neurotoxic compound BIA 10-2474. ACS Chem Biol. 2019;14(2):192–7.

Hughes J, Smith TW, Kosterlitz HW, Fothergill LA, Morgan BA, Morris HR. Identification of two related pentapeptides from the brain with potent opiate agonist activity. Nature. 1975;258(5536):577–80.

Humphrey PP. The discovery and development of the triptans, a major therapeutic breakthrough. Headache. 2008;48(5):685–7.

Indo Y. NGF-dependent neurons and neurobiology of emotions and feelings: lessons from congenital insensitivity to pain with anhidrosis. Neurosci Biobehav Rev. 2018;87:1–16.

Jarvis MF. The neural-glial purinergic receptor ensemble in chronic pain states. Trends Neurosci. 2010;33(1):48–57.

Julius D, Basbaum AI. Molecular mechanisms of nociception. Nature. 2001;413(6852):203–10.

Kambur O, Mannisto PT. Catechol-O-methyltransferase and pain. Int Rev Neurobiol. 2010;95:227–79.

Kang K, Pulver SR, Panzano VC, Chang EC, Griffith LC, Theobald DL, et al. Analysis of Drosophila TRPA1 reveals an ancient origin for human chemical nociception. Nature. 2010;464(7288):597–600.

Kannampalli P, Sengupta JN. Role of principal ionotropic and metabotropic receptors in visceral pain. J Neurogastroenterol Motil. 2015;21(2):147–58.

Kays J, Zhang YH, Khorodova A, Strichartz G, Nicol GD. Peripheral synthesis of an atypical protein kinase C mediates the enhancement of excitability and the development of mechanical hyperalgesia produced by nerve growth factor. Neuroscience. 2018;371:420–32.

Khalid S, Tubbs RS. Neuroanatomy and neuropsychology of pain. Cureus. 2017;9(10):e1754.

Khuong TM, Wang QP, Manion J, Oyston LJ, Lau MT, Towler H, et al. Nerve injury drives a heightened state of vigilance and neuropathic sensitization in Drosophila. Sci Adv. 2019;5(7):eaaw4099.

Kidd BL, Urban LA. Mechanisms of inflammatory pain. Br J Anaesth. 2001;87(1):3–11.

Kissin I. The development of new analgesics over the past 50 years: a lack of real breakthrough drugs. Anesth Analg. 2010;110(3):780–9.

Kleiber B, Jain S, Trivedi MH. Depression and pain: implications for symptomatic presentation and pharmacological treatments. Psychiatry. 2005;2(5):12–8.

Koboldt DC, Steinberg KM, Larson DE, Wilson RK, Mardis ER. The next-generation sequencing revolution and its impact on genomics. Cell. 2013;155(1):27–38.

Kuczynska K, Grzonkowski P, Kacprzak L, Zawilska JB. Abuse of fentanyl: an emerging problem to face. Forensic Sci Int. 2018;289:207–14.

Latremoliere A, Costigan M. Combining human and rodent genetics to identify new analgesics. Neurosci Bull. 2018;34(1):143–55.

Latremoliere A, Woolf CJ. Central sensitization: a generator of pain hypersensitivity by central neural plasticity. J Pain. 2009;10(9):895–926.

Lee YC. Effect and treatment of chronic pain in inflammatory arthritis. Curr Rheumatol Rep. 2013;15(1):300.

Lemmela S, Solovieva S, Shiri R, Benner C, Heliovaara M, Kettunen J, et al. Genome-wide meta-analysis of sciatica in Finnish population. PLoS One. 2016;11(10):e0163877.

Li L, Vlisides PE. Ketamine: 50 years of modulating the mind. Front Hum Neurosci. 2016; 10:612.

Li XY, Wan Y, Tang SJ, Guan Y, Wei F, Ma D. Maladaptive plasticity and neuropathic pain. Neural Plast. 2016;2016:4842159.

Liang S, Xu C, Li G, Gao Y. P2X receptors and modulation of pain transmission: focus on effects of drugs and compounds used in traditional Chinese medicine. Neurochem Int. 2010;57(7):705–12.

Lim TK, Macleod BA, Ries CR, Schwarz SK. The quaternary lidocaine derivative, QX-314, produces long-lasting local anesthesia in animal models in vivo. Anesthesiology. 2007;107(2):305–11.

Lin YT, Ro LS, Wang HL, Chen JC. Up-regulation of dorsal root ganglia BDNF and trkB receptor in inflammatory pain: an in vivo and in vitro study. J Neuroinflammation. 2011;8:126.

Lisowska B, Lisowski A, Siewruk K. Substance P and chronic pain in patients with chronic inflammation of connective tissue. PLoS One. 2015;10(10):e0139206.

Lodge D, Mercier MS. Ketamine and phencyclidine: the good, the bad and the unexpected. Br J Pharmacol. 2015;172(17):4254–76.

Lopes DM, Denk F, Chisholm KI, Suddason T, Durrieux C, Thakur M, et al. Peripheral inflammatory pain sensitisation is independent of mast cell activation in male mice. Pain. 2017;158(7):1314–22.

Luo ZD, Chaplan SR, Higuera ES, Sorkin LS, Stauderman KA, Williams ME, et al. Upregulation of dorsal root ganglion (alpha)2(delta) calcium channel subunit and its correlation with allodynia in spinal nerve-injured rats. J Neurosci. 2001;21(6):1868–75.

Macone AE, Perloff MD. Triptans and migraine: advances in use, administration, formulation, and development. Expert Opin Pharmacother. 2017;18(4):387–97.

Malecki JM, Willemen H, Pinto R, Ho AYY, Moen A, Kjonstad IF, et al. Lysine methylation by the mitochondrial methyltransferase FAM173B optimizes the function of mitochondrial ATP synthase. J Biol Chem. 2019;294(4):1128–41.

Mallet C, Dubray C, Duale C. FAAH inhibitors in the limelight, but regrettably. Int J Clin Pharmacol Ther. 2016;54(7):498–501.

Manolio TA. Genomewide association studies and assessment of the risk of disease. N Engl J Med. 2010;363(2):166–76.

Matsuoka Y, Yamashita A, Matsuda M, Kawai K, Sawa T, Amaya F. The NLRP2 inflammasome in dorsal root ganglion as a novel molecular platform that produces inflammatory pain hypersensitivity. Pain. 2019;160:2149.

McDermott LA, Weir GA, Themistocleous AC, Segerdahl AR, Blesneac I, Baskozos G, et al. Defining the functional role of NaV1.7 in human nociception. Neuron. 2019;101(5):905–19. e8.

Meade EA, Smith WL, DeWitt DL. Differential inhibition of prostaglandin endoperoxide synthase (cyclooxygenase) isozymes by aspirin and other non-steroidal anti-inflammatory drugs. J Biol Chem. 1993;268(9):6610–4.

Meltzer H, Marom E, Alyagor I, Mayseless O, Berkun V, Segal-Gilboa N, et al. Tissue-specific (ts)CRISPR as an efficient strategy for in vivo screening in Drosophila. Nat Commun. 2019;10(1):2113.

Meng W, Adams MJ, Hebert HL, Deary IJ, McIntosh AM, Smith BH. A genome-wide association study finds genetic associations with broadly-defined headache in UK biobank (N=223,773). EBioMedicine. 2018;28:180–6.

Miller RJ, Jung H, Bhangoo SK, White FA. Cytokine and chemokine regulation of sensory neuron function. Handb Exp Pharmacol. 2009;194:417–49.

Miller RE, Malfait AM, Block JA. Current status of nerve growth factor antibodies for the treatment of osteoarthritis pain. Clin Exp Rheumatol. 2017;35(Suppl 107):85–7.

Montell C, Rubin GM. Molecular characterization of the Drosophila trp locus: a putative integral membrane protein required for phototransduction. Neuron. 1989;2(4):1313–23.

Moore RA, Derry S, Aldington D, Cole P, Wiffen PJ. Amitriptyline for neuropathic pain and fibromyalgia in adults. Cochrane Database Syst Rev. 2012;12:CD008242.

Nackley AG, Tan KS, Fecho K, Flood P, Diatchenko L, Maixner W. Catechol-O-methyltransferase inhibition increases pain sensitivity through activation of both beta2- and beta3-adrenergic receptors. Pain. 2007;128(3):199–208.

Nagakura Y. Challenges in drug discovery for overcoming 'dysfunctional pain': an emerging category of chronic pain. Expert Opin Drug Discovery. 2015;10(10):1043–5.

Nair AS, Poornachand A, Kodisharapu PK. Ziconotide: indications, adverse effects, and limitations in managing refractory chronic pain. Indian J Palliat Care. 2018;24(1):118–9.

Navratilova E, Porreca F. Substance P and inflammatory pain: getting it wrong and right simultaneously. Neuron. 2019;101(3):353–5.

Neely GG, Hess A, Costigan M, Keene AC, Goulas S, Langeslag M, et al. A genome-wide Drosophila screen for heat nociception identifies alpha2delta3 as an evolutionarily conserved pain gene. Cell. 2010;143(4):628–38.

Noble F, Marie N. Management of opioid addiction with opioid substitution treatments: beyond methadone and buprenorphine. Front Psych. 2018;9:742.

Onutu AH. Duloxetine, an antidepressant with analgesic properties - a preliminary analysis. Rom. J. Anaesth. Intensive Care. 2015;22(2):123–8.

Pacula RL, Smart R. Medical marijuana and marijuana legalization. Annu Rev Clin Psychol. 2017;13:397–419.

Pecze L, Blum W, Schwaller B. Mechanism of capsaicin receptor TRPV1-mediated toxicity in pain-sensing neurons focusing on the effects of Na(+)/Ca(2+) fluxes and the Ca(2+)-binding protein calretinin. Biochim Biophys Acta. 2013;1833(7):1680–91.

Pertwee RG. Cannabinoid pharmacology: the first 66 years. Br J Pharmacol. 2006;147(Suppl 1):S163–71.

Pertwee RG. The diverse CB1 and CB2 receptor pharmacology of three plant cannabinoids: delta9-tetrahydrocannabinol, cannabidiol and delta9-tetrahydrocannabivarin. Br J Pharmacol. 2008;153(2):199–215.

Pitsikas N. Duloxetine Eli Lilly & Co. Curr Opin Investig Drugs. 2000;1(1):116–21.

Powis TG, Gallaga Murrieta E, Lesure R, Lopez Bravo R, Grivetti L, Kucera H, et al. Prehispanic use of chili peppers in Chiapas, Mexico. PLoS One. 2013;8(11):e79013.

Preuss CV, Kalava A, King KC. Prescription of controlled substances: benefits and risks. Treasure Island: StatPearls; 2019.

Ramachandraih CT, Subramanyam N, Bar KJ, Baker G, Yeragani VK. Antidepressants: from MAOIs to SSRIs and more. Indian J Psychiatry. 2011;53(2):180–2.

Ramos-Matos CF, Lopez-Ojeda W. Fentanyl. Treasure Island: StatPearls; 2019.

Rao PP, Kabir SN, Mohamed T. Nonsteroidal anti-inflammatory drugs (NSAIDs): progress in small molecule drug development. Pharmaceuticals. 2010;3(5):1530–49.

Reardon C, Murray K, Lomax AE. Neuroimmune communication in health and disease. Physiol Rev. 2018;98(4):2287–316.

Rico-Villademoros F, Slim M, Calandre EP. Amitriptyline for the treatment of fibromyalgia: a comprehensive review. Expert Rev Neurother. 2015;15(10):1123–50.

Riquelme G, Sepulveda JM, Al Ghumgham Z, Del Campo M, Montero C, Lagos N. Neosaxitoxin, a paralytic shellfish poison toxin, effectively manages bucked shins pain, as a local long-acting pain blocker in an equine model. Toxicon. 2018;141:15–7.

Roberson DP, Binshtok AM, Blasl F, Bean BP, Woolf CJ. Targeting of sodium channel blockers into nociceptors to produce long-duration analgesia: a systematic study and review. Br J Pharmacol. 2011;164(1):48–58.

Rosenblum A, Marsch LA, Joseph H, Portenoy RK. Opioids and the treatment of chronic pain: controversies, current status, and future directions. Exp Clin Psychopharmacol. 2008;16(5):405–16.

Rupniak NM, Boyce S, Williams AR, Cook G, Longmore J, Seabrook GR, et al. Antinociceptive activity of NK1 receptor antagonists: non-specific effects of racemic RP67580. Br J Pharmacol. 1993;110(4):1607–13.

Sansone RA, Sansone LA. Pain, pain, go away: antidepressants and pain management. Psychiatry. 2008;5(12):16–9.

Schank JR, Heilig M. Substance P and the neurokinin-1 receptor: the new CRF. Int Rev Neurobiol. 2017;136:151–75.

Schou WS, Ashina S, Amin FM, Goadsby PJ, Ashina M. Calcitonin gene-related peptide and pain: a systematic review. J Headache Pain. 2017;18(1):34.

Schug SA, Zech D, Grond S. Adverse effects of systemic opioid analgesics. Drug Saf. 1992;7(3):200–13.

Shang X, Wang Z, Tao H. Mechanism and therapeutic effectiveness of nerve growth factor in osteoarthritis pain. Ther Clin Risk Manag. 2017;13:951–6.

Sieberg CB, Taras C, Gomaa A, Nickerson C, Wong C, Ward C, et al. Neuropathic pain drives anxiety behavior in mice, results consistent with anxiety levels in diabetic neuropathy patients. Pain Rep. 2018;3(3):e651.

Smith ES, Lewin GR. Nociceptors: a phylogenetic view. J Comp Physiol A Neuroethol Sens Neural Behav Physiol. 2009;195(12):1089–106.

Sneddon LU. Comparative physiology of nociception and pain. Physiology. 2018;33(1):63–73.

Sun L, Lutz BM, Tao YX. The CRISPR/Cas9 system for gene editing and its potential application in pain research. Translat Perioper Pain Med. 2016;1(3):22–33.

Szczot M, Pogorzala LA, Solinski HJ, Young L, Yee P, Le Pichon CE, et al. Cell-type-specific splicing of piezo2 regulates mechanotransduction. Cell Rep. 2017;21(10):2760–71.

Tammimaki A, Mannisto PT. Catechol-O-methyltransferase gene polymorphism and chronic human pain: a systematic review and meta-analysis. Pharmacogenet Genomics. 2012;22(9):673–91.

Tawfik GM, Hashan MR, Abdelaal A, Tieu TM, Huy NT. A commentary on the medicinal use of marijuana. Trop Med Health. 2019;47:35.

Tepper SJ. History and review of anti-calcitonin gene-related peptide (CGRP) therapies: from translational research to treatment. Headache. 2018;58(Suppl 3):238–75.

Thour A, Marwaha R. Amitriptyline. Treasure Island: StatPearls; 2019.

Toussaint K, Yang XC, Zielinski MA, Reigle KL, Sacavage SD, Nagar S, et al. What do we (not) know about how paracetamol (acetaminophen) works? J Clin Pharm Ther. 2010;35(6):617–38.

Tracey I, Woolf CJ, Andrews NA. Composite pain biomarker signatures for objective assessment and effective treatment. Neuron. 2019;101(5):783–800.

Treede RD. The International Association for the Study of Pain definition of pain: as valid in 2018 as in 1979, but in need of regularly updated footnotes. Pain Rep. 2018;3(2):e643.

Treede RD, Rief W, Barke A, Aziz Q, Bennett MI, Benoliel R, et al. A classification of chronic pain for ICD-11. Pain. 2015;156(6):1003–7.

Tsantoulas C. Emerging potassium channel targets for the treatment of pain. Curr Opin Support Palliat Care. 2015;9(2):147–54.

Tsantoulas C, McMahon SB. Opening paths to novel analgesics: the role of potassium channels in chronic pain. Trends Neurosci. 2014;37(3):146–58.

Tsuda M, Shigemoto-Mogami Y, Koizumi S, Mizokoshi A, Kohsaka S, Salter MW, et al. P2X4 receptors induced in spinal microglia gate tactile allodynia after nerve injury. Nature. 2003;424(6950):778–83.

Tsuda M, Tozaki-Saitoh H, Inoue K. Pain and purinergic signaling. Brain Res Rev. 2010;63(1–2):222–32.

Vink S, Alewood PF. Targeting voltage-gated calcium channels: developments in peptide and small-molecule inhibitors for the treatment of neuropathic pain. Br J Pharmacol. 2012;167(5):970–89.

Viswanatha R, Li Z, Hu Y, Perrimon N. Pooled genome-wide CRISPR screening for basal and context-specific fitness gene essentiality in Drosophila cells. elife. 2018;7:e36333.

Walitt B, Ceko M, Gracely JL, Gracely RH. Neuroimaging of central sensitivity syndromes: key insights from the scientific literature. Curr Rheumatol Rev. 2016;12(1):55–87.

Wallace JL, McCafferty DM, Sharkey KA. Lack of beneficial effect of a tachykinin receptor antagonist in experimental colitis. Regul Pept. 1998;73(2):95–101.

Wang WB, Cao YJ, Lyu SS, Zuo RT, Zhang ZL, Kang QL. Identification of a novel mutation of the NTRK1 gene in patients with congenital insensitivity to pain with anhidrosis (CIPA). Gene. 2018;679:253–9.

Watson JC, Sandroni P. Central neuropathic pain syndromes. Mayo Clin Proc. 2016;91(3):372–85.

Wemmie JA, Taugher RJ, Kreple CJ. Acid-sensing ion channels in pain and disease. Nat Rev Neurosci. 2013;14(7):461–71.

Wiffen PJ, Derry S, Moore RA, McQuay HJ. Carbamazepine for acute and chronic pain in adults. Cochrane Database Syst Rev. 2011;1:CD005451.

Woolf CJ. What is this thing called pain? J Clin Invest. 2010;120(11):3742–4.

Wright CL, Mist SD, Ross RL, Jones KD. Duloxetine for the treatment of fibromyalgia. Expert Rev Clin Immunol. 2010;6(5):745–56.

Xie YF, Huo FQ, Tang JS. Cerebral cortex modulation of pain. Acta Pharmacol Sin. 2009;30(1):31–41.

Xu Q, Yaksh TL. A brief comparison of the pathophysiology of inflammatory versus neuropathic pain. Curr Opin Anaesthesiol. 2011;24(4):400–7.

Yang F, Zheng J. Understand spiciness: mechanism of TRPV1 channel activation by capsaicin. Protein Cell. 2017;8(3):169–77.

Yasuda H, Hotta N, Nakao K, Kasuga M, Kashiwagi A, Kawamori R. Superiority of duloxetine to placebo in improving diabetic neuropathic pain: results of a randomized controlled trial in Japan. J Diab Invest. 2011;2(2):132–9.

Young JW, Juurlink DN. Tramadol. CMAJ. 2013;185(8):E352.

Zarghi A, Arfaei S. Selective COX-2 inhibitors: a review of their structure-activity relationships. Iran J Pharm Res. 2011;10(4):655–83.

Zhang S, Zhao J, Meng Q. AAV-mediated siRNA against TRPV1 reduces nociception in a rat model of bone cancer pain. Neurol Res. 2019;41:972–9.

Zhao L, Huang J, Fan Y, Li J, You T, He S, et al. Exploration of CRISPR/Cas9-based gene editing as therapy for osteoarthritis. Ann Rheum Dis. 2019;78(5):676–82.

Zhu YF, Wu Q, Henry JL. Changes in functional properties of A-type but not C-type sensory neurons in vivo in a rat model of peripheral neuropathy. J Pain Res. 2012;5:175–92.

Zhuo M. Neural mechanisms underlying anxiety-chronic pain interactions. Trends Neurosci. 2016;39(3):136–45.

Enabling Precision Health Approaches for Symptom Science Through Big Data and Data Science

17

Suzanne Bakken, Theresa A. Koleck, Caitlin Dreisbach, and Kathleen T. Hickey

Contents

17.1   Introduction

Big data and data science are essential to enabling precision health approaches for symptom science (Hickey et al. 2019). Big data can be conceptualized in terms of four "Vs"—volume, variety, velocity (e.g., sensor or streaming data), and veracity (i.e., level of uncertainty associated with the collection of data sources) (IBM 2015). These multiple attributes of big data reflect an additional complexity beyond that of simple volume. Consistent with this definition, the types and volume of data related to the symptom experience, symptom management strategies, and outcomes are increasingly accessible for research. Traditional data streams for symptom science

S. Bakken (✉)
School of Nursing, Department of Biomedical Informatics, and Data Science Institute, Columbia University, New York, NY, USA
e-mail: sbh22@cumc.columbia.edu

T. A. Koleck
School of Nursing, University of Pittsburgh, Pittsburgh, PA, Pennsylvania

C. Dreisbach
Data Science Institute, Columbia University, New York, NY, USA

K. T. Hickey
School of Nursing, Columbia University, New York, NY, USA

© Springer Nature Switzerland AG 2020   239
S. G. Dorsey, A. R. Starkweather (eds.), *Genomics of Pain and Co-Morbid Symptoms*, https://doi.org/10.1007/978-3-030-21657-3_17

such as patient reports and electronic health record (EHR) data are now complemented by consumer-generated (i.e., quantified self), environmental data, and *-omic* data streams (Bakken and Reame 2016).

In contrast to big data, which is defined by the attributes of the data, data science explicitly focuses on extraction of value from data. For example, the National Consortium for Data Science has emphasized value in its definition of data science "as the systematic study of the organization and use of digital data in order to accelerate discovery, improve critical decision-making processes, and enable a data-driven economy" (p.iii) (Ahalt et al. 2014). Data science draws upon theories and techniques from multiple fields including mathematics, statistics, and information technology and includes methods such as probability models, machine learning, statistical learning, computer programming, pattern recognition and learning, visualization, predictive analytics, uncertainty modeling, data warehousing, and high performance computing (Wikipedia 2015).

Consequently, data science requires a technical pipeline that is informed by a research question and supported by data governance policies. The pipeline typically includes sets of tools that support: (1) extraction/ingestion, (2) wrangling (pre-processing using semi-automated tools), (3) computation and analysis, (4) modeling and application, and (5) reporting and visualization (Tesla Institute 2018). Of note, the pipeline required for data science is often significantly different from the data management and analytic pipelines typically available to nurse scientists (e.g., SAS, STATA) due to the complexity of data and the required processing power. Tools such as Apache Hadoop Map-Reduce, Apache Mahout (machine learning algorithms), Sparks Machine Learning Library, and R-Hadoop are needed to support reduction and analysis of multi-dimensional data through methods such as K-means, neural network backpropagation, support vector machines, and Gaussian discriminative analysis. Visualization of the data for analysis, interpretation, and reporting can be supported through general tools such as Tableau and tools for special purposes (e.g., Sentiment Viz for visualization of Tweet contents, ORA for visualization of network structures) (Bakken and Koleck 2019).

The nursing science community has recognized the relevance of data science (Bakken and Reame 2016; Brennan and Bakken 2015). For example, a recent systematic review of application of data science in nursing evaluated 17 studies conducted in 2009–2015 that focused on nursing practice and systems that affect nurses (Westra et al. 2017). The majority of studies were in acute care settings, used the EHR as their data source, and had study purposes categorized as knowledge discovery, prediction, and evaluation. Since the timeframe of the review, additional nursing studies have been conducted that reflect data sources beyond EHRs including *-omics*, (Koleck and Conley 2016), social media (Odlum and Yoon 2015), and sensors (Rantz et al. 2015a). From a policy perspective, Brennan and Bakken argued that nursing policy statements inform a principled and ethical approach to big data and data science, (Brennan and Bakken 2015) and a nursing science perspective on health policy considerations for data science has been delineated (Bakken 2017).

However, there has been limited research explicitly identified as data science that has focused on symptom science.

In this chapter, we describe how big data and data science can advance precision health approaches for symptom science, summarize current challenges to use of big data and data science methods, and present future research directions.

17.2 How Can Big Data and Data Science Advance Precision Approaches for Symptom Science?

The Nursing Science Precision Health (NSPH) Model (Fig. 17.1) as applied to the National Institutes of Health (NIH) Symptom Science Model (Hickey et al. 2019) is used as the organizing framework to summarize the literature on how big data and data science can advance precision approaches for symptom science. In addition, four case studies illustrate the application of data science methods to support the precision health components of the NSPH model: (a) measurement, (b) characterization of phenotype including lifestyle and environmental factors, (c) characterization of genotype and other biomarkers, and (d) identification of intervention targets, design of interventions, and delivery of interventions. These components are supported by an information and data science infrastructure.

17.2.1 Precision in Characterization of Phenotype Including Lifestyle and Environmental Factors

A variety of data sources are relevant to characterization of symptom phenotypes including lifestyle and environmental factors and can be used to increase the precision of the phenotype. While patient reports from standardized surveys, including those considered National Institute of Nursing Research (NINR) common data elements (CDEs), (Redeker et al. 2015) and ecological momentary assessments remain important sources of phenotypic data for symptom science, other data sources amenable to data science methods include EHRs, social media, wearables, and sensors.

Based on the premise that the volume of longitudinal symptom data available in EHR free-text clinical narratives offers an unprecedented opportunity to study the biological and behavioral foundations of symptom occurrence as well as symptom documentation practices, Koleck et al. synthesized the literature on the use of natural language processing (NLP) to process and/or analyze symptom information documented in EHR free-text narratives (Koleck et al. 2019). The 27 studies reviewed encompassed a wide variety of symptoms, including shortness of breath, pain, nausea, dizziness, disturbed sleep, constipation, and depressed mood and applied a variety of NLP approaches. The authors concluded that current focus is on the development of methods to extract symptom information and the use of symptom information for disease classification tasks rather than the examination of symptoms themselves.

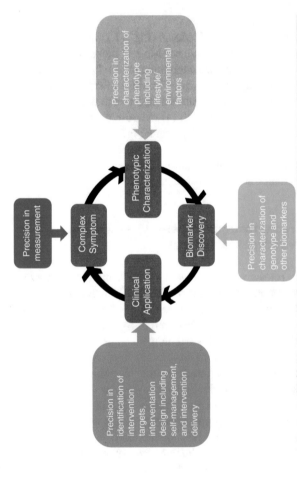

Fig. 17.1 The Nursing Science Precision Health (NSPH) Model Applied to Stages of the National Institutes of Health (NIH) Symptom Science Model. The four precision components in the boxes with arrows and the information and data science infrastructure comprise the NSPH model. The components can be applied to different phenomena as illustrated by the stages of the NIH Symptom Science Model. Reprinted from Nursing Outlook, 67 (4), Kathleen T. Hickey, Suzanne Bakken, Mary W. Byrne, et al., Precision health: Advancing symptom and self-management science, 462–475, Copyright (2019), with permission from Elsevier

Dreisbach et al. critically examined and summarized the literature on the use of NLP and text mining for symptom extraction and processing in electronic patient-authored text (ePAT) (Dreisbach et al. 2019). The 21 studies reviewed accessed ePAT from a variety of sources including Twitter and online community forums or patient portals focused on diseases (e.g., diabetes, cancer, and depression). Pain and fatigue and sleep disturbance were the most frequently evaluated symptom clinical content categories with both appearing in 18 studies. Individual studies of social media have focused on particular symptom areas. For example, Marshall et al. compared symptom clusters from an online breast cancer forum with a symptom checklist finding some differences and some overlap between the two data sources (Marshall et al. 2016). Other researchers have developed tools to improve extraction of such information from online forums (Zhang et al. 2017).

Although polysomnography and accelerometers have been widely used in symptom studies for measurement of sleep and physical activity, recent systematic reviews have focused on comparison among consumer devices or between consumer devices and those that have traditionally been used in research. In a narrative review focused on sleep, Kolla et al. concluded that when compared to polysomnography/actigraphy, consumer devices tend to underestimate sleep disruptions and overestimate total sleep times and sleep efficiency in normal subjects, applications (apps) perform worse than the wearable devices, and both devices and apps are better at detecting sleep than wake patterns (Kolla et al. 2016). A review of 22 studies focused on comparisons among consumer devices (Fitbit and Jawbone) for sleep and physical activity monitoring found higher validity for step counts, lower validity for energy expenditure and sleep, and high inter-device reliability for steps, distance, energy expenditure, and sleep for some Fitbit models (Evenson et al. 2015). In terms of home sensors, Rantz et al. have examined the use of sensor technology to enable aging in place through a series of studies (Rantz et al. 2015a, b). Sensors continuously monitor functional status including: (a) respiration, pulse, and restlessness during sleep; (b) gait speed, stride length, and stride time for calculation of fall risk; and (c) fall detection and algorithms are applied to these data to generate automated health alerts to healthcare staff for possible action (Liu et al. 2014). A multitude of databases provide data on built environment (e.g., healthy food availability) (Co Jr. and Bakken 2018) and data gathered through environmental sensors focused on aspects that influence symptoms such as pollutants, allergens, and noise levels. The increase in breadth and depth of phenotypic data related to symptom status, behaviors, and the context of the symptom experience provides new opportunities in terms of research questions that can be addressed.

17.2.2 Precision in Characterization of Genotype and Other Biomarkers

As described in detail elsewhere in this book, nurse scientists have systematically integrated -omic (e.g., genomics, epigenomics, transcriptomics, proteomics, metabolomics) and other biomarker discovery into their programs of research in symptom management (Koleck and Conley 2016; Cashion and Grady 2015; Alfaro et al. 2014; Maki et al. 2018). Recognizing the importance of biomarkers, the NINR Center Directors Network proposed a minimum set of biomarker CDEs for symptom and self-management science, including pro- and anti-inflammatory cytokines, a hypothalamic-pituitary-adrenal axis marker, cortisol, neuropeptide brain-derived neurotrophic factor, and DNA polymorphisms (Page et al. 2018). Out of this set, DNA polymorphisms, in particular, represent a high volume of complex data when studied using a genome wide methodology (i.e., genome wide association study [GWAS]). Specialized toolsets (e.g., PLINK, R) are required to manage, check the quality of, statistically summarize, and visualize genome wide data. Single assay – *omics* association studies like GWAS and related methodologies (e.g., transcriptome wide association study, methylome wide association study) are common "big data" approaches undertaken by nurse scientists to uncover biomarkers. Fortunately, it may not be necessary for nurse scientists to collect biological specimens and generate primary data to incorporate –*omics* into their symptom work. Existing epidemiological data sets, where symptoms are well-characterized but not the main phenotype of interest, can be repurposed for symptom science (Osier et al. 2017). These data are readily available and accessible to researchers in resources such as the database of genotypes and phenotypes (dbGaP) (www.ncbi.nlm.nih.gov/gap) or Gene Expression Omnibus (www.ncbi.nlm.nih.gov/geo). Nevertheless, it is important to consider that while –*omic* association studies generate data-driven hypotheses, statistically significant findings require extensive follow-up and validation to pinpoint causal variants, genes, and/or pathways prior to clinical implementation. Data science techniques, such as data mining or machine learning, allow us to move beyond association-level statistics. However, the biological complexity of symptom and symptom management mechanisms necessitates investigation of multiple levels of biological information and changes in dynamic systems over time. More recent bioinformatics tools (e.g., Multi-Omics Factor Analysis [https://github.com/bioFAM/MOFA]) and publicly funded studies (e.g., Trans-Omics for Precision Medicine [TOPMed]) are aimed at integration from multiple assays. With robust symptom characterization in large, longitudinal data sets, research aimed at diverse data integration are poised to make a significant impact on precision symptom management.

17.2.3 Precision in Identification of Intervention Targets, Intervention Design Including Self-management, and Intervention Delivery

A challenge to precision in identification of intervention targets and design of interventions is the lack of a common method for describing and organizing intervention components. To address this issue for evidence-based behavioral mechanisms, Michie et al. delineated a taxonomy (Michie et al. 2013). Of particular relevance to precision in intervention design for symptom management, this taxonomy not only provides an important foundation for building the knowledge base on evidence-based behavior change mechanisms within a population, but also facilitates comparisons across populations to determine whether efficacy differs.

In terms of precision in intervention delivery, technology-enabled tailored symptom management interventions based on a limited set of tailoring variables are fairly common and are most often tested in an RCT or, alternatively, using Multiphase Optimization Strategy (MOST) if the intervention is multi-component in nature. A newer precision approach is just-in-time adaptive interventions (JITAIs) that are mobile technology-enabled and evaluated through micro-randomized trials in which an individual is randomized to intervention options multiple times during the course of the study based upon decision points (Klasnja et al. 2015). Because individuals receive intervention components multiple times over the course of the trial, the proximal causal effects of the randomized intervention can be modeled and the time-varying moderation of the effects of the intervention can be assessed. This results in a more in-depth understanding of the intervention effects and contributes to the design of more effective JITAIs.

17.2.4 Case Studies

Four case studies illustrate application of data science approaches in support of precision in symptom science. Table 17.1 delineates the components of the NSPH model for each case study. Table 17.2 summarizes the key data science challenges experienced in the case studies.

17.2.4.1 Electronic Health Record Phenotyping for Postoperative Nausea and Vomiting

Secondary use of symptom data captured and stored within EHRs has the potential to facilitate symptom science. However, in order to exploit this big data resource, researchers must first create accurate and reproducible *computable phenotypes* for symptoms. A computable phenotype is a "characteristic that can be ascertained via a computerized query to an EHR system or clinical data repository using a defined set of data elements and logical expressions" (Richesson and Smerek 2014). Koleck et al. are working to develop a computable phenotype for nausea and its accompanying sign, vomiting, in women undergoing gynecologic surgical procedures (Koleck et al. 2018). After obtaining data from their institution's clinical data repository through formal data acquisition procedures (governance) and storing the data on a Health Insurance Portability and Accountability Act (HIPAA) compliant server (infrastructure), the team decided which data type and sources were most appropriate and practical for the phenotyping task. EHRs are comprised of two main data types—unstructured (e.g., discharge summaries, progress notes) and structured (e.g., diagnosis codes, medications, laboratory test results). NLP has been used to extract a wide variety of symptom terms, including nausea, pain, shortness of breath, disturbed sleep, and depressed mood, from clinical narratives (Koleck et al. 2019). However, because unstructured data introduces additional pre-processing challenges (wrangling), the team evaluated three structured data sources: ICD-9

Table 17.1 NSPH model focus of case studies

Case study	Characterization		Intervention		
	Phenotype including lifestyle and environment	Genotype and other biomarkers	Target discovery	Design	Delivery
Electronic health record phenotyping for postoperative nausea and vomiting	x				
Gut–brain axis and symptom perception in irritable bowel syndrome	x	x			
Twitter and symptom science in dementia caregiving	x		x	x	
Mobile electrocardiograms and text messages in atrial fibrillation	x	x			x

Table 17.2 Summary of key data science challenges for case studies

Case example	Governance	Infrastructure	Capture/ Extraction/ Ingestion	Wrangling	Computation and analysis	Modeling and application	Reporting and visualization
Electronic health record phenotyping for postoperative nausea and vomiting	x	x	x	x			
Gut–brain axis and symptom perception in irritable bowel syndrome					x	x	x
Twitter and symptom science in dementia caregiving		x	x				
Mobile electrocardiograms and text messages in atrial fibrillation			x	x			

diagnosis codes, antiemetic medication orders, and antiemetic medication administration records. In particular, the medication order and administration record extraction algorithms required iterative refinement directed by an expert-curated list of antiemetic drugs derived from professional practice guidelines. After algorithm development, the essential next steps and challenges are assessment of data quality (including completeness, correctness, concordance, plausibility, and currency) (Weiskopf et al. 2013) and validation of the computable phenotype against a *gold standard* (Richesson and Smerek 2014). Because a gold standard for the true phenotype was not available, the team compared estimated postoperative nausea and/or vomiting prevalence rates to those reported in the literature. Estimates based on antiemetic documentation in administration records generated rates most comparable. The team's analysis calls attention to the thoughtful use of medication order versus administration information as a data element for EHR phenotyping, especially for drugs like antiemetics that are frequently prescribed *PRN* (i.e., as needed). With additional refinement and validation, a postoperative nausea and vomiting computable phenotype has utility in future EHR-based risk factor and treatment response studies.

17.2.4.2 Gut–Brain Axis and Symptom Perception in Irritable Bowel Syndrome

The gut microbiome is thought to play a crucial role in normal physiological functioning; and yet, relatively little is known about its specific contributions to health and disease (D'Argenio and Salvatore 2015). The emerging study of the connection between the gut microbiome and the brain, termed the gut–brain axis, is especially relevant for the understanding of symptoms. Specifically, symptoms from irritable bowel syndrome (IBS) account for over 50% of primary care clinical visits for gastrointestinal (GI) distress annually (Wilson et al. 2004). Differences in microbial composition and reduction in diversity of species have been shown in IBS patients compared with healthy subjects (Distrutti et al. 2016). Further reinforcing the heterogeneous etiology of IBS, a study by Molinder et al. showed that previously undiagnosed individuals with IBS use words outside of the current diagnostic criteria to describe their symptoms (Molinder et al. 2015). A conceptual framework by Distrutti et al. illustrates how dysregulation of the autonomic nervous system in the gut from GI symptomology influences the sensory perception and activation of the hypothalamic-pituitary-adrenal axis (Distrutti et al. 2016). The combination of patient-authored description, symptom perception, measurement precision, and lifestyle/environment factors emphasizes the complexity in understanding the biological mechanisms of IBS through precision health advances. Measurement considerations include the types of sampling and storage procedures (i.e., rectal swabbing, stool sample), varying sequencing methods (i.e., 16S RNA sequencing and shotgun metagenomics) on multiple branded platforms (i.e., Illumina, Ion Torrent) and the explosion of new bioinformatics tools to process the raw sequencing data. Data science techniques that address the volume and veracity of data, including dimensionality reduction algorithms and cloud computing, can help to mitigate the complexity of microbial biology and mechanisms for pertinent symptoms and clinical outcomes (Kelsey et al. 2018). Expert phenotyping of diet, lifestyle measures, and clinical symptoms in combination with rigorous genomic sequencing can contribute to knowledge discovery and therapeutic advances. To guide nursing research, Maki et al. outlined the methodological considerations for designing a microbiome focused study (Maki et al. 2018). They argued that nurses are uniquely positioned to contribute to microbiome research based on their holistic perspective of the patient experience, environment, and genomics underpinnings.

17.2.4.3 Twitter and Symptom Science in Dementia Caregiving

Although very limited in character length, Tweets have associated metadata which results in more than 20 data elements per Tweet including explicit and extractable characteristics of the user and the Tweet (Sinnenberg et al. 2017). In a series of studies, Yoon et al. have applied a set of data science methods to Twitter to examine symptom self-management topics through topic modeling and sentiment analysis and the social network through analyzing macro-, meso-, and micro-network structures. In terms of the NSPH model, these studies have focused on characterization of the phenotype including lifestyle and environment and for discovery of intervention targets and intervention design. A variety of existing tools were combined to create a pipeline to support data extraction and analysis: extraction/ingestion (NODEXL, nCapture), wrangling (Notepad++, Tableau), structural analyses including visualization (ORA, Pajek), and content analysis including visualization (Weka, Sentiment Viz). The earliest study focused on types of physical activity and the associated positive and negative sentiments. For example, biking, dancing, and hiking were consistently associated with positive sentiments, while running was a mixture, and yard work and wood chopping were frequently associated with negative sentiments (Yoon et al. 2013). To examine the mechanisms of social support for caregivers of those with Alzheimer's disease or other dementias in a later study, the investigators created corpora containing Tweets about Alzheimer's disease, dementia, and caregiving. Through the methods of topic modeling, sentiment analysis, and network analysis, they found that (a) frequently occurring dementia topics were related to mental health and caregiving, (b) the sentiments expressed in the Tweets were more negative than positive, and (c) network patterns demonstrated lack of social connectedness (Yoon 2016; Yoon et al. 2016). A subsequent analysis of a subset of the corpora explicitly focused on symptoms, applied a new technique for sentiment analysis based on the concept of transfer learning, i.e., the algorithm was trained on an unrelated data set and then applied to the Tweet corpus to generate a continuous rather than dichotomous sentiment score (unpublished). Across the studies, there were challenges related to data science infrastructure and pipeline. These included institutional policies that limited data storage, lack of graphical user interfaces to high performance computing resources, and labor-intensive extraction of publically available Tweets on a daily basis due to lack of budget for purchasing a retrospective Twitter database. To address the last issue, relevant Tweets were downloaded on a daily basis, pre-processed, and then combined to form the analytic Tweet corpus.

17.2.4.4 Integrating Mobile Electrocardiograms and Text Messages in Atrial Fibrillation

Atrial fibrillation is estimated to affect 2.7–6.1 million individuals in the USA and can lead to significant morbidity including stroke (Centers for Disease Control and Prevention 2015). Yet, atrial fibrillation, including recurrence after treatment, is often undetected because it is episodic, requires electrocardiography for documentation, and is poorly correlated with patient-reported symptoms. Moreover, modifiable lifestyle factors including obesity and heavy alcohol use influence the occurrence of atrial fibrillation. The iPhone® Helping Evaluate Atrial Fibrillation Heart Rhythm Through Technology (iHEART) symptom management intervention combines the AliveCore® mobile electrocardiogram (ECG) and related Kardia mobile app for capture of the ECG and subsequent provider review with educational and motivational text messages aimed at modifiable lifestyle factors (Hickey et al. 2016). iHEART addresses three precision components of the NSPH model: (a) characterization of phenotype including lifestyle and environment, (b) characterization of *-omics* and other biomarkers, and (c) intervention delivery. Participants captured ECGs using AliveCore and uploaded the ECGs to the Cloud via the Kardia app. Participants also recorded symptoms and triggers via an app and received text messages. Providers and research staff viewed data via a web portal. There were several challenges related to the data science pipeline. Some participants had difficulties in capturing or transmitting the ECG due to poor connectivity between fingertips and the device or between the device and the app resulting in tracings that could not be classified or were not uploaded (Reading et al. 2018). In addition, some participants found the app to capture symptoms and triggers difficult to use, and the capture of symptoms via voice was sometimes obscured by background noise limiting the utility of the data for analysis. The resulting data set is high volume and complex including >30,000 ECGs, patient-reported triggers and symptoms, as well as traditional standardized surveys and clinical measures. Analyses to determine the efficacy of the iHEART intervention focus on disease-specific knowledge, clinical outcomes, quality of life, and quality-adjusted life years.

For the case studies, the challenges varied by the type of data needed for the research project. For example, use of EHR data to create computable phenotypes was challenging from the perspective of governance and infrastructure as well as data extraction and wrangling. In contrast, data collected for research purposes had less governance issues, but can experience challenges later in the data science pipeline as in the IBS example researchers. Moreover, publically available data like Tweets have no governance issues, but institutions may lack sufficient infrastructure to support efficient extraction of large volumes of data.

17.3 Future Directions

Future directions for the use of data science methods in symptom science should address the key data challenges that are outlined in this chapter, including governance, infrastructure, extraction/ingestion, wrangling, computation and analysis, modeling and application, and reporting and visualization. Further, an emphasis on health equity, implementation and evaluation, and open science should be strongly considered in future directions given their relevance to the nursing perspective on symptom science.

In relation to health equity, the Belmont Report delineated principles for ethical conduct of research that must be considered for use of big data streams and data science methods: respect for persons (i.e., autonomy), beneficence, and justice (The National Commission for the Protection of Human Subjects of Biomedical and Behavioral Research 1979). Informed consent is the primary mechanism for protection of autonomy. Biobanks have explicit opt-in or opt-out consent processes. Use of protected health information (PHI) from EHRs and other electronic clinical data resources for research has ethical and regulatory oversight from institutional review boards and national regulations such as HIPAA. In contrast, terms of agreement of data use from social network sites and other quantified-self technologies may not be read or fully comprehended by users including those with low health literacy or limited English proficiency. Consequently, data may be used for multiple purposes in the absence of truly informed consent.

Poor methodological rigor and loss of confidentiality pose threats to the principle of beneficence. Methodological rigor is needed to select appropriate data streams as well as at each stage of the data science pipeline to ensure appropriate decision making based on study findings. Loss of confidentiality and subsequent commodification of data can occur as digital content is produced and consumed by patients and consumers as they access websites, use mobile health apps, and post and respond to social network messages. Since individuals do not typically reap financial benefits from commodification of their data, they may vary in their willingness to have their data used for public health versus commercial purposes (Vayena et al. 2015; Lupton 2014).

For data science, the principle of justice means careful consideration of characteristics of the individuals or populations comprising the data streams that will be used to address the research question. For example, the literature suggests that (a) patient severity of illness and sociodemographic characteristics and, consequently, EHR data vary by type and location of the healthcare organization, (b) Latinos are less likely than Whites or Blacks to use health tracking apps (Fox and Duggan 2013), and (c) racial and ethnic minorities are less likely than non-minorities to participate in biobanks (Dang et al. 2014; Shaibi et al. 2013). Such biases in the data streams may limit the relevance of discoveries and predictions to those at greatest risk for health disparities. Additionally, machine learning algorithms that are trained using historical data may be susceptible to poor representation from vulnerable populations who are most at risk for health disparities. When implemented clinically or used to guide health policy, these algorithms have the potential to perpetuate social and structural biases by reinforcing poor public health and neighborhood outcomes and inequitable healthcare costs (Rajkomar et al. 2018). In order to diminish the potential for harm in symptom science research, researchers must carefully match their selection of data streams to their research questions. Modeling strategies need to include diverse representation and demonstrate adequate performance across all populations of interest. Moreover, while the Belmont principles are relevant to data science, additional guidance is needed to inform their operationalization.

In terms of implementation and evaluation, deployment of predictive or prescriptive analytic technologies is not sufficient to make meaningful change and reduce adverse health outcomes. Further qualitative research is needed to explore the experiences of clinicians and patients as they implement, use, and troubleshoot new products resulting from the analytics such as clinical dashboards, clinical decision support systems, and mobile apps. It is clear that nurses and other health professionals at the point of care are key stakeholders as users of such technologies to improve quality of care and health outcomes for patients and their families.

Finally, efforts consistent with the tenets of open science should be undertaken to remove barriers to symptom-related data and resource sharing and to make research products (e.g., data sets, data processing pipelines, clinically actionable predictive algorithms, software) openly available (Watson 2015; McKiernan et al. 2016). As a foundational element of open science, the NINR Center Directors proposed CDEs for symptom science (Redeker et al. 2015; Page et al. 2018), and Centers submit these data to a nursing science central data repository. In terms of data access for open science, a key strategy for implementing the FAIR (findable, accessible, interoperable, and reusable) principles (Wilkinson et al. 2016) at the federal level is the NIH Commons Framework (National Institutes of Health 2015). The Commons comprises three layers: (a) bottom layer—a computational platform comprising high performance and cloud computing, (b) middle layer—reference data sets and user-defined data, and (c) top layer—services (e.g., application programming interfaces [APIs], containers, indexing such as PubMed and DataMed) and tools (e.g., scientific analysis, workflows). Nurse scientists should engage in such open science initiatives to advance interdisciplinary symptom science research as well as to increase the influence of their data sets for care and policy.

17.4   Conclusions

The recent big data and data science movement affords researchers and clinicians an unprecedented opportunity to approach questions regarding precision health for symptom science from a new perspective. The symptom science domain is primed for the application of data science methods to phenotype characterization, biomarker discovery, and precision interventions across a variety of big data sources such as EHRs, -*omics*, social media, wearables, and sensors. Moreover, nurse scientists, working with interdisciplinary collaborators, are well-suited to advance symptom science through data science.

References

Ahalt SC, Bizen C, Evans J, Erlich Y, Ginsburg GS, Krishnamurthy A, et al. Data to discovery: genes to health - a white paper from the National Consortium for Data Science. The National Consortium for Data Science; 2014.

Alfaro E, Dhruva A, Langford DJ, Koetters T, Merriman JD, West C, et al. Associations between cytokine gene variations and self-reported sleep disturbance in women following breast cancer surgery. Eur J Oncol Nurs. 2014;18(1):85–93.

Bakken S. Data science. In: Hinshaw AS, Grady PA, editors. Shaping health policy through nursing research. Cham: Springer; 2017.

Bakken S, Koleck TA. Big data challenges from a nursing perspective. In: Kushniruk A, Borycki E, Househ M, editors. Big data big challenges: a healthcare perspective. Cham: Springer; 2019.

Bakken S, Reame N. The promise and potential perils of big data for advancing symptom management research in populations at risk for health disparities. Annu Rev Nurs Res. 2016;34(1):247–60.

Brennan PF, Bakken S. Nursing needs big data and big data needs nursing. J Nurs Scholarsh. 2015;47(5):477–84.

Cashion AK, Grady PA. The National Institutes of Health/National Institutes of nursing research intramural research program and the development of the National Institutes of Health symptom science model. Nurs Outlook. 2015;63(4):484–7.

Centers for Disease Control and Prevention. Atrial Fibrillation Fact Sheet. 2015 [cited 2018 December 8]. Available from: https://www.cdc.gov/dhdsp/data_statistics/fact_sheets/fs_atrial_fibrillation.htm.

Co MC Jr, Bakken S. Influence of the local food environment on Hispanics' perceptions of healthy food access in New York City. Hispanic Health Care Int. 2018;16(2):76–84.

D'Argenio V, Salvatore F. The role of the gut microbiome in the healthy adult status. Clin Chim Acta. 2015;451(Pt A):97–102.

Dang JH, Rodriguez EM, Luque JS, Erwin DO, Meade CD, Chen MS Jr. Engaging diverse populations about biospecimen donation for cancer research. J Community Genet. 2014;5(4):313–27.

Distrutti E, Monaldi L, Ricci P, Fiorucci S. Gut microbiota role in irritable bowel syndrome: new therapeutic strategies. World J Gastroenterol. 2016;22(7):2219–41.

Dreisbach C, Koleck TA, Bourne PE, Bakken S. A systematic review of natural language processing and text mining of symptoms from electronic patient-authored text data. Int J Med Inform. 2019;125:37–46.

Evenson KR, Goto MM, Furberg RD. Systematic review of the validity and reliability of consumer-wearable activity trackers. Int J Behav Nutr Phys Act. 2015;12:159.

Fox S, Duggan M. Tracking for health. Pew Internet and American Life Project; 2013.

Hickey KT, Hauser NR, Valente LE, Riga TC, Frulla AP, Masterson Creber R, et al. A single-center randomized, controlled trial investigating the efficacy of a mHealth ECG technology intervention to improve the detection of atrial fibrillation: the iHEART study protocol. BMC Cardiovasc Disord. 2016;16:152.

Hickey K, Bakken S, Byrne M, Bailey D Jr, Demiris G, Docherty S, et al. Precision health: advancing symptom and self-management science. Nurs Outlook. 2019;67(4):462–75.

IBM. IBM big data & analytics hub. 2015. Available from: http://www.ibmbigdatahub.com/infographic/four-vs-big-data.

Kelsey C, Dreisbach C, Alhusen J, Grossmann T. A primer on investigating the role of the microbiome in brain and cognitive development. Dev Psychobiol. 2018;61(3):341–9.

Klasnja P, Hekler EB, Shiffman S, et al. Microrandomized trials: an experimental design for developing just-in-time adaptive interventions. Health Psychol. 2015;34S:1220–8.

Koleck TA, Conley YP. Identification and prioritization of candidate genes for symptom variability in breast cancer survivors based on disease characteristics at the cellular level. Breast Cancer. 2016;8:29–37.

Koleck T, Bakken S, Tatonetti N. Comparison of electronic medication orders versus administration records for identifying prevalence of postoperative nausea and vomiting. In: AMIA Annu Symp, San Francisco, CA; 2018.

Koleck TA, Dreisbach C, Bourne PE, Bakken S. Natural language processing of symptoms documented in free-text narratives of electronic health records: a systematic review. J Am Med Inform Assoc. 2019;26(4):364–79.

Kolla BP, Mansukhani S, Mansukhani MP. Consumer sleep tracking devices: a review of mechanisms, validity and utility. Expert Rev Med Devices. 2016;13(5):497–506.

Liu L, Popescu M, Skubic M, Rantz M. An automatic fall detection framework using data fusion of Doppler radar and motion sensor network. Conf Proc IEEE Eng Med Biol Soc. 2014;2014:5940–3.

Lupton D. The commodification of patient opinion: the digital patient experience economy in the age of big data. Sociol Health Illn. 2014;36(6):856–69.

Maki KA, Diallo AF, Lockwood MB, Franks AT, Green SJ, Joseph PV. Considerations when designing a microbiome study: implications for nursing science. Biol Res Nurs. 2018;21(2):125–41. 1099800418811639.

Marshall SJ, Livingstone KM, Celis-Morales C, et al. Reproducibility of the online Food4Me food-frequency questionnaire for estimating dietary intakes across Europe. J Nutr. 2016;146(5):1068–75.

McKiernan EC, Bourne PE, et al. How open science helps researchers succeed. elife. 2016;5:e16800.

Michie S, Richardson M, Johnston M, et al. The behavior change technique taxonomy (v1) of 93 hierarchically clustered techniques: building an international consensus for the reporting of behavior change interventions. Ann Behav Med. 2013;46(1):81–95.

Molinder H, Agreus L, Kjellstrom L, et al. How individuals with the irritable bowel syndrome describe their own symptoms before formal diagnosis. Ups J Med Sci. 2015;120(4):276–9.

National Institutes of Health. The NIH commons. 2015. Available from: https://datascience.nih.gov/commons.

Odlum M, Yoon S. What can we learn about the Ebola outbreak from tweets? Am J Infect Control. 2015;43(6):563–71.

Osier N, Imes C, Khalil H, Zelazny J, Johansson A, Conley Y. Symptom science: repurposing existing omics data. Biol Res Nurs. 2017;19(1):19–27.

Page GG, Corwin EJ, Dorsey SG, et al. Biomarkers as common data elements for symptom and self-management science. J Nurs Scholarsh. 2018;50(3):276–86.

Rajkomar A, Hardt M, Howell MD, Corrado G, Chin MH. Ensuring fairness in machine learning to advance health equity. Ann Intern Med. 2018;169(12):866–72.

Rantz MJ, Skubic M, Popescu M, et al. A new paradigm of technology-enabled 'Vital Signs' for early detection of health change for older adults. Gerontology. 2015a;61(3):281–90.

Rantz M, Lane K, Phillips LJ, et al. Enhanced registered nurse care coordination with sensor technology: impact on length of stay and cost in aging in place housing. Nurs Outlook. 2015b;63(6):650–5.

Reading M, Baik D, Beauchemin M, Hickey KT, Merrill JA. Factors influencing sustained engagement with ECG self-monitoring: perspectives from patients and health care providers. Appl Clin Inform. 2018;9(4):772–81.

Redeker NS, Anderson R, Bakken S, et al. Advancing symptom science through use of common data elements. J Nurs Scholarsh. 2015;47(5):379–88.

Richesson R, Smerek M. Electronic health records-based phenotyping. Rethinking clinical trials: a living textbook of pragmatic clinical trials. Durham: Duke University; 2014.

Shaibi GQ, Coletta DK, Vital V, Mandarino LJ. The design and conduct of a community-based registry and biorepository: a focus on cardiometabolic health in Latinos. Clin Trans Sci. 2013;6(6):429–34.

Sinnenberg L, Buttenheim AM, Padrez K, Mancheno C, Ungar L, Merchant RM. Twitter as a tool for health research: a systematic review. Am J Pub Health. 2017;107(1):e1–8.

Tesla Institute. Understanding the data science pipeline. 2018. Available from: http://www.tesla-institute.com/index.php/using-joomla/extensions/languages/278-understanding-the-data-science-pipeline.

The National Commission for the Protection of Human Subjects of Biomedical and Behavioral Research. The Belmont Report: ethical principles and guidelines for the protection of human subjects of research. Washington, DC; 1979.

Vayena E, Salathe M, Madoff LC, Brownstein JS. Ethical challenges of big data in public health. PLoS Comput Biol. 2015;11(2):e1003904.

Watson M. When will 'open science' become simply 'science'? Genome Biol. 2015;16:101.

Weiskopf NG, Hripcsak G, Swaminathan S, Weng C. Defining and measuring completeness of electronic health records for secondary use. J Biomed Inform. 2013;46(5):830–6.

Westra BL, Sylvia M, Weinfurter EF, Pruinelli L, Park JI, Dodd D, et al. Big data science: a literature review of nursing research exemplars. Nurs Outlook. 2017;65(5):549–61.

Wikipedia. Data science. 2015. Available from: http://en.wikipedia.org/wiki/Data_science.

Wilkinson MD, Dumontier M, Aalbersberg IJ, et al. The FAIR guiding principles for scientific data management and stewardship. Sci Data. 2016;3:160018.

Wilson A, Longstreth GF, Knight K, Wong J, Wade S, Chiou CF, et al. Quality of life in managed care patients with irritable bowel syndrome. Manag Care Interface. 2004;17(2):24–8, 34.

Yoon S. What can we learn about mental health needs from tweets mentioning dementia on World Alzheimer's day? J Am Psychiatr Nurses Assoc. 2016;22(6):498–503.

Yoon S, Elhadad N, Bakken S. A practical approach for content mining of tweets. Am J Prev Med. 2013;45(1):122–9.

Yoon S, Co MC Jr, Bakken S. Network visualization of dementia tweets. Stud Health Technol Inform. 2016;225:925.

Zhang S, Grave E, Sklar E, Elhadad N. Longitudinal analysis of discussion topics in an online breast cancer community using convolutional neural networks. J Biomed Inform. 2017;69:1–9.